Thermodynamics: Four Laws That Move the Universe

Jeffrey C. Grossman, Ph.D.

PUBLISHED BY:

THE GREAT COURSES
Corporate Headquarters
4840 Westfields Boulevard, Suite 500
Chantilly, Virginia 20151-2299
Phone: 1-800-832-2412
Fax: 703-378-3819
www.thegreatcourses.com

Printed in the United States of America

Jeffrey C. Grossman, Ph.D.

Professor in the Department of Materials
Science and Engineering
Massachusetts Institute of Technology

Professor Jeffrey C. Grossman is a Professor in the Department of Materials Science and Engineering at the Massachusetts Institute of Technology. He received his B.A. in Physics from Johns Hopkins University in 1991 and his M.S. in Physics from the University of Illinois at Urbana-Champaign in 1992. After receiving his Ph.D. in Physics in 1996 from the University of Illinois at Urbana-Champaign, he performed postdoctoral work at the University of California, Berkeley, and was one of five selected from 600 to be a Lawrence Fellow at the Lawrence Livermore National Laboratory. During his fellowship, he helped to establish their research program in nanotechnology and received both the Physics Directorate Outstanding Scientific Achievement Award and the Science and Technology Award.

Professor Grossman returned to UC Berkeley as director of a nanoscience center and head of the Computational Nanoscience research group, which he founded and which focuses on designing new materials for energy applications. He joined the MIT faculty in the summer of 2009 and leads a research group that develops and applies a wide range of theoretical and experimental techniques to understand, predict, and design novel materials with applications in energy conversion, energy storage, and clean water.

Examples of Professor Grossman's current research include the development of new, rechargeable solar thermal fuels, which convert and store the Sun's energy as a transportable fuel that releases heat on demand; the design of new membranes for water purification that are only a single atom thick and lead to substantially increased performance; three-dimensional photovoltaic panels that when optimized deliver greatly enhanced power per area footprint of land; new materials that can convert waste heat directly into electricity; greener versions of one of the oldest and still most widely used building materials in the world, cement; nanomaterials for storing hydrogen safely and at high densities; and the design of a new type of solar cell made entirely out of a single element.

As a teacher, Professor Grossman promotes collaboration across disciplines to approach subject matter from multiple scientific perspectives. At UC Berkeley, he developed two original classes: an interdisciplinary course in modeling materials and a course on the business of nanotechnology, which combined a broad mix of graduate students carrying out cutting-edge nanoscience research with business students eager to seek out exciting venture opportunities. At MIT, he developed two new energy courses, taught both undergraduate and graduate electives, and currently teaches a core undergraduate course in thermodynamics.

To further promote collaboration, Professor Grossman has developed entirely new ways to encourage idea generation and creativity in interdisciplinary science. He invented speedstorming, a method of pairwise idea generation that works similarly to a round-robin speed-dating technique. Speedstorming combines an explicit purpose, time limits, and one-on-one encounters to create a setting where boundary-spanning opportunities can be recognized, ideas can be generated at a deep level of interdisciplinary specialty, and potential collaborators can be quickly assessed. By directly comparing speedstorming to brainstorming, Professor Grossman showed that ideas from speedstorming are more technically specialized and that speedstorming participants are better able to assess the collaborative potential of others. In test after test, greater innovation is produced in a shorter amount of time.

Professor Grossman is a strong believer that scientists should teach more broadly—for example, to younger age groups, to the general public, and to teachers of varying levels from grade school to high school to community college. To this end, he has seized a number of opportunities to perform outreach activities, including appearing on television shows and podcasts; lecturing at public forums, such as the Exploratorium, the East Bay Science Cafe, and Boston's Museum of Science; developing new colloquia series with Berkeley City College; and speaking to local high school chemistry teachers about energy and nanotechnology.

The recipient of a Sloan Research Fellowship and an American Physical Society Fellowship, Professor Grossman has published more than 110 scientific papers on the topics of solar photovoltaics, thermoelectrics, hydrogen storage, solar fuels, nanotechnology, and self-assembly. He has appeared on a number of television shows and podcasts to discuss new materials for energy, including PBS's *Fred Friendly Seminars*, the *Ecopolis* program on the Discovery Channel's Science Channel, and NPR's *On Point* with Tom Ashbrook. He holds 18 current or pending U.S. patents.

Professor Grossman's previous Great Course is *Understanding the Science for Tomorrow: Myth and Reality*. ■

Table of Contents

INTRODUCTION

LECTURE GUIDES

Table of Contents

Table of Contents

Thermodynamics: Four Laws That Move the Universe

Scope:

In this course on thermodynamics, you will study a subject that connects deep and fundamental insights from the atomic scale of the world all the way to the highly applied. Thermodynamics has been pivotal for most of the technologies that have completely revolutionized the world over the past 500 years.

The word "thermodynamics" means "heat in motion," and without an understanding of heat, our ability to make science *practical* is extremely limited. Heat is, of course, everywhere, and putting heat into motion was at the core of the industrial revolution and the beginning of our modern age. The transformation of energy of all forms into, and from, heat is at the core of the energy revolution and fundamentally what makes human civilization thrive today.

Starting from explaining the meaning of temperature itself, this course will cover a massive amount of scientific knowledge. Early on, you will learn the crucial variables that allow us to describe any system and any process. Sometimes these variables will be rather intuitive, such as in the case of pressure and volume, while other times they will be perhaps not so intuitive, as in the case of entropy and the chemical potential. Either way, you will work toward understanding the meaning of these crucial variables all the way down to the scale of the atom and all the way up to our technologies, our buildings, and the planet itself. And you will not stop at simply learning about the variables on their own; rather, you will strive to make connections between them.

Once you have a solid understanding of the different variables that describe the world of thermodynamics, you will move on to learn about the laws that bind them together. The first law of thermodynamics is a statement of conservation of energy, and it covers all of the different forms of energy possible, including thermal energy. When you learn the first law, you will discover the many different ways in which energy can flow into or out of a material. By taking a look under the hood of materials and peering across vast length and time scales, you will learn that at the fundamental level, for the energy stored in the material itself, all energy flows are equivalent.

You will learn that internal energy represents a kind of energy that is stored inside of a material and that it can tell us how that material holds onto energy from heat and responds to changes in temperature. In addition, you will learn how a change in temperature represents a crucial driving force for heat to flow—a concept also known as the zeroth law of thermodynamics.

From the first law and your knowledge of the internal energy, you will understand that performing work on a system—for example, applying a pressure or initiating a reaction—is entirely equivalent to adding heat. This is one of the fundamental pillars of thermodynamics. And armed with this knowledge, you will see that there are many different ways to manipulate the properties of materials and that you can adapt your processing techniques to whatever is most convenient given the tools you have on hand.

You will learn that vast amounts of heat can be generated not just by lighting fire, but also by other means, such as the friction of two pieces of metal rubbing together. And by understanding the relationships between the basic variables pressure, temperature, volume, and mass, you will learn how heat can go the other way, converting into some form of useful work. For example, taking advantage of the fact that heat makes a gas expand, which in turn can be used to push on a piston, you will learn about how the very first engines were made.

As you explore thermodynamics beyond the driving force and corresponding response of a material, you will learn what it is that governs when and why the system stops being driven. In other words, you will learn why a system comes to equilibrium. This means that, given any set of external conditions—such as temperature, pressure, and volume—on average, the system no longer undergoes any changes. The concept of equilibrium is so important in thermodynamics that a good portion of the early lectures of the course will be dedicated to explaining it. But it is not until the lecture on entropy and the second law of thermodynamics that you discover just how to find this special equilibrium place.

As you learn about the second law, you will understand just why it is that thermodynamics is the subject that treats thermal energy on an equal footing with all of the other forms: The reason it is able to do so has to do with entropy. You will be learning a lot about entropy throughout this course because it is of absolutely fundamental importance to this subject. Entropy is the crucial link between temperature and thermal energy. It is a way to quantify how many different ways there are to distribute energy, and as you will learn, it is the foundation for the second law of thermodynamics.

By understanding entropy, you will gain an intuitive sense for the connectedness of thermal energy to all other forms of energy, and you will understand why a perpetual-motion machine can never exist. Entropy offers nothing less than the arrow of time, because it is the thermodynamic variable that provides a way to predict which direction all processes will occur. And you will learn that like temperature, entropy can have an absolute minimum value, as stated in the third law of thermodynamics.

Once you have learned these laws, you will be able to apply them to the understanding of a wide range of properties and processes—for example, how materials change their phase between gas, liquid, and solid in many different ways and sometimes even possess completely new phases, as in the case of a supercritical fluid. You will learn that when two different materials are mixed together, the phase diagram provides the ultimate materials map and can help navigate through the immense complexity of mixtures.

You will learn about heat engines and discover that while work can make heat with 100 percent efficiency, heat cannot do work with 100 percent efficiency. It's the Carnot limit that tells us that the absolute best a heat engine can ever achieve is dependent on a temperature difference.

You will learn how other types of work—including magnetism, surface tension, phase change, and entropy itself—can be used to make stuff move. And beyond these highlights, you will learn many other foundational thermodynamic concepts that are critical pillars of the science and engineering of modern times, including eutectic melting; osmotic pressure; the electrochemical cell; and that beautiful moment for a material where all three phases coexist, called the triple point; along with many, many other important phenomena.

In this course, you will learn how the four laws of thermodynamics will continue to provide the foundation of science technology into the future, for any single aspect of our lives where a material is important. ■

Thermodynamics—What's under the Hood

Lecture 1

The discipline of materials science and engineering is the connection between the science of materials and the technology those materials can influence—it's the study of what makes a material behave the way it does. What's under the hood of a given material? And once you know that, how can you engineer it to be something better or do something different? Some aspect of thermodynamics is always under that hood. By the end of this course, you will gain the knowledge of the laws of thermodynamics: four laws that move the universe.

What Is Thermodynamics?

- Thermodynamics is a subject that connects deep and fundamental insights from the atomic scale of the world all the way to the highly applied. And it is a topic that has been pivotal for the technologies that have completely revolutionized the world as we know it over the past 500 years.

Thermodynamics helps you understand what's under the hood of a given material and how you can engineer it to be something better or do something different.

- Perhaps the single most important differentiator for this subject is the prefix "thermo," which means "heat"; it is the fundamental understanding of heat and temperature and, ultimately, entropy that separates this subject from all others.

- The word "thermodynamics" means "heat in motion," and without an understanding of heat, our ability to make science practical is extremely limited. Heat is everywhere, and putting heat into motion is at the core of the industrial revolution and the beginning of our modern age. The transformation of energy of all forms into, and from, heat is at the core of the energy revolution and fundamentally what makes human civilization thrive today.

Rebuilding the World

- What if all the buildings were leveled, the cars and roads were gone, hospitals disappeared, and electricity was only found in lightning strikes? Basically, what if nothing were here? How would we rebuild our world? What would the challenges be? What tools would we need? What knowledge would we need?

- The place to start, in order to answer such questions, is to ask what the variables are in the first place. What are the "givens"? That is, what are the things that we know about, the things that we have? What are the things that we may not have but need to know about? These are the variables.

- In our effort to rebuild the world, we would ask what raw materials we could find, plus the different forms of energy we would be able to harness to manipulate those ingredients. These energies could include heat, or mechanical forces, or chemical reactions, as examples.

- Once we knew what the variables were, next we would need to know how those variables interact with one another. In other words, how do materials respond to their environment? How does one form of energy change when another form of energy is introduced? What is the interplay between energy of all forms and matter?

- The answers to these questions can be found in thermodynamics, which is the subject that holds the key to understanding how energy in all of its forms—including heat—changes the nature of the materials that make up the world.

- Without understanding these relationships, we could not rebuild our world. We would not be able to make an engine that turns heat into mechanical work; we would not be able to make electricity from spinning turbines.

- So, how would we begin? First, we would need to get a sense of what the variables are that we can control. These variables include temperature and pressure, the volume or mass of a substance, the phase of the material (which would mean working with the material as a gas, liquid, or solid), and the chemistry itself.

- Then, we would start varying the properties with respect to one another to see what happens. We would try to vary just a few variables at a time, while holding as many of the other ones fixed. This is what scientists and engineers did to lay the foundations of thermodynamics.

- We'll develop a strong sense for each of these crucial variables and learn the relationships between them. We'll build our knowledge base starting with the properties of materials when they're in a single phase, like a gas, and then we'll learn how and why materials can change phase as a function of their environment.

- But just knowing the variables is not enough to build something useful, which is why, once we have a solid understanding of the different variables of thermodynamics, we will move on to learn about the law that binds them all together. The first law of thermodynamics is a statement of conservation of energy, and it covers all of the different forms of energy possible, including thermal energy.

- At that stage of knowledge needed in our attempt to rebuild the world, we would have a fantastic grasp of the different ways in which energy can flow into or out of a material. By taking a look under the hood of materials and peering across vast length and time scales, we'll learn that at the fundamental level, for the energy stored in the material itself, all energy flows are equivalent.

- We would see that performing work, such as applying a pressure or initiating a reaction, is entirely equivalent to adding heat. This is one of the fundamental pillars of thermodynamics, and it is captured in the first law.

- And armed with that knowledge, we would learn that there are many ways to manipulate the properties of materials and that we can adapt our processing techniques to whatever is most convenient, given the tools we have on hand.

- We would also be able to experiment with the trading back and forth between heat and work. We would see that vast amounts of heat can be generated not just by lighting fire, but also by other means, including the friction of two pieces of metal rubbing together.

- And by playing with the relationships between those basic variables of pressure, temperature, volume, and mass, we'd start to figure out how to make heat go the other way and convert into some useful form.

Entropy and the Second Law of Thermodynamics

- At this point, in terms of our rebuilding process, we could do a whole lot. We'd be able to make an efficient form of concrete that pours easily and dries quickly, because we would have learned what temperatures are needed to make the calcium silicate particles of just the right composition and size.

- We would probably have figured out that the properties of steel depend very closely not just on how much carbon is added to iron, but also on how rapidly the molten metal is cooled into a solid.

- We would then be able to take advantage of that knowledge to make the different types of steels needed—whether for buildings, railroad tracks, or plumbing pipes. And we would be making some of our first combustion engines, turning the heat from fire into mechanical motion. In short, with the topics of thermodynamics described so far, we would be able to reenter the industrial revolution.

- However, it would be difficult to move past this point, because there would still remain enormous gaps, including the ability to make and control many more kinds of materials, such as plastics; the development of modern medicine; the ability to create highly tailored chemistries, and the entire information technology revolution.

- In terms of the knowledge we would need, we would have a thorough sense of the different driving forces in nature and how to control them to change the variables we care about. But the driving force and corresponding response alone do not tell the whole story. For that, we need to know what it is that governs when and why the system stops being driven.

- In other words, we need to understand why a system comes to equilibrium. This means that, given any set of external conditions, such as temperature, pressure, and volume, on average the system no longer undergoes any changes. But it is not until we learn about entropy and the second law of thermodynamics that we find the missing knowledge we need.

- The second law of thermodynamics allows us to understand and predict the equilibrium point for any set of processes involving any type of energy transfer, including thermal energy. By combining the first and second laws of thermodynamics, we can predict the equilibrium point for a system under any conditions.

- In addition, by understanding entropy, we will learn how to determine when and why a system reaches equilibrium. We will gain an intuitive sense for the connectedness of thermal energy to all other forms of energy.

- Our attempt to rebuild the world will involve a constant transformation of energy, and heat will always be in the mix during these transformations. What we will need to do with this energy very often is to create order from disorder.

- At this stage, we will understand why it takes energy to create order and why things don't sometimes just become ordered on their own. The concept of entropy answers these questions, because it provides a crucial bridge between heat and temperature.

- We will tightly connect the first and second laws to make the most use of our thermodynamic understanding as we rebuild. And by doing so, we'll be in a position to create maps of materials—maps that tell us precisely what a material will do for any given set of conditions. These maps are called phase diagrams, and they represent one of the single most important outcomes of thermodynamics.

- With phase diagrams, we unlock the behavior of a given material or mixture of materials in a manner that lets us dial in just the right processing conditions to achieve just the right behavior. With phase diagrams, we are able to make lighter machines, more efficient engines, myriad new materials, and completely new ways to convert and storage energy.

- After learning about phase diagrams, we will learn about a number of practical examples of how the knowledge of thermodynamics leads to crucial technologies. We will fill in the gaps and learn key elements that it would take to put the world back.

- Fundamentally, at the core of this subject is the knowledge of what makes a material behave the way it does under any circumstances. And all along the way, as we gain this core knowledge of thermodynamics, we will see countless examples of how it impacts our lives.

- Through it all, the underlying processes will involve some form of an energy transfer—either to, from, or within a given material, from one form to another, and to another still. Thermodynamics is the subject that governs all possible forms of energy transfer and tells us how to predict the behavior of materials as a result of these energy ebbs and flows.

Suggested Reading

Anderson, *Thermodynamics of Natural Systems*, Chap. 1.

DeHoff, *Thermodynamics in Materials Science*, Chap. 1.

Questions to Consider

1. If the modern era had to be rebuilt from scratch, what kinds of information would we need to know in order to get started? What forms of energy would you need to understand and control?

2. How many elements are in your cell phone?

Thermodynamics—What's under the Hood

Lecture 1—Transcript

Hello, and welcome to our first lecture on thermodynamics. In this lecture I hope to share with you my excitement and passion for this topic and to convey to you why the subject of thermodynamics is so crucial. I started out studying theoretical physics, and I got both my bachelors and Ph.D. degrees in that subject. Now, I'm a materials scientist and engineer, and for me, it's the connection of fundamental understanding to actual, working, relevant devices and technologies that strikes at the heart of my passion. What drives my work is the connection between the science of materials, and the technology those materials can influence, and that is what the discipline of materials science and engineering is. It is, at its core, the study of what makes a material behave the way it does. What's under the hood of a given material? And once I know that, how can I engineer it to be something better or to do something different? And I'm excited to teach you this course, because it's always some aspect of thermodynamics that is, in fact, "under that hood."

Consider the following question. You've probably heard of the periodic table of the elements; 118 of them exist officially, but there are around a hundred elements that are stable and found naturally. So, how many of those 100 elements do you think there are in your cell phone? It's quite remarkable to think about; my cell phone has 64 different elements in it. How did we do that? How did we take around $^2/_3$ of all of the known stable elements and make them perform all of the different functions that a cell phone does?

We literally dug up chunks of the earth's crust, extracted out this broad range of elements, and then processed them in a whole bunch of particular ways, and voila, I can shoot a high definition video and text it anywhere on the globe, check sports scores, the weather, or movie times, listen to millions of songs, or simply make a phone call, all because I figured out how to coax those 64 different elements into just the right positions and combinations.

But how is this actually done? It has to do with the way materials interact with each other and how they respond to processing conditions. If you've ever cooked a meal, then you know what I'm talking about. Let's think of the elements in my cell phone as the ingredients we're using to cook with. The processing conditions for those chocolate chip cookies might be a temperature of 375ºF and a time of 10 minutes. But for the brownies, it's a temperature of 300ºF and 45 minutes of cook time. And the oven is not under any extra pressure, so those are both examples of cooking at the same pressure as the rest of the kitchen. But, if you've ever used a pressure cooker, then you know that pressure is another way to change the processing conditions, and just like temperature, changing the pressure makes a big difference. So we have some initial set of raw materials—the flour, vanilla extract, eggs, sugar, and so on. And when we mix them together in just the right proportions and then process them under just the right conditions, then we can achieve a certain desired outcome.

It's the same thing with the elements and materials science and engineering. We start with our initial ingredients, which correspond to the elements of the periodic table, process them in just the right ways, and we get a material that has just the right properties for some desired application. At the heart of this example is the fundamental nature of energy and all of its forms. I put mechanical energy into the eggs when I beat them. Chemical energy is rearranged when sugar is dissolved, and heat energy is inserted when the food is baked. One of the key questions you are answering when you cook a meal is, how does energy in all of its different forms change the nature of a material? And how can we understand these energy flows so that we can control them to make materials take on certain properties or behave in a certain way?

Taking this a bit further, how can we become master chefs of not just food, but of all known elements? This, in its essence, is the topic of thermodynamics, and it is absolutely essential in science and technology. And thermodynamics is not just a foundational topic for materials science and engineering; it's also at the core of many other disciplines, like chemical, mechanical engineering, or biology, and chemistry, and physics. Each discipline may take a different view of thermodynamics, some emphasizing more how it applies to engines, while others, delving more into the theoretical, statistical nature of the subject. But in the end, they all place this topic at the top of their academic food chain. And for very good reason. As Einstein famously said about thermodynamics, "It is the only physical theory of universal content that, within the framework of applicability of its basic concepts, will never be overthrown." Now, that's a powerful statement, given that he had literally just overthrown the laws of classical physics by helping to develop the quantum theory.

Thermodynamics is a subject that connects deep and fundamental insights from the atomic scale of the world all the way to the highly applied. And it is a topic that has been pivotal for the technologies that have completely revolutionized the world as we know it over the past 500 years. And perhaps, the single most important differentiator for this subject, and the aspect that makes thermodynamics so incredibly important to my own research in energy, is that very word thermo itself. It is the fundamental understanding of heat and temperature, and ultimately entropy, that separates this subject from all others.

Thermodynamics means, literally, "heat in motion," and without an understanding of heat, our ability to make science practical is extremely limited. Heat is, of course, everywhere, and putting heat into motion? Well, that alone is at the core of the industrial revolution and the beginning of our modern age. The transformation of energy of all forms into and from heat is at the core of the energy revolution and fundamentally what makes human civilization thrive today.

So, in answer to the question of why thermodynamics is such a pivotal topic, let's pose another question. What if we had to start all over? What if the buildings were leveled, the cars and roads were gone, hospitals disappeared, and electricity was only found in lightning strikes? Basically, what if nothing were here? How would we rebuild our world? What would the challenges be? What tools would we need? What knowledge would we need?

The place to start, in order to answer such a question, is to ask what the variables are in the first place. What are the givens? That is, what are the things that we know about, the things that we have? What are the things that we may not have but need to know about? These are the variables. And as you can imagine, these would be the ingredients at our disposal, as well as what our kitchen can do. In the example of cooking a meal, it would come down to different plants and animals and different ways to process those ingredients. In our effort to rebuild the world, we make this more general. We ask what raw materials we could find, plus the different forms of energy we would be able to harness to manipulate those ingredients. These energies could include heat, or mechanical forces, or chemical reactions as examples. And once we knew what the variables were, next we'd need to know how those variables interact with one another. In other words, how do materials respond to their environment? How does one form of energy change when another form of energy is introduced? What is the interplay between energy of all forms, and matter?

The answers to these questions can be found in thermodynamics. Thermodynamics is the subject that holds the key to understanding how energy in all of its forms—including heat—changes the nature of the materials that make up the world. Without understanding these relationships, we could not rebuild our world. We would not be able to make an engine that turns heat into mechanical work. We would not be able to make electricity from spinning turbines and then use that electricity to perform useful tasks, like produce light or communicate across great distances. We would have no understanding of what ingredients to mix in what proportions and at what temperatures to make steel strong enough to build buildings and airplanes, and we would certainly not be able to make those airplanes fly.

How would we begin? Well, first, we'd need to get a sense of what the variables are that we can control. I mentioned a few already, but let's list them all together here. There would be temperature and pressure, since I could make fire to control heat and squeezing or banging on a material applies a force, or pressure, to it. I could also control the volume or mass of a substance simply by changing the amount, and I can think about controlling, at least in some cases, the phase of the material. This means working with the material as a gas, liquid, or a solid, which as you probably already know, is going to be a function of the temperature. And then there would be the chemistry itself, which goes back to those raw ingredients and how I mix different relative amounts together.

Then I'd start varying the properties with respect to one another to see what happens. I'd try to vary just a few variables at a time while holding as many of the other ones fixed. For example, I'd change the temperature and see if the volume changes, or hold the temperature fixed and add more material and see if the pressure changes, and so on. And this is exactly what scientists and engineers did to lay the foundations of thermodynamics, and it's what we'll do in the early lectures of this course. We'll develop a strong sense for each of these crucial variables, and we'll learn the relationships between them. We'll build our knowledge starting with the properties of materials when they're in a single phase, like a gas, and then, we'll learn how and why materials can change phase as a function of their environment. And since heat is one the single most important kinds of energy, we'll begin our course in the very next lecture with the way in which thermal energy is measured—that would be temperature.

But just knowing the variables is not enough to build something useful. It would be like knowing all of the ingredients in a chocolate chip cookie recipe but not knowing how much of each one to use or how to cook them. And even more importantly, we'd have to figure out how one cooking process is related to another. Would I achieve the same results by baking in an oven, frying on a pan, or boiling in water? That's why, once we have a solid understanding of the different variables of thermodynamics, we'll move on to learn about the laws that bind them all together.

The first law of thermodynamics is a statement of conservation of energy, and it covers all of the different forms of energy possible, including thermal energy. At that stage of our knowledge, needed in our attempt to rebuild the world, we would have a fantastic grasp of the different ways in which energy can flow into or out of a material. By taking a look under the hood of materials and peering across vast length and time scales, we'll learn that at the fundamental level, for the energy stored in the material itself, all energy flows are equivalent. We'd see that performing work, like say, applying a pressure or initiating a reaction, is entirely equivalent to adding heat. This is one of the fundamental pillars of thermodynamics, and it's captured in the first law.

And armed with that knowledge, we would learn that there are many ways to manipulate the properties of materials, and that we can adapt our processing techniques to whatever is most convenient given the tools we have on hand. We would also be able to experiment with the trading back and forth between heat and work. We'd see that vast amounts of heat can be generated not just by lighting fire, but also by other means, like the friction of two pieces of metal rubbing together. As chefs, we're learning that there are more ways to process our ingredients than we might have originally thought. And by playing with the relationships between those basic variables—pressure, temperature, volume, and mass, we'd start to figure out how to make heat go the other way and convert into some form of useful work. For example, taking advantage of the fact that heat makes a gas expand, which in turn can be used to push on a piston, we'd make our very first engines.

Now, at this point, in terms of our rebuilding process, we could do a whole lot. We'd be able to make an efficient form of concrete that pours easily and dries quickly, since we'd have learned what temperatures are needed to make the calcium silicate particles of just the right composition and size. We'd probably have figured out that the properties of steel depend very closely not just on how much carbon is added to iron, but also on how rapidly the molten metal is cooled into a solid. We'd then be able to take advantage of that knowledge to make the different types of steels needed, whether for buildings, railroad tracks, or plumbing pipes. And we'd be making some of our very first combustion engines, turning the heat from fire into mechanical motion. In short, with the topics of thermodynamics I have described to you so far, we'd be able to reenter the industrial revolution.

And yet, it would be hard to move past this point. Yes, we've put back some of the basic building blocks of the world, but there would still remain enormous gaps, like the ability to make and control many more kinds of materials, like plastics, or the development of modern medicine, or the ability to create highly tailored chemistries, or for that matter, the entire information-technology revolution, to name just a few examples.

You see, in terms of the knowledge we'd need, we would have a thorough sense of the different driving forces in nature and how to control them to change the variables we care about—put energy into a system and that system will respond. This is the nature of a thermodynamic driving force; it's a way to push or pull on a material with any type of energy. Apply a pressure, and see the volume change. Increase the temperature, and watch a material melt. A force is applied, and a response is observed.

But the driving force and corresponding response alone do not tell the whole story. For that, we need something else. We need to know what it is that governs when and why the system stops being driven. In other words, we need to understand why a system comes to equilibrium. This means that given any set of external conditions, like temperature, pressure, and volume, on average, the system no longer undergoes any changes. This concept of equilibrium is so important in thermodynamics that I'll spend a good amount of time discussing it in the early lectures of the course. But it is not until our lecture on entropy and the second law of thermodynamics that we find the missing knowledge I just described.

Consider an analogy of a ball rolling down a track. If the track points down, then we know the ball will experience a driving force—due to gravity, in this case—to go down the track. So, that's the driving force and the response. But we still have the crucial question of where the ball will come to rest. Of course, we have an intuitive sense here that it will stop where there is some sort of minimum point in the track. At that point, the ball might go back and forth for a while, but if there's friction so it loses a little energy as it goes, eventually it will come to rest there—That's the equilibrium point.

The second law of thermodynamics allows us to understand and predict this minimum point for any set of processes involving any type of energy transfer, including thermal energy. By combining the first and second laws of thermodynamics, we can predict the equilibrium point for a system under any conditions. That's pretty powerful stuff! And the thing is that, at least at first, it's not as intuitive as the simple ball-rolling-on-a-track example. That's because there are a whole lot more possible forms of energy that may be present than just gravitational energy. For example, there's the mechanical energy, which I already mentioned, and is due to pressure times a change in volume. There's chemical energy, which can lead to changes in the material composition itself, and many other forms, like energy due to magnetism, surface tension, and electric fields, to name a few. So the track of thermodynamics can get pretty complicated.

But there's one particular kind of energy that leads to the biggest departure from an intuitive picture, and that would be thermal energy. Thermodynamics is the subject that treats this energy on an equal footing with all of the other terms. And the reason it's able to do so has to do with another variable, called entropy. We'll be talking a lot about entropy throughout this course, because it's of absolutely fundamental importance to our subject. Entropy gives us the crucial link between temperature and thermal energy. It's a way to quantify how many different ways there are to distribute energy, and as we'll see, it's the foundation for the second law of thermodynamics.

By understanding entropy, we will learn how to determine when and why a system reaches equilibrium. We'll gain an intuitive sense for the connectedness of thermal energy to all other forms of energy, and we'll understand why a perpetual motion machine can never exist. Entropy gives us nothing less than the arrow of time, since it's the thermodynamic variable that provides a means to predict which direction all processes will occur. It tells us how to find where the minimum of the track is, not just for a ball and gravity, but for any set of materials and any kind of energy terms.

So, armed with entropy and the second law, we will have filled the knowledge gap we needed to tackle the next step in our effort to rebuild the world. We'll be able to gain a deeper understanding of how materials both hold onto thermal energy, as well as how they respond to it, something that we'll discuss in our lectures on heat capacity and thermal expansion. Our attempt to rebuild the world will involve a constant transformation of energy, and heat will always be in the mix during these transformations. And what we will need to do with this energy very often is to create order from disorder. We'll make a car move by blowing up little carbonaceous molecules in an engine; we'll make chemicals pure from their raw, mixed-up natural sources; we'll make computer chips out of sand; and turn limestone into liquid to make roads and houses. Our bodies themselves are continuously taking energy from food to maintain ordered arrangements of trillions of cells, and we would begin to understand these processes more deeply.

At this stage, we'll understand why it takes energy to create order and why things don't just sometimes become ordered on their own. The concept of entropy answers these questions, as it provides a crucial bridge between heat and temperature. As we'll see, unlike energy, entropy is not conserved, and yet, it does play a key role in energy conservation, since entropy allows us to bring heat into our fundamental equation for energy. We will tightly connect the first and second laws together to make the most use of our thermodynamic understanding as we rebuild.

And by doing so, we'll be in a position to create maps of materials, maps that tell us precisely what a material will do for any given set of conditions. These maps are called phase diagrams, and they represent one of the single most important outcomes of thermodynamics. They can be constructed from the knowledge of the first and second laws for any material. With phase diagrams, we unlock the behavior of a given material, or mixture of materials, in a manner that lets us dial in just the right processing conditions to achieve just the right behavior.

With this type of control, we can begin making again the materials needed for electronics. We can dictate precisely when reactions will happen, and in which direction, allowing us to make many more materials, as well as control over chemistry accurate enough to lead to modern biotechnology and medicine. With phase diagrams, we are able to make lighter machines, more efficient engines, myriad new materials, and completely new ways to convert and store energy itself.

Because of how important they are, we'll spend five whole lectures to fully cover phase diagrams. And after that, we'll talk about a number of practical examples of how the knowledge of thermodynamics leads to crucial technologies. We'll fill in the gaps and learn key elements that it would take to put the world back. Fundamentally, at the core of this subject, is the knowledge of what makes a material behave the way it does under any circumstances. And all along the way, as we gain this core knowledge of thermodynamics, we'll see countless examples of how it impacts our lives. Back to the analogy of cooking in the kitchen, we'll see why you might burn your mouth on the cheese, rather than the crust of a piece of pizza—that's the heat capacity, and it tells us how and why materials can hold on to different amounts of heat.

When we learn about phase transitions and latent heat, we'll see how much energy it takes to make a material transition from one phase, like a solid, into another, like a liquid. We'll see why ice floats, rather than sinks, and how pressure plays a role in determining what phase a material takes on. We'll learn why it's possible to both boil and freeze a liquid at exactly the same time, and we'll learn what governs the behavior of mixtures of materials. We'll learn about the limits of making heat do mechanical work, like in a car engine, and we'll also explore a number of other ways that materials can be used to make things move. We'll learn about electrochemistry and the basic principles of a battery, and we'll put to rest, once and for all, the question of whether a potato can really power a clock.

And through it all, the underlying processes will involve some form of an energy transfer, either to, from, or within a given material—from one form to another, and to another still. Thermodynamics is the subject that governs all possible forms of energy transfer and tells us how to predict the behavior of materials as a result of these energy ebbs and flows.

Thermodynamics is crucially important to our understanding of the world, and it lies at the core of our understanding of materials. And yet, it's perhaps the least intuitive introductory subject of them all. When we see a ball bounce, or not, we immediately "feel" the physics concepts of gravity, or energy and momentum. But when we add a dash of salt into our soup, does the increase in the salt's configurational entropy, which is, after all, what makes it dissolve, does that come to us with the same intuition? The answer is, in my experience, that the same basic intuition is missing from thermodynamics.

A number of textbooks, and now many websites, do a fantastic job at teaching the key concepts, like entropy, temperature, heat, phases, the conversion of energy from one form into another, and the seemingly endless permutations of fundamental equations and possible thermodynamic work terms, and of course, the four laws that form the basis of the theory of thermodynamics—Those are the ones Einstein said "will never be overturned." But you see, all these concepts often become stuck in many students' minds as either only sets of mathematical operations, or else, ideas that only apply to highly contrived examples, as opposed to real-life problems. The concepts of thermodynamics badly need to be brought to life. I want to bring thermodynamics to that same level as other subjects where intuition is more natural, to get you to feel that same kind of intuition for the key concepts, which at first can appear quite a bit more elusive—Entropy, I'm talking to you.

So, I'm going to be sure to get across the excitement of real problems that we can interact with and see that relate the key concepts directly to experiments. To accomplish this, this course will be demo-driven, meaning that in nearly every lecture, at some point or another I will take a break, put on my lab coat and goggles, and go and smash, break, light on fire, or otherwise convert energy from one form into another.

I'll pour water into a bowl that looks just like regular water but becomes ice as soon as it hits the bowl. I'll light a hundred dollar bill on fire without it charring a single bit. I'll break the bottom of a glass bottle cleanly off by striking it at the top, and I'll boil water by cooling it, instead of heating it. I'll show you what is perhaps the world's simplest motor involving a battery, a magnet, and a single wire. And I'll give many other examples of mechanical work. We'll light a piece of cotton on fire using a hammer and the laws of the ideal gas, and I'll show you how to make a balloon not pop, even when you light it on fire. I'll demonstrate that matter can exist in all three phases at the same time—gas, liquid, and solid. We'll run all sorts of engines, from ones you may know about already, that run on heat, to ones that are less common, like those that run on surface tension or none other than entropy itself. And of course, this list wouldn't be complete if I didn't tell you that I'll also use a very simple chemical reaction to create a massive 10-foot-high eruption.

Now, I don't plan to make this a course with a whole lot of derivations, but I won't shy away from the math either. To learn thermodynamics means learning some of the basic math as needed, so that you can go back and forth between the key concepts and the variables and relationships that represent them. It would be a disservice to the topic if I didn't get into some of the math in order to show you how the important relationships between thermodynamics can be quantified. That is, after all, what engineers would need to do to rebuild the world. But I will always strive for a balance, never getting too far away from the key conceptual pictures. I'll be giving many analogies and many comparisons along the way, and as much as possible, I'll ground our discussions in real problems.

By the end of this course, you'll be a master chef of materials who possesses not just the recipes, but a deep understanding of why a recipe gives the results it does. You may not be able to combine 64 elements to make a cell phone, but you'll have the knowledge that it takes to see how such materials could be processed in just the right ways to do so. You'll learn about the most important variables and how they depend on one another. And you'll gain the knowledge of the laws of thermodynamics—four laws that govern the behavior of these variables. These are the four laws that move the universe. So, what are we waiting for? Let's get started!

Variables and the Flow of Energy

Lecture 2

Thermodynamics is not always an intuitive subject. This course is going to make the concepts of thermodynamics come alive and teach you how they relate to essentially everything you do. In this lecture, you will ease your way into some of the important history of thermodynamics, and you will learn some underlying concepts. By the end of this lecture, you will have a good grasp of some of the key historical milestones of thermodynamics and the difference between a macroscopic and microscopic view of the world.

The Rise of Thermodynamics

- It is interesting to note that the time periods of human history correspond to a given material, from the Stone Age to the Bronze Age to the Iron Age and beyond. In fact, humanity's ability to thrive throughout history has often relied on its material of choice and, importantly, the properties that this material possesses—or that humans figured out how to control.

- In most of human history, it took centuries to master a particular material. For example, over a whole lot of time and trial and error, people realized that iron could be made to be many things—weak, strong, ductile, or brittle—depending on where the material is in something called its phase diagram, which is a kind of map. In thermodynamics, a phase diagram provides the key knowledge to predict what material you've made and what properties it will have once you've made it.

- For centuries, even millennia, engineers recognized that properties of materials were determined by their nature, or the composition of the material and what it looks like. These engineers saw that the nature of a material could be intentionally modified or controlled by processing it in an appropriate way. For example, different materials melted at different temperatures, and mixtures of materials would have different, sometimes surprising, melting points.

- But for all of these observations about the ways materials can be engineered by processing them in different ways, the rules that were derived over thousands of years were entirely empirical.

- It took a long time after the Iron Age for this to change, although in the 1600s, a great advance was made by the German scientist Otto von Guericke, who built and designed the world's first vacuum pump in 1650. He created the world's first large-scale vacuum in a device he called the Magdeburg hemispheres, named after the town for which von Guericke was the mayor.

- The Magdeburg hemispheres that von Guericke designed were meant to demonstrate the power of the vacuum pump that he had invented and the power of the vacuum itself. One of the two hemispheres had a tube connection to it with a valve so that when the air was sucked out from inside the hemispheres, the valve could be closed, the hose from the pump detached, and the two hemispheres were held together by the air pressure of the surrounding atmosphere.

- One of the misunderstandings that Guericke helped to clarify is that it's not the vacuum itself that has "power" but, rather, the difference in pressure across the boundary creates the power of the vacuum. Guericke's pump was a true turning point in the history of thermodynamics.

- After learning about Guericke's pump, just a few years later British physicist and chemist Robert Boyle worked with Robert Hooke to design and build an improved air pump. Using this pump and carrying out various experiments, they observed the correlation between pressure, temperature, and volume.

- They formulated what is called Boyle's law, which states that the volume of a body of an ideal gas is inversely proportional to its pressure. Soon after, the ideal gas law was formulated, which relates these properties to one another.

- After the work of Boyle and Hooke, in 1679 an associate of Boyle's named Denis Papin built a closed vessel with a tightly fitting lid that confined steam until a large pressure was generated inside the vessel. Later designs implemented a steam-release valve to keep the machine from exploding.

- By watching the valve rhythmically move up and down, Papin conceived of the idea of a piston and cylinder engine, and based on his designs, Thomas Savery built the very first engine 20 years later.

- This, in turn, attracted many of the leading scientists of the time to think about this work, one of them being Sadi Carnot, who many consider the father of thermodynamics. In 1824 Carnot published a paper titled "Reflections on the Motive Power of Fire," a discourse on heat, power, and engine efficiency. This marks the start of thermodynamics as a modern science.

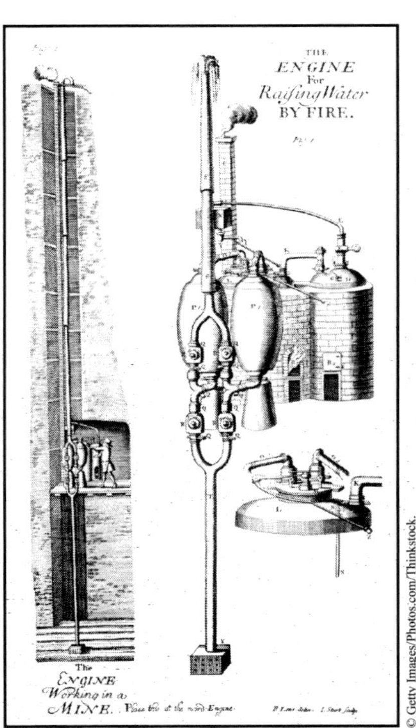

Thomas Savery built the first steam engine in the late 1600s.

- In 1876, Josiah Willard Gibbs published a paper called "On the Equilibrium of Heterogeneous Substances." Gibbs and his successors further developed the thermodynamics of materials, revolutionized materials engineering, and turned it into materials science.

- A crucial development was in Gibbs's derivation of what we now call the fundamental equation that governs the properties of a material, as well as in his demonstration that this fundamental equation could be written in many different forms to define useful thermodynamic potentials. These potentials define the way energy, structure, and even information can be turned from one form into another.

- Gibbs's formulation for the conditions of equilibrium—that is, when energy flows find a stability—was an absolutely critical turning point that enabled scientists to determine when changes can occur in a material and when they cannot.

Thermodynamics Today

- Scientists and engineers across all disciplines are continually working to control the structure of materials and synthesize them with properties that provide optimum performance in every type of application. They want to identify the relationship among the structure, properties, and performance of the material they're working with.

- Thermodynamics addresses the question that lies at the heart of these relationships—namely, how a material changes in response to its environment. The response of materials to the environment determines how we can synthesize and fabricate materials, build devices from materials, and how those devices will operate in a given application.

- If we change the processing conditions, then the structure of the material may change. And if the structure of the material changes, then its properties change. A change in the properties of a given material will in turn change the performance of a device that is made from this material.

- The variety of ways in which materials can be changed or affected are called thermodynamic forces. These can be mechanical, chemical, electrical, magnetic, or many other kinds of forces that have the ability to act on the material. All of these forces are what push or pull on the material in different ways.

- Thermodynamics addresses the question of how a material changes in response to some condition or some force that is acting on it. And controlling the response of a material to its environment has formed the basis for many, many technologies.

- Quite broadly speaking, thermodynamics is a framework for describing what processes can occur spontaneously in nature. It provides a formalism to understand how materials respond to all types of forces in their environment, including some forces you may have not thought about or even recognized as a "force."

Classical versus Statistical Thermodynamics

- There are two different points of view of thermodynamics that have been developed over the past several hundred years. First, there is classical thermodynamics, which is the focus of this course and provides the theoretical framework to understand and predict how materials will tend to change in response to forces of many types on a macroscopic level, which means that we are only interested in the average behavior of a material at large length and time scales.

- The second point of view takes the opposite approach—namely, the calculation of thermodynamic properties starting from molecular models of materials. This second viewpoint takes a microscopic view of the subject, meaning that all thermodynamic phenomena are reduced to the motion of atoms and molecules.

- The challenge in taking this approach is that it involves such a large number of particles that a detailed description of their behavior is not possible. But that problem can be solved by applying statistical mechanics, which makes it possible to simplify the microscopic view and derive relations for the average properties of materials. That's why this microscopic approach is often called statistical thermodynamics.

- In classical thermodynamics, the fundamental laws are assumed as postulates based on experimental evidence. Conclusions are drawn without entering into the microscopic mechanisms of the phenomena. The advantage of this approach is that we are freed from the simplifying assumptions that must be made in statistical thermodynamics, in which we end up having to make huge approximations in order to deal with so many atoms and molecules.

- Statistical thermodynamics takes almost unimaginable complexity and begins to simplify it by making many assumptions. This type of approximation is difficult and often only works for one type of problem but then not for another, different type of problem.

- Taking instead the macroscopic view, we simply use our equations of thermodynamics to describe the interdependence of all of the variables in the system. We do not derive these equations and relationships from any models or microscopic picture; rather, they are stated as laws, and we go from there. The variables are related to one another, and the flow of energy is described without having to know anything about the detailed atomic arrangements and motion.

Suggested Reading

Callister and Rethwisch, *Materials Science and Engineering*, Chap. 1.

DeHoff, Thermodynamics in Materials Science, Chap. 1.

Questions to Consider

1. Think of an example of a technology you use, and place it in the context of the property-structure-performance-characterization tetrahedron. How do you think thermodynamics plays a role in each of the correlations?

2. Who are some of the most important "founders" of thermodynamics, and what were their contributions?

3. What experiments were performed to test whether nature truly "abhors a vacuum"?

Variables and the Flow of Energy

Lecture 2—Transcript

Welcome to our second lecture on thermodynamics. In this lecture, we'll ease our way into some of the important history of thermodynamics and discuss some underlying concepts, which will help set the context for the course. By the end of this lecture I'd like you to have a good grasp of the following concepts: first, the structure-property-performance triangle; second, the difference between a macroscopic and microscopic view of the world; and third, some of the key historical milestones of thermodynamics. Now, most of us are familiar with at least some of the various archeological ages of human civilization—from the Stone Age, to the Bronze Age, to the Iron Age, to the age of plastics, and recently, what many consider to be the age of silicon.

Notice that the way we like to define the time periods of human history correspond to a given material. In fact, humanity's ability to thrive throughout history has often relied upon its material of choice and importantly, the properties that this material possesses or, I should say, that humans figured out how to control.

And in most of human history, it took centuries to master a particular material. For example, the iron that appeared just at the end of the Bronze Age was quite inferior in its salient mechanical properties to bronze. In fact, over a whole lot of time and trial and error, people realized that iron could be made to be many things; it could be weak, strong, ductile, or brittle; it all depends on where you are in something called the phase diagram of the material. This phase diagram is a kind of map. And much like any map, once we learn how to read it, it unlocks important knowledge.

In thermodynamics, a phase diagram provides the key knowledge to predict what material you've made and what properties it will have once you've made it. Let's take a look at how our ability to create these kinds of materials maps—these phase diagrams—came about. You see, for centuries, even millennia, engineers recognized that properties of materials were determined by their nature. What I mean by nature here is the composition of the material and what it looks like—what's in it, how do those things that are in it come together, and what structure do they ultimately form.

Now, these engineers saw that the nature of a material could be intentionally modified or controlled by processing it in an appropriate way. If they heated water, for example, they could repeatedly make it boil. If they banged on a piece of metal, they saw that its mechanical properties changed and could in some cases get stronger the more they deformed it. Different materials melted at different temperatures. And mixtures of materials would have different, sometimes surprising, melting points.

But for all of these wonderful observations about the way materials can be engineered by processing them in different ways, the rules that were derived over thousands of years were entirely empirical. In a way, a materials engineer resembled more a master chef than a modern technologist. Recipes were passed on from engineer to engineer, from generation to generation, to be memorized rather than understood.

It took a long time after the Iron Age for this to change. Although still more than 400 years ago, way back in the 1600s, a great advance was made by the German scientist Otto von Guericke. He built and designed the world's first vacuum pump in 1650 and created the world's first large-scale vacuum in a device he called the Magdeburg hemispheres, named after the town for which Otto von Guericke was the mayor.

Now, why would someone want to build a vacuum pump? Well, for literally thousands of years, since the time when Aristotle famously claimed that "Nature abhors a vacuum," people had been trying to create a vacuum to see if that statement was really true. Guericke's goal was to disprove Aristotle's long-held supposition. The Magdeburg hemispheres that Otto von Guericke designed were meant to demonstrate the power of the vacuum pump that he had invented, and more to the point, the power of the vacuum itself.

One of the two hemispheres had a tube connected to it with a valve so that when the air was sucked out from inside of the hemispheres, the valve could be closed, the hose from the pump detached, and the two hemispheres were held together by the air pressure of the surrounding atmosphere.

In one demonstration, back in the mid-1600s, Guericke used 30 horses, 15 of them attached to each hemisphere, and trying as hard as they could to run in opposite directions. No matter how hard the horses pulled, they could not pull the pieces apart after the vacuum was set up. This was quite a demonstration of the power of the vacuum!

Now, that phrase itself was one of the misunderstandings that Guericke helped to clarify. That is, it's not the vacuum itself that has and power, but rather, the difference in pressure across the boundary which creates the power of the vacuum. What we are really witnessing in the hemispheres experiment is the power of one atmosphere of pressure, which is pushing on the two hemispheres to keep them together. This pushing force is due to the pressure difference between inside and outside of the sphere.

Let's take a look, a first-hand look, at how powerful indeed the creation of a vacuum can be. We're not going to be bringing horses into the studio here, but I've got a smaller, horse-free demonstration of the power of the vacuum that I'd like to show you now.

Okay, so, as I promised, there are not going to be any horses in this demo for me to show you the power of the vacuum. Instead, what I'm going to do is a pretty cool trick that almost seems impossible, and all I need is an empty bottle and some water.

Now, here I have three bottles. Two of them are filled with water; one of them is empty. You know, I could do this trick just with my bare hand, but to make it a little bit more comfortable, I'm going to use a rubber mallet. So first, let's start with the empty bottle. If I strike this from the top with a mallet, you can see that nothing happens at all. Right? Nothing's happening. But, now, I'm going to take the same kind of bottle, exactly the same, but now it's half-way filled with water. I'm going to do the same thing; I'm going to strike the top with the mallet. I'm going to do it over here, over a bucket. Okay? So in this case, I've got the same bottle, but it's half-filled with water.

There, what you saw, is that the bottom broke out when I hit it in the same way. What's going on? Well, one thing you might think is that, when I hit the top, I kind of covered the top perfectly and created a sort of pressure wave inside of the glass. To test that idea, let's use a different bottle that has a wider opening at the top, and what I'm going to do now is I'm going to be sure not to hit it in the middle, but I'm going to hit it on the side. So let's go over to the side here. And now, I'm going to hit the side of the top, so there's no way that a pressure wave could form inside of this experiment.

And there you see, we get the same effect. So, what is going on? Even though in this case it was open to the air when I hit it, the bottom still broke out. Well, it's because of a vacuum. What happens is that when I hit the top of the bottle, for just a slight millisecond, the glass moves down before the water inside had a chance to move with it. So for that brief instant in time, a vacuum is formed at the bottom.

Now, at the top, we have just open air, so it's one atmosphere. Below, we have almost zero atmospheres, because we have that vacuum that formed, and that is a huge pressure difference. That pressure difference means that you have a whole lot of force being exerted on the water, and the water accelerates like a bullet into the bottom of the glass. So that is really showing the power of the vacuum. And I hope that you know from the lecture that it's not the vacuum that has the power, it's the pressure difference. So what we really saw here is the power of the pressure of one atmosphere.

Now, finally, if you do this at home, please do be careful, since it involves broken glass. But, anyway, the most important point is that it's the pressure differences that we saw, and that these were a crucial part of the history and the foundation of thermodynamics.

So there you saw the power of the vacuum. Or, as I've mentioned, really, the power of a pressure differential in action. Guericke's pump was a true turning point in the history of thermodynamics. After learning about Guericke's pump just a few years later, British physicist and chemist Robert Boyle worked with Robert Hooke to design and build an improved air pump. Using this pump and carrying out various experiments, they observed the correlation between pressure, temperature, and volume. They formulated what is called Boyle's law, which states that the volume of a body of an ideal gas is inversely proportional to its pressure. Soon after, the ideal gas law was formulated, which relates these properties to one another. We'll come back to these relationships in another lecture.

After the work of Boyle and Hooke, in 1679, an associate of Boyle's named Denis Papin, built a closed vessel with a tightly fitting lid that confined steam until a large pressure was generated inside the vessel. Later designs implemented a steam release valve to keep the machine from exploding. And by watching the valve rhythmically move up and down, Papin conceived of the idea of a piston and a cylinder engine, and based on his designs, twenty years later, Thomas Savery built the very first engine. This in turn attracted many of the leading scientists of the time to think about this work, one of them being Sadi Carnot, whom many consider the "father of thermodynamics." In 1824 Carnot published a paper titled, "Reflections on the Motive Power of Fire," a discourse on heat, power, and engine efficiency. And this really marks the start of thermodynamics as a modern science.

And 50 years later, along came Josiah Willard Gibbs, who in 1876 published a paper called, "On the Equilibrium of Heterogeneous Substances." Gibbs and his successors further developed the thermodynamics of materials, revolutionized materials engineering, and turned it into materials science. Recipes were replaced by full understanding and theory—actually laws—which were found to govern the behavior of materials. A crucial development was in the Gibbs derivation of what we call today the fundamental equation that governs the properties of a material, as well as in his demonstration that this fundamental equation could be written in many different forms to define useful thermodynamic potentials. And these potentials lie at the heart of what we will learn in this course, as they define nothing less than the way energy, structure, and even information can be turned from one form into another.

Gibbs' formulation for the conditions of equilibrium, that is, when energy flows find a stability, that was an absolutely critical turning point that enabled scientists to determine when changes can occur in a material and when they cannot. Many years later, another scientist you may have heard of, Albert Einstein, weighed in on thermodynamics, and famously said, "[Thermodynamics] is the only physical theory of universal content which, within the framework of the applicability of its basic concepts, I am convinced will never be overthrown." A pretty powerful statement, especially considering it comes from one of the greatest scientific minds of all time.

So what is it that thermodynamics does for us today? What impact does it have on our lives? Scientists and engineers across all disciplines are continually working to control the structure of materials and synthesize them with properties that provide optimum performance in every type of application. They want to identify the relationship between the structure, properties, and performance of the material they're working with. Thermodynamics addresses the question that lies at the heart of these relationships, namely, how a material changes in response to its environment. The response of materials to the environment determines how we can synthesize and fabricate materials, build devices from materials, and how those devices will operate in a given application.

Armed with knowledge of thermodynamics, we can explain many phenomena in the natural world. With the thermodynamics we'll be learning in this course, we will answer specific questions such as the following:

How do I predict a material's behavior?

What sets the boundaries for the behavior of a material?

What processes make a material unstable?

How is heat related to mechanical, chemical, electrical, magnetic, and other types of work?

How do I predict whether and how a material will change in time?

What are the limitations of a material?

And, what is the maximum response of a material to a given force?

So the answer to the question, "How does thermodynamics impact our lives?" is simply, "in any and every instance in which we have needed to control the properties of a material or process." That covers quite a bit!

So to sum this up graphically, we can visualize a tetrahedron with four corners that we'll label property, performance, structure, and processing. Because in the shape of a tetrahedron each of these labels has a relationship to each other, it helps us see how they are interconnected. If we change the processing conditions, say by increasing the temperature, then the structure of the material may change. And if the structure of the material changes, then its properties change; it could be the electric properties, or optical properties, or simply how stable the materials are and under what conditions they stay stable or not.

A change in the properties of a given material will, in turn, change the performance of a device that is made from this material. For example, if I make silicon using lower temperature processing conditions, then its structure will contain a higher proportion of impurities, which in turn, degrade the ability of the material to transport electricity. This leads to a lower efficiency solar cell. But in every type of application, at least every single one in which a material is present, the optimization and understanding of this tetrahedron is critical.

Now, in the example I just gave, I referred to temperature as a way to change the processing conditions of the material. But beyond temperature, there are many, many different ways in which a material can be made to undergo a change. The variety of ways in which a material can be changed or affected are what we call thermodynamic forces. These can be mechanical forces, chemical, electrical, magnetic, or many other kinds of forces that have the ability to act on a material. All of these forces are what push or pull on the material in different ways.

Thermodynamics addresses the question of how a material changes in response to some condition or some force that is acting upon it. It tells us how to navigate through that structure-property-processing-performance tetrahedron. And, as I already mentioned, controlling the response of a material to its environment has formed the basis for many, many technologies.

Here are a few examples of specific thermodynamic driving forces and the technologies they have been used to create. With the driving force of temperature, we've been able to make internal combustion engines; with the driving force of the electrostatic potential, we make batteries; with the driving force of mechanical stress, we make, essentially, any and all materials that are made for load-bearing applications; with the driving force of concentration gradients, we make dialysis machines; with the driving force of chemical reactions, we make what are called piezoelectric materials; with the driving force of surfaces, surface forces, we create composite materials; and with magnetic fields as the driving force, we've made disk drives. And the list goes on, and on, and on!

Quite broadly speaking, thermodynamics is a framework for describing what processes can occur spontaneously in nature. It provides a formalism to understand how materials respond to all types of forces in their environment, including some forces you may have not thought about or even recognized as a force.

There are two different points of view of thermodynamics that have been developed over the past several hundred years. First, there is classical thermodynamics, which is the focus of this course, and it provides the theoretical framework to understand and predict how materials will tend to change in response to forces of many types on a macroscopic level. That's the important word here—macroscopic. This means that we are only interested in the average behavior of a material at large length and time scales.

The second point of view takes the opposite approach, namely, the calculation of thermodynamic properties starting from molecular models of materials. This second viewpoint takes a microscopic view of the subject, meaning that all thermodynamic phenomena are reduced to the motion of atoms and molecules. The challenge in taking this approach is that it involves such a large number of particles that a detailed description of their behavior is simply not possible. But that problem can be solved by applying statistical mechanics to the problem, which makes it possible to simplify the microscopic view and derive relations for the average properties of materials. That's why this microscopic approach is often called statistical thermodynamics.

In classical thermodynamics, the fundamental laws are assumed as postulates, based on experimental evidence. Conclusions are drawn without entering into the microscopic mechanisms of the phenomena. The advantage of this approach is that we are freed from the simplifying assumptions that must be made in statistical thermodynamics, in which we wind up having to make huge approximations in order to deal with so many atoms and molecules.

Here, I think it's helpful to show a simple example that illustrates these two different types of thermodynamics—classical and statistical. We'll use a piston in this example, and by the way, a piston is perhaps the most classic example used for thermodynamics. It allows us to talk about the relationship between those very same properties that Guericke, Boyle, Hooke, Papin, and Carnot were all trying to understand nearly 500 years ago, namely, the interdependence of pressure, volume, and temperature.

So in this piston, we have the usual chamber filled with gas and the piston, which can move in to compress the gas, or move out to uncompress it. But in this case, I've also attached a spring around the handle, so that when we compress the piston, we also compress the spring, which wants to oppose the force pushing in. If I pull the handle out, the spring will tend to pull it back in.

Now, let's assume that this whole apparatus is packed in some type of thermal insulation. So, no heat can get into the system, and no heat can leave it. If the piston is displaced from its original rest position to some new position and then released, you can see in this animation that it will oscillate. After a bunch of oscillations, the piston will finally come to rest in a new position. And here is the key point, trying to analyze the motion of the piston in terms of classical Newtonian mechanics, that is, force equals mass times acceleration, will fail. The acceleration of the piston, and thus, the force, is not a unique function of its position and velocity. So what's going on here?

As you may have guessed, a measure of the temperature in the gas would show that its temperature rose during the process. The mechanical motion of the piston gave some energy to the gas molecules as it pushed back and forth on them, and this trading of energy led to an increase in the gas temperature. But when the gas is at a different temperature, as Boyle would have attested, its volume will change. It will, in fact, expand, and that expansion pushes on the piston, and that is the reason why the piston comes to a new resting place, one that is different than its starting point.

Now, we could eliminate the gas completely from this picture, and then we would simply have a piston oscillating on a spring, which is a standard problem in classical physics. In that case, the piston would come to a stop at the same place where it started, and we'd be exploring the exchange of the potential energy in the spring with the kinetic energy of the piston. But with the gas there, and in particular, remember, we said that the gas could not give any of its heat away, it will heat up, and the picture is quite different. The piston comes to a new resting place.

So, what would happen then if we took away that thermal insulation? In that case, the piston is allowed to exchange thermal energy with its environment, and let's also suppose that we immerse the whole piston into some sort of an ice bath, so that it would very quickly have its heat removed. In that case, the energy that was going from the motion of the piston into the heating of the gas, that energy would now leave the gas and flow out in the form of heat into the ice bath. So in that case, even with the gas present, the piston would return to its original starting position, since the temperature of the gas would remain constant and there would be no volume expansion.

So now we come to the whole point of my going through this piston example. And it comes down to this simple question, could microscopic physics be used to solve this problem when the gas is present and expanding with temperature? To do so, we'd have to track all of the forces on each and every gas molecule as they collide with the container walls, the piston, and one another. If we could do this, then have all of the forces in the system as a function of time, which would give us all the positions and velocities and should, in principle, describe the system.

However, if we have a normal-sized piston, say, one we can hold in our hands, then this amount of gas means that we'll have about 10^{23} particles—That's a 10 with 23 zeros after it. So that's the problem and challenge with a microscopic view of thermodynamics, namely, that it's simply not possible to solve for.

What statistical thermodynamics does is to take this almost unimaginable complexity and begin to simplify it by making many assumptions. For example, one could attempt to use the laws of Newtonian physics, like force equals mass times acceleration, but now include temperature as a kind of new variable that determines force. But this type of approximation is difficult and often only works for one type of problem but then not for another, different type of problem.

Taking instead the macroscopic view, we simply use our equations of thermodynamics, ones we'll learn throughout this course, to describe the interdependence of all of the variables in the system. We do not derive these equations and relationships from any models or microscopic picture; rather, they're stated as laws, and we go from there. The variables are related to one another, and the flow of energy is described without having to know anything about the detailed atomic arrangements and motion.

As I already mentioned, most of this class will be devoted to this macroscopic view of thermodynamics, and that is, therefore, where we shall begin. In the next lecture we'll cover one of the single, most important macroscopic variables of this whole course, namely temperature. And in a few lectures from now, we'll return to this very same piston example, but by that point, we'll be armed with the first law of thermodynamics, which directly relates heat to mechanical motion.

Let me end this lecture with one last quote from another famous founding father of thermodynamics, Arnold Sommerfeld. He said,

> "Thermodynamics is a funny subject. The first time you go through it, you don't understand it at all. The second time you go through it, you think you understand it, except for one or two small points. The third time you go through it, you know you don't understand it, but by that time you are so used to it, it doesn't bother you any more."

Of course, Sommerfeld was being a bit facetious here, but the point is well taken; thermodynamics is not always an intuitive subject. We're going to take up the challenge of feeling our oneness with thermodynamics, of making the concepts come alive, and of understanding how they relate to essentially everything we do.

Temperature—Thermodynamics' First Force

Lecture 3

This lecture addresses what is perhaps the most intuitive thermodynamic concept out of all the ones covered in this course: temperature. This variable is of utmost importance in thermodynamics, but as you will learn in this lecture, the very notion of temperature might surprise you. In order to truly understand temperature and the fundamental reasons for it, you will need to head all the way down to the scale of the atom.

The Zeroth Law of Thermodynamics

- The zeroth law of thermodynamics is related at its core to temperature. Imagine that we have three objects: object A, object B, and object C. Let's place object A in contact with object C, and let's also place object B in contact with object C. The zeroth law of thermodynamics states that if objects A and B are each in thermal equilibrium with object C, then A and B are in thermal equilibrium with each other—where equilibrium is the point at which the temperatures become the same.

- The zeroth law of thermodynamics allows us to define and measure temperature using the concept of thermal equilibrium, and importantly, it tells us that changes in temperature are thermodynamic driving forces for heat flow. From this law, we can define temperature as "that which is equal when heat ceases to flow between systems in thermal contact," which is the foundational framework for thinking about energy in its many forms.

- In the most general terms, temperature is a measure of how "hot" or "cold" a body is. Of course, "hot" and "cold" are completely relative terms. So, temperature is meaningful only when it is relative to something.

- This very simple but powerful statement leads us to a particular need: We need to define a temperature scale, such as the Celsius (used throughout the world) and Fahrenheit (used predominantly in the United States) scales. Kelvin is another scale used to measure temperature, and it is known as an absolute scale—the standard temperature units of thermodynamics.

- A temperature scale allows us to measure relative changes. Both the Celsius and Fahrenheit scales use numbers, but the relative differences between them are different. In other words, in this comparison, one scale is not just a rigid shift from the other, but rather, each degree means something different.

- For example, the 100 Celsius degrees between water freezing (at 0) and water boiling (at 100) are equivalent to the 180 Fahrenheit degrees between 32 and 212. To convert from Fahrenheit to Celsius, we take the Fahrenheit reading, subtract 32 degrees, and multiply by 5/9.

- Energy due to motion is called kinetic energy, and the average kinetic energy of an object is defined as $1/2mv^2$, where m is the mass of the object and v is its velocity. If we have a bunch of objects, then this definition still holds: The average kinetic energy of the collection of objects would simply be the average of $1/2mv^2$ over all of the different objects.

- All matter—whether solid, liquid, or gas—is made up of atoms, which are the building blocks for all of the things we know and love. By measuring temperature, we're measuring how fast the atoms in a material are moving. The higher the average velocity of the atoms, the higher the temperature of the material.

The Kelvin Scale

- The Kelvin scale is the temperature scale of thermodynamics. In general, it's the temperature scale used in science and engineering. It makes a lot of sense and also makes solving problems in thermodynamics easier.

- From our definition of temperature as the average kinetic energy, the lowest possible temperature value is zero: This happens only if the average velocity of a material is zero, which would mean that every single atom in the material is completely motionless.

- We know that the square of velocity can never be negative, and because the mass cannot be negative either, the kinetic energy is always a positive value. So, zero is the lower bound on temperature, and it's the critical reference point in the Kelvin temperature scale (often referred to as "absolute zero" because there's no possibility for negative values).

William Thomson, Baron Kelvin (1824–1907), also known as Lord Kelvin, was a Scottish physicist who invented the Kelvin scale.

- The Kelvin scale, named after Lord Kelvin of Scotland, who invented it in 1848, is based on what he referred to as "infinite cold" being the zero point. The Kelvin scale has the same magnitude as the Celsius scale, just shifted by a constant of about 273 degrees. So, zero degrees kelvin—corresponding mathematically to the lowest possible value of temperature—is equivalent to −273.15 degrees Celsius.

- We know that the average speed of particles that make up a system is related to the temperature of the system. This is an example of connecting a macroscopic property to a microscopic property. In this case, "macro" relates to the single value we measure to describe the system—its temperature—that corresponds to an average over all those "micro" quantities—the particle velocities.

- A Maxwell-Boltzmann distribution gives us a distribution of particle velocities that depends on the temperature and pressure of the system, as well as the masses that make up the particles. These average distributions of velocities of atoms form a basis for understanding how often they collide with one another, which gives us another crucial macroscopic thermodynamic variable: pressure.

Thermometers and Thermal Equilibrium

- A thermometer, which is what we use to determine how hot or cold something is, works using the principle of thermal equilibrium. Remember that equilibrium means that the system is no longer changing on average. So, if we measure the temperature of the system, then even though the various atoms could be changing their individual velocities—either slowing down or speeding up—the average over all of the velocities stays the same. This is the state of equilibrium at the macroscopic scale.

- Thermal equilibrium is important for measuring temperature, because if we place a thermometer in contact with an object, after some time, the reading on the thermometer reaches a steady value. This value is the temperature of the object, and it is the value at which the thermometer is in thermal equilibrium with the object.

- One of the earliest records of temperature being measured dates all the way back to the famous Roman physician Galen of Pergamon, who in the year 170 AD proposed the first setup with a temperature scale. Although Galen attempted to quantify and measure temperature, he did not devise any sort of apparatus for doing so.

- In the year 1592, Galileo proposed the thermoscope, a device that predicted temperature fluctuations by using a beaker filled with wine plus a tube open at one end and closed with a bulb shape at the other. If you immerse the open end of the tube into the beaker of wine, then the bulb part is sticking up at the top.

- With this instrument, if the temperature gets warmer, the air inside the bulb at the top expands, and the wine is forced back down the tube, while if the air temperature gets colder, the air contracts, making the wine rise up in the tube.

- If we were to engrave marks onto the side of the glass tube, something that was not done until 50 years later, then the thermoscope becomes a thermometer—that is, it becomes a way to measure and quantify changes in temperature.

- A little later, a more sophisticated kind of thermometer was developed, called a Galilean or Florentine thermometer. The key concept in such a thermometer is that the density of water depends on temperature, where density is the mass of water divided by volume. This density changes as the temperature changes.

- If we placed an object in water that has a greater density than the water, then it would sink to the bottom, while an object with density less than the water would float to the top. If the object's density were exactly equal to the water density, then it would rest in the middle of the container. The Florentine thermometer was constructed using this principle.

- Compared to previous thermometer designs, the advantage of this new design was that because it was based on a fully sealed liquid, the pressure of the air outside didn't change the temperature reading, as it did with the older air thermoscopes.

- Taking the idea of a fully sealed liquid even further, the next type of thermometer that was developed was still based on the change in density of a fluid with temperature, but now the effect was exaggerated by the geometry of the design.

- The fluid fills a small glass bulb at one end and is allowed to expand or contract within a tiny, fully sealed channel stemming out from the bulb. Mercury was the fluid of choice for a long time, because it expands and contracts in quite a steady manner over a wide temperature range.

- However, various alcohols maintain the same steady expansion and contraction versus temperature as mercury, and because mercury is poisonous, it has largely been phased out as the thermometer's working fluid of choice.

- These types of fluid-filled glass thermometers, based essentially on the original Florentine thermometer designs, remained the dominant temperature measurement devices for centuries. More recently, though, new temperature measurement techniques have emerged, ones that are even more accurate and can span a wider range of temperatures.

- One such type of thermometer is also based on volume expansion, but in this case, the material being expanded is a solid metal as opposed to a liquid. By placing two thin sheets of different metals together, this bimetal strip works as a way to measure temperature because, as the temperature changes, the two metals expand and contract differently.

- Because one strip expands more than the other, and because the two strips are glued strongly together, they will tend to bend. By calibrating this bending, you can measure temperature. Because bimetal strips are made out of solid metal, they can go to very low as well as very high temperatures.

- Regardless of the type of thermometer, it must always be calibrated. A device that is not calibrated will simply show an increase in numbers when heated and a decrease in numbers when cooled—but these numbers have no physical meaning until reference points are used to determine both what the actual temperature as well as what changes in temperature mean.

- We don't necessarily always need to have a thermometer that gives us the most accurate temperature and over the largest range possible. It depends on how we plan to use the thermometer. Liquid crystal thermometers, a low-cost option, respond to temperature change by changing their phase and, therefore, color.

- The key principle of an infrared (IR) thermometer is that by knowing the infrared energy emitted by an object, you can determine its temperature. To use this thermometer, you simply point it where you want to measure, and you get an almost instant reading of the temperature.

- In contrast to the super-cheap liquid crystal thermometers, IR thermometers are quite expensive. However, they do possess many advantages, such as high accuracy over a very wide temperature range, the possibility to collect multiple measurements from a collection of spots all near one another, and the ability to measure temperature from a distance.

Suggested Reading

Halliday, Resnick, and Walker, *Fundamentals of Physics*, Chap. 18.

http://hyperphysics.phy-astr.gsu.edu/hbase/thermo/thereq.html.

http://hyperphysics.phy-astr.gsu.edu/hbase/thermo/temper.html#c1.

http://www.job-stiftung.de/pdf/skripte/Physical_Chemistry/chapter_2.pdf.

Questions to Consider

1. What happens during the process of thermal equilibrium?

2. Is thermal equilibrium attained instantaneously, or is there a time lag?

3. If two discs collide on an air hockey table, is energy conserved? Is velocity conserved in the collision?

Temperature—Thermodynamics' First Force
Lecture 3—Transcript

In this lecture we'll be discussing what is perhaps the most intuitive thermodynamic concept out of all the ones we cover in this course. It's a concept that you know and feel every minute of every day, and you probably check its value multiple times per day. It's the thermodynamic conversation starter, since many of us bring it up when we're not sure what to say, like, for example, "It's cold out there today!".

Of course, we're talking here about temperature. This variable is of utmost importance in thermodynamics, After all, how could we study "heat in motion" without a way to know if something is hot? And yet, as we'll see in this lecture, the very notion of temperature may surprise us. In order to truly understand temperature and the fundamental reasons for it, we'll need to head all the way down to the scale of the atom.

But, before I go down to that scale, I want to motivate this understanding of temperature even more by introducing to you our very first thermodynamic law for this course, called the zeroth law of thermodynamics. Notice that it's the first law we're learning, but those wacky thermodynamicists wanted to start counting their laws at zero, so they call this one the zeroth law. The law is related at its core to temperature, and it goes as follows. Imagine we have three objects; let's call them object A, object B, and object C. Now, let us place object A in contact with object C. And let us also place object B in contact with object C. The zeroth law of thermodynamics states that if objects A and B are each in thermal equilibrium with object C, then A and B are in thermal equilibrium with each other.

Equilibrium here means the point at which the temperatures become the same, and it's a concept I'll go into much more deeply in the next lecture. Now, I know that you're probably thinking, well, of course, isn't that law kind of intuitive? Well, yes it is, but it's important—and foundational—to state it as a law. It allows us to define and measure temperature using the concept of thermal equilibrium, and importantly, it tells us that changes in temperature are thermodynamic driving forces for heat flow. From this zeroth law of thermodynamics, one can define temperature as "that which is equal when heat ceases to flow between systems in thermal contact." This is the foundational framework for thinking about energy in its many forms!

Okay, so now let's discuss how we experience temperature, and more importantly, how it is measured. In the most general terms, temperature is a measure of how "hot" or how "cold" a body is. Of course, you may be thinking that "hot" and "cold" are completely relative terms, and you'd be absolutely right. Ice cubes are colder than boiling water, but they're a whole lot hotter than liquid nitrogen. And that first warm spring day of the year somehow feels much warmer than the rest of spring.

So temperature is meaningful only when it is relative to something. That's a really important statement, so I'd like to repeat it—Temperature is meaningful only when it is relative to something. And this very simple but powerful statement leads us immediately to a particular need, that is, we need to define a temperature scale. We've all probably heard of the Celsius and Fahrenheit scales. Here in the U.S., for some reason, we still use Fahrenheit, while in just about all of the rest of the world the Celsius scale is used. Kelvin is another scale used to measure temperature, and this is known as an "absolute scale." It's the standard temperature units of thermodynamics, but we'll come back to that in a moment.

A temperature scale allows us to measure relative changes. Now, if you take a look at Celsius vs. Fahrenheit, you can see that both the numbers and the relative differences between them are different. In other words, in this comparison, one scale is not just a rigid shift from the other, but rather, each degree means something different. So the 100 Celsius degrees between water freezing at 0° and water boiling at 100° are equivalent to those 180 Fahrenheit degrees between 32° and 212°. To convert between Fahrenheit to Celsius, we take the Fahrenheit reading, subtract 32°, and multiple by $^5/_9$, as many of us learn early on in school.

Now, as I mentioned, in the U.S.—ok, I suppose I should also include the Cayman Islands and the Bahamas—we're all used to the Fahrenheit scale. So when we take our body temperature, a reading of 98.6 is normal, and a chilly day would be around 50°. For the rest of the world, since they're used to the Celsius scale, a normal body temperature is 37°, and that same chilly day would be around 10°. Same exact meanings, just different scales and therefore different numbers.

The history of the Fahrenheit scale goes back to Daniel Fahrenheit, who in 1724 was trying to develop a systematic way to measure differences in temperatures. What he needed were reference points, things that seemed to happen at the same temperature with pretty high consistency. He decided water was a really good candidate for this, so he picked three reference points related to water. First, because he had observed that salty water has a different freezing point than fresh water, he chose the temperature at which salt water freezes, that is, when it transforms from a liquid into a solid. He called this point 0°. Second, he chose the point at which just plain water without the salt freezes; he called that 32°. And third, he picked the temperature of the human body, which at the time he called "blood heat" and referred to it as 96°. The reason for this particular value of 96 was that it meant there would be 64° between that reference point and the freezing point of fresh water. The choice of 64 intervals was not arbitrary; he picked it so that it would be easy to divide by two up to six times (2 to the 6th power is 64). Simply put, it made for easier math, and he also wanted this difference to be exactly twice the difference between the freezing points of salty and fresh water. Later, other scientists decided they wanted the boiling point of water to be exactly 180° higher than the freezing point, so they set that as a reference point, which in turn brought up the body temperature reading to 98 instead of 96°.

Now, why am I giving you this history? Well, first of all, to give old Fahrenheit his due. But also, so that you can see that these temperature scales are really quite arbitrary; choices are made regarding reference points, and so a scale is defined. And perhaps not surprisingly, historically, a number of different scales have been introduced. For example, in addition to the three I already mentioned—Fahrenheit, Celsius, and Kelvin—there's also the Rankine, Newton, Delisle, and Romer scales, to name just a few. By the early 18th century, as many as 35 different temperature scales had been devised!

The Celsius, or Centigrade, scale was created by Anders Celsius in 1742, and it was presented in a paper he wrote titled, "Observations of two persistent degrees on a thermometer." Notice the word "persistent." Again, because a good temperature scale should have as its reference points phenomena that repeatedly happen at the same temperatures. In this case, the phenomena he chose were the freezing and boiling points of water, although in his original version freezing was at 100 and boiling was at 0. Later, others decided to reverse these two reference points to the Celsius scale we know today, where 100° is boiling and 0° is freezing.

Okay, so those are the two scales the world uses—Fahrenheit for the roughly 320 million people living in the United States, Cayman Islands, the Bahamas, Palau, and Belize; and Celsius for the 6.7 billion people, give or take a few, living everywhere else. But, regardless of the scale we use, what is it we're actually feeling when we feel, or really, I should say, measure, a temperature? In order to answer this, we'll have to turn our attention to a definition some of you may have already learned from an introductory physics course, namely, the definition of the energy of a moving body.

This energy, that is, the energy due to motion, is called the kinetic energy, and the average kinetic energy of an object is defined as $\frac{1}{2} mv^2$, where that m is the mass of the object, and the v is its velocity. If we have a bunch of objects, then this definition still holds; the average kinetic energy of the collection of objects would simply be the average of $\frac{1}{2} mv^2$ over all of the different objects.

Now, let's remember that all matter, whether solid, liquid, or gas, is made up of atoms. We'll be talking about atoms at various points in this course, and I'm sure many of you have seen the periodic table of the elements, which shows all of the various atoms that we know about. These atoms are the building blocks for all of the things we know and love, including ourselves by the way! We're made up of a whole lot of carbon, plus oxygen, hydrogen, nitrogen, calcium, phosphorous, sulfur, sodium, and magnesium, among a bunch of other things. So, when we go back to our definition of kinetic energy, and we consider those moving objects to be the atoms that make up a material, well, that's where we get to a crucial part of our understanding of temperature. By measuring temperature, what we're actually doing is measuring how fast the atoms in a material are moving.

Okay, that was so important, I'd like to repeat it. When we measure temperature, we're measuring how fast the atoms in a material are moving.

The higher the average velocity of the atoms, the higher the temperature of the material. Just to be sure we really feel our oneness with this idea, let's take a look at my Brownian motion table to help bring home the point.

So, as I talked about, temperature arises fundamentally from the random vibrations of atoms. Now, this is related to the concept of Brownian motion, which means, just the motion, the random motion of particles. And since we're talking about the scale of the atom, it means the random motion of the atoms. And that's what gives rise to temperature.

So, to visualize this, what I have here is called a Brownian motion table. Now, it's a really cool thing, as a thermodynamicist, to have one of these things, because, basically, what it does, is it reproduces the effects of those random motions. So this is basically like an air-hockey table that you might have seen in an arcade, but it's a much smaller size.

And what I'm going to do is I'm going to turn on this pump, which blows air through tiny, little holes on the top. And then, I can imagine that these are atoms, and I put them on top of the table, and we watch them moving around randomly, okay. So that's random motion bumping around. But as I put more on, they don't just bump off the sides of the box, they actually bump off each other, and as I put more and more on, you can see them moving around with different speeds, bumping around off of each other and off of the container walls. This is a very nice, intuitive picture of what's happening to atoms as they bump around off of each other and off of the container walls. And you can see that I can put different size atoms on this Brownian motion table, and they'll have different speeds and different trajectories.

But notice something really important. The speeds are not all the same, so everyone of these is moving at a different speed, and then the speed changes as they get bumped into by others. The temperature is only the average over all of these particles, and that is the constant for this table. If I let it go like this, the temperature is staying the same, even though, the particles are changing their speeds, because temperature is only an average over all the particles.

Even when I add more particles to the table, you can see that what happens is the speed of the particles does not change as I add more. But what does change is how often they bump into each other. That is the collision rate, but that's not related to temperature. That is related to pressure.

So there we saw an example of particles moving around randomly with different individual speeds, but for the whole table, the average speed was roughly a constant. I hope that gives you a feel for the meaning of temperature, which corresponds to that macroscopic average. And notice that if I had turned off the air for the table, all of the particles would have come to rest. What would the temperature in that case be? This question brings us to another temperature scale, called the Kelvin scale, and it is the temperature scale of thermodynamics. For that matter, it's the temperature scale used in science and engineering in general, and it's the one that we'll use throughout this course. It's a scale that makes a lot of sense and also makes solving problems in thermodynamics easier.

Notice from our definition of temperature as the average kinetic energy, that the lowest possible temperature value is zero; this happens only if the average velocity of a material is zero, which would mean that every single atom in the material is completely motionless—the Brownian motion table was turned off. We know that the square of velocity can never be negative, and since the mass cannot be negative either, the kinetic energy is always a positive value. So zero is the lower bound on temperature, and it's the critical reference point in the Kelvin temperature scale, often referred to as absolute zero, since there's no possibility for negative values.

So the Kelvin scale, named after Lord Kelvin of Scotland. who invented it in 1848, is based on what he referred to as "infinite cold" being the zero point. The Kelvin scale has the same magnitude as the Celsius scale, just shifted by a constant of about 273°. So zero degrees Kelvin, remember, corresponding mathematically to the lowest possible value of temperature, is equivalent to about negative 273°C. Well, −273.15 to be precise.

So, a nice comfortable temperature for us humans would lie at around 298 Kelvin, corresponding to 25°C, or 77°F. Now, I know that 298 is a strange value to think of as feeling like a warm, sunny day for many of us not used to the Kelvin scale. But the Kelvin scale is fundamental because of the fact that it has built into it the notion of an absolute zero. In fact, the Kelvin scale is so central in thermodynamics that it's often referred to as the thermodynamic temperature scale.

Okay, so we know that the average speed of particles that make up a system is related to the temperature of the system. This is an example of connecting a macroscopic property to a microscopic property, something that we discussed in our last lecture. In this case macro relates to the single value we measure to describe the system—its temperature. That corresponds to an average over all those micro quantities, the particle velocities.

For example, here's a distribution of the velocities of different atoms at room temperature, or, to be precise, 298.15 Kelvin. The x-axis corresponds to the velocity of the particle, and the y-axis is the probability that the particle has that speed. Note that all of these atoms are gases at this temperature. Also notice that the lightest atom, Helium, has a more spread out distribution of velocities and can reach much faster speeds, while the heavier atoms, like Argon and Xenon, have a sharper peak, so less of a spread in velocities and tend to be slower on average. This type of relationship is known as a Maxwell-Boltzmann distribution, which gives us a distribution of particle velocities that depends on the temperature and pressure of the system, as well as the masses that make up the particles.

By the way, did you notice how fast those atoms are moving at room temperature? Five hundred meters per second, just for reference, is 1,118 miles per hour! As we'll see, these average distributions of velocities of atoms form a basis for understanding how often they collide with one another, which gives us another crucial macroscopic thermodynamic variable, namely pressure. But, I'm getting ahead of myself here. For the rest of this lecture, I want to stick with temperature.

So, now that we know what temperature is all the way down to the scale of the atom, let's take a look at how we actually measure it. How is it that we determine how hot or cold something is? As you all know, of course, the answer is to use a thermometer! A thermometer works using the principle of thermal equilibrium. And there's that really important thermodynamics concept I mentioned earlier when I told you about the zeroth law. Equilibrium means that the system is no longer changing on average. So, if we measure the temperature of the system, then, even though the various atoms could be changing their individual velocities—either slowing down or speeding up—the average over all of the velocities stays the same. This is the state of equilibrium at the macroscopic scale, and it's a concept we will use a lot in this course.

So, thermal equilibrium is important for measuring temperature, since, if I place a thermometer in contact with an object, after some time, the reading on the thermometer reaches a steady value. This value is the temperature of the object, and it is the value at which the thermometer is in thermal equilibrium with the object.

One of the earliest records of temperature being measured dates all the way back to the famous Roman Physician, Galen of Pergamon, who, in the year 170 A.D., proposed the first-ever setup with a temperature scale. He used equal quantities of boiling water and ice, which he called neutral temperature. And on either side of this point he had 4° of heat and 4° of cold. Although Galen attempted to quantify and measure temperature, he did not devise any sort of apparatus for doing so. For that, we have to fast forward more than 1,400 years to the time of Galileo.

In the year 1592, Galileo proposed the thermoscope, a device that predicted temperature fluctuations by using a beaker filled with wine, plus a tube open at one end and closed with a bulb shape at the other. If you immerse the open end of the tube into the beaker of wine, as shown here, then the bulb part is sticking up at the top. With this instrument, if the temperature gets warmer, the air inside the bulb at the top expands, and the wine is forced back down the tube, while if the air temperature gets colder, the air inside the bulb contracts, making the wine rise up in the tube. Now, if we were to engrave marks onto the side of the glass tube, something that was not done until 50 years later, then this thermoscope becomes a thermometer. That is, it becomes a way to measure and quantify changes in temperature.

A little bit later, a more sophisticated kind of thermometer was developed, called a Galilean Thermometer, although, there is compelling evidence that it was not actually invented by Galileo himself, but rather by his students at the Accademia del Cimento, or, Academy of the Experiment. Although most people still refer to it as a Galilean thermometer, the term Florentine thermometer is perhaps more accurate. Let's take a look

The key concept in such a thermometer is that the density of water depends on temperature. By density, I mean the mass of water divided by its volume. At room temperature, this value is just under 1 gram per centimeter cubed—0.997044 to be exact. The point is that this density changes as the temperature changes.

Now, if I place an object in water that has a greater density than the water, then it will sink to the bottom, while an object with density less than the water floats to the top. If the object's density is exactly equal to the water density, then it rests in the middle of the container. Using this principle, the Florentine thermometer was constructed. And it looks something like this.

Five bulbs of different densities are used in this case, each one carefully calibrated to be exactly equal to the density of water at a particular temperature. The bulbs are then placed in a tube filled with water. So, if the temperature is, say, 76°F, the bulb representing 76, that is, the one whose density matches the density of water at exactly 76°F, that bulb floats around in the middle, as shown, while denser bulbs, that would be 64°, 68°, and then 72° in this case, sink down to the bottom, and the lighter bulb corresponding to the density of water at 80°, floats to the top. The bulb floating in between gives the temperature reading.

Now, let's change the temperature by applying heat to the outside wall. This changes the temperature of the water inside, which in turn changes its density, making it decrease, which in turn causes the less dense bulbs to sink further. Notice that the reaction of this thermometer to temperature change is quite slow, not exactly ideal for measuring temperature. This thermometer also requires a rather large piece of equipment; imagine if you wanted to know the temperature to an accuracy of a single degree in a range from 0° to 100°; you'd need 100 balls inside a pretty big container!

And yet, compared to previous thermometer designs, the advantage if this new design was that, since it was based on a fully sealed liquid, the pressure of the air outside didn't change the temperature reading, as it did with the older air thermoscopes. Taking the idea of a fully sealed liquid even further, the next type of thermometer that was developed was still based on the change in density of a fluid with temperature, but now the effect was exaggerated by the geometry of the design.

As you can see here, the fluid fills a small glass bulb at one end and is allowed to expand or contract within a tiny fully sealed channel stemming out from the bulb. Many of us are quite familiar with this type of thermometer and have, in fact, had our temperature taken with it many times. Mercury was the fluid of choice for a long time, since it expands and contracts in quite a steady manner over a wide temperature range. This is a good thing for a temperature measurement apparatus, since we'd like the measurement of a degree to be the same, whether it's taken at 60° or 80°. But, various alcohols maintain the same steady expansion and contraction versus temperature, as mercury does. And since mercury is poisonous, it has largely been phased out as the thermometer's working fluid of choice.

And you may be thinking, why not use water as the fluid? Well, an advantage of alcohol over water is the larger temperature range for which alcohol remains a fluid. For example, ethanol, which is the fluid inside the thermometer I just showed you, can be used to measure temperatures all the way down to −70°C, while water, of course, would have frozen at 0°.

These types of fluid-filled glass thermometers, based on, essentially, the same original Florentine thermometer designs, remained the dominant temperature measurement devices for centuries. And in fact, it is these types of thermometers that were used to develop the most important temperature scales that we've already discussed. More recently, though, new temperature measurement techniques have emerged, ones that are even more accurate and can span a wider range of temperatures.

One such type of thermometer some of you may be familiar with is also based on volume expansion, but in this case, the material being expanded is a solid metal, as opposed to a liquid. By placing two thin sheets of different metals together, this bi-metal strip works as a way to measure temperature, since, as the temperature changes, the two metals expand and contract differently. Since one strip expands more than the other, and since the two strips are glued together strongly, they will tend to bend. By calibrating this bending, one can measure temperature. And bi-metal strips are useful because, since they're made out of solid metal, they can go to very low as well as very high temperatures. It's the kind of thermometer you might find in your thermostat at home, or, like this one here, which you can safely stick into your turkey in the oven.

Note that regardless of the type of thermometer, it must always be calibrated. A device that is not calibrated will simply show an increase in numbers when heated and a decrease in numbers when cooled. But these numbers have no physical meaning until reference points are used to determine both what the actual temperature, as well as changes in temperature, mean. In the case of the bimetal strip, a certain bend may be calibrated with the boiling point of water, while a change from this bend to another reference point, say the freezing point of water, would indicate how much bend corresponds to each degree.

Now, it's important to keep in mind that we don't necessarily always need to have a thermometer that gives us the most accurate temperature and over the largest possible range. It depends on how we plan to use the thermometer. For example, in many third-world countries, low cost is the most important factor for a technology to become useful, so if one could measure temperature within only a very narrow temperature range but for extremely low cost, it would be quite valuable.

That's exactly the case for these for these liquid crystal thermometers, which respond to temperature change by changing their phase and, therefore color. I simply take one of these strips, which costs less than a penny, and place it on my forehead to check my temperature.

And the last type of thermometer we'll examine here is called an infrared—or IR—thermometer. The key principle of this type of temperature measurement is that by knowing the infrared energy emitted by an object, one can determine its temperature using a theory related to what is called blackbody radiation. Now, I'm not planning to cover blackbody radiation here in any sort of detail, but I just want you to know that from this theory we know that any material with a certain temperature emits energy.

So an IR thermometer captures this energy and then converts it into an electrical signal. From this, a corresponding temperature value can be derived using one of the equations from the theory of blackbody radiation.

And so, to use this thermometer, I simply point it where I want to measure, and I get an almost instant reading of the temperature as you can see. Notice that this IR thermometer points a red spot to where it's measuring the temperature. That's just coming from a laser pointer built into the thermometer to make sure we know where it's measuring, but it has nothing to do with the measurement itself. People have been confused at times by this part of the thermometer, and even refer to it as a laser thermometer. But rest assured, this is not measuring temperature using a laser.

Now, in contrast to the super-cheap liquid crystal thermometers, these IR thermometers are quite expensive. However, they do possess many advantages, such as high accuracy over a very wide temperature range, the possibility to collect multiple measurements from a collection of spots all near one another, creating an effective heat map, and the ability to measure temperature from a distance, so not in direct contact with the object.

And speaking of being in thermal contact, that brings me back to the very final point of this lecture. Namely, a reminder of the zeroth law of thermodynamics. Remember, if we have two objects, A and B, in thermal contact, then A and B will come to thermal equilibrium, that is, they will have the same temperature. If a third object, C, is then brought into contact with either one of these, then this object will have the same temperature as both objects A and B. All three of the objects will come to thermal equilibrium with one another. This law seems kind of intuitive, but importantly, it tells us that changes in temperature are thermodynamic driving forces for heat flow. The concept of temperature itself is based upon the zeroth law of thermodynamics.

As I said before, from this zeroth law of thermodynamics, one can define temperature as "that which is equal when heat ceases to flow between systems in thermal contact." This is the foundational framework for thinking about energy in its many forms!

Salt, Soup, Energy, and Entropy

Lecture 4

This lecture will lay the foundation for the "dynamics" part of the term "thermodynamics"—that is, a framework for understanding how heat changes and flows from one object to another and from one form of energy to another. In this lecture, you will explore some basic concepts that are critical to thermodynamics and that will serve as a kind of language for the rest of the course. Many of these involve some simple but crucial definitions and terms, while other concepts will involve a bit of discussion.

General Definitions

- The definition of a system is a group of materials or particles, or simply a space of interest, with definable boundaries that exhibit thermodynamic properties. More specifically, the four different systems that are possible in thermodynamics have unique boundary conditions that limit the exchange of energy or atoms/molecules with their surroundings.
  - An isolated system is one in which the boundaries do not allow for passage of energy or particles or matter of any form. Think of this type of system as one with the absolute ultimate walls around it: a barrier that permits nothing at all, not even energy, to pass through it.
  - In a closed system, the boundaries do not allow for the passage of matter or mass of any kind. However, unlike the isolated system, in a closed system, the boundaries do allow for the passage of energy.
  - In an adiabatic system, the boundaries do not allow passage of heat. Matter can pass through a boundary in this type of system, but only if it carries no heat. No heat can enter or leave this type of system.

- In an open system, the boundaries allow for the passage of energy and particles or any type of matter. Pretty much anything can flow both in and out of an open system.

- When we solve for problems in thermodynamics, the first thing we must do is define our system and how it interacts with the rest of the universe.

- Thermodynamics is a theory for predicting what changes will happen to a material or a system. A key part of making correct predictions about a system is identifying what processes can happen within the system. To do this, we have to introduce the language for the different types of processes.
    - An adiabatic process is one in which no heat is transferred during the process.

    - In an isothermal process, the temperature during the process remains constant.

    - In an isobaric process, the pressure remains constant.

    - In an isochoric process, the volume remains constant during the process.

    - Finally, we can combine these types of processes together. For example, in an isobarothermal process, both temperature and pressure are held constant.

- This way of thinking about different thermodynamic processes is essential because these are just the types of variables we know how to control when we manipulate a material.

- The broadest kind of classification of a variable is whether it is intensive or extensive. An intensive variable is one whose magnitude does not vary with the size of the system but can vary from point to point in space. Pressure and temperature are intensive variables. In the case of extensive variables, the magnitude of the variable varies linearly with the size of the system. Volume is an extensive variable.

- Intensive and extensive variables can form coupled pairs that when multiplied together give a form of thermodynamic work, which means a kind of energy. Pressure and volume are one example of such a pair.

States and Equilibrium

- The state of a system is a unique set of values for the variables that describe a material on the macroscopic level. A macroscopic variable is something you can measure related to the entire system.

- Examples of macroscopic variables are pressure, temperature, volume, and amount of the material—and all of these variables are also called state variables. It doesn't matter if the variable is intensive or extensive for it to be considered a state variable. What does matter is whether the variable depends on the sequence of changes it went through to get to where it is.

- Thermodynamics only makes predictions for equilibrium states. Thermodynamics does not make predictions for the manner in which the system changes from one state to another. Rather, it takes the approach that eventually it will happen, and it is a science that tells us what the system will be like once it does happen.

- Equilibrium is a state from which a material has no tendency to undergo a change, as long as external conditions remain unchanged.
    - An unstable state is actively going to tend toward a new equilibrium state.

    - In a metastable equilibrium state, the system can change to a lower total energy state, but it may stay stable in a higher energy state as well.

- In an unstable equilibrium state, any perturbation or small change will cause the material to change to a new equilibrium state.

- In a stable equilibrium state, the material does not change its state in response to small perturbations.

- In physics, you learn that stable mechanical equilibrium is achieved when the potential energy is at its lowest level—for example, the potential energy is minimized when the ball rolls down to the bottom of the track. Similar principles will come into play in reaching internal equilibrium in materials: We may look for the maximum or minimum of a thermodynamic function to identify equilibrium states.

Internal Energy and Entropy

- In physics, you learn that the total energy is equal to the sum of two terms. First, the potential energy is related to how much energy the system has to potentially be released. Second, the kinetic energy is due to the motion of the system.

- A third term is the thermodynamic concept of internal energy, where the total energy is now a sum over all three of these individual energy terms (potential, kinetic, and internal). The additional energy term represented by the internal energy is of central importance. The internal energy of a system is a quantity that measures the capacity of the system to induce a change that would not otherwise occur.

- Internal energy in a material can be thought of as its "stored" energy. But it's not the same kind of stored energy as in the potential energy. In this case, we're talking about energy storage in the basic building blocks of the material itself—namely, the atoms.

- Energy is transferred to a material through all of the possible forces that act upon it—pressures, thermal energy, chemical energy, magnetic energy, and many others—and this internal energy is stored within the random thermal motions of the molecules, their bonds, vibrations, rotations, and excitations.

- These are the things that lead to the difference between the two states of the same material—for example, water—and this is quantified by the concept of internal energy. The internal energy of the ice is different than the internal energy of the liquid, and that is the only energy term that captures the difference between these two phases.

- Using the laws of thermodynamics, we're going to be able to predict the phase of a material for any given set of thermodynamic state variables, and these predictions will lead to phase diagrams, which are one of the holy grails of thermodynamics, like having GPS navigation for materials.

- Entropy is a nonintuitive but absolutely critical parameter of materials—along with the more common parameters like volume, pressure, temperature, and number of molecules. This concept is different than these other concepts because it's not as straightforward to grasp, at least at first.

- In fact, there has been quite a bit of confusion over what entropy really is, and unfortunately, it gets misrepresented or misunderstood often, even among scientists. Entropy has been used synonymously with words such as "disorder," "randomness," "smoothness," "dispersion," and "homogeneity."

- However, we do know what entropy is, and today we can measure it, calculate it, and understand it. In simple conceptual terms, entropy can be thought of as an index or counter of the number of ways a material can store internal energy.

- For a given set of state variables—such as the temperature, pressure, and volume of a cup of water—it is equally likely to be in any of the possible microstates consistent with the macrostate that gives that temperature, pressure, and volume. Entropy is directly related to the number of those microstates that the system can have for a given macrostate.

- Basically, entropy is a measure of the number of degrees of freedom a system has. Sometimes people confuse this with randomness or things being somehow more spread out and smooth. And you might hear people say that if something is more uniform, or less ordered, then it must have more entropy. But that's not quite right.

- Finding the balance between energy and entropy is what defines equilibrium states. And the connection between energy, entropy, and those equilibrium states is governed by four thermodynamic laws.

Suggested Reading

Anderson, Thermodynamics of Natural Systems, Chap. 2.

Moran, Shapiro, Boettner, and Bailey, *Fundamentals of Engineering Thermodynamics*, Chap. 1.

http://www.felderbooks.com/papers/dx.html

Questions to Consider

1. What is the difference between intensive and extensive variables?

2. What does it mean for a variable to be a state function?

3. Why do some chemical reactions proceed spontaneously while others do not, and why does heat affect the rate of a chemical reaction?

Salt, Soup, Energy, and Entropy
Lecture 4—Transcript

In the last lecture, we talked about the most basic thermodynamic concept of all, namely, temperature. Since the "thermo" part of thermodynamics means "heat," you can imagine how important our understanding of temperature is for this course; it serves as the foundational concept to measure and compare how hot or cold a particular object gets. But to complete the title of our subject, we need to lay the foundation for the "dynamics" part of thermodynamics, that is, a framework for understanding how heat changes and flows from one object to another and from one form of energy to another.

In this lecture, we're going to explore the other basic concepts that are critical to thermodynamics and that will serve as a kind of language for the rest of the course. Many of these involve some simple but crucial definitions and terms, while other concepts will involve a bit of discussion. One that falls into this latter category is entropy, which I'm very excited to introduce to you today. But I don't want to get ahead of myself.

So, first of all, we need to define the different types of systems that are possible in thermodynamics. There are four important kinds I will discuss, but before I do that, perhaps I should clarify what I even mean by the word "system" itself. The definition of a system is a group of materials or particles, or simply a space of interest with definable boundaries that exhibit thermodynamic properties. Okay, so the key word there is boundary, and that's really what I want to distinguish here. More specifically, the four different systems that are possible in thermodynamics have unique boundary conditions that limit the exchange of energy or atoms and molecules with their surroundings.

First, an isolated system is one in which the boundaries do not allow for passage of energy or particles or matter of any form. Think of this type of system as one with the absolute ultimate walls around it; a barrier that permits nothing at all, not even energy, to pass through it.

Second, we have a closed system. Here, the boundaries do not allow for the passage of matter or mass of any kind. However, unlike the isolated system, in a closed system, the boundaries do allow for the passage of energy.

Third, there is an adiabadic system, where the boundaries do not allow passage of heat. Matter can flow through a boundary in this type of system, but only if it carries no heat. The key here is that no heat can enter or leave this type of system.

And last, we have an open system, which is pretty much how it sounds. In this case, the boundaries allow for the passage of energy and particles or any type of matter. Pretty much anything can flow both in and out of an open system.

Now, don't worry about memorizing these definitions, or for that matter, all of the rest of the definitions I'm going to set forth in this lecture. More important than the actual words we use—I'll remind you of them as we continue to use them throughout the course—I want you to start thinking about the basic concepts we're discussing. In this case, the point is that when we solve for problems in thermodynamics, the first thing we must do is define our system and how it interacts with the rest of the universe.

And now that we've defined types of systems, the next thing we need to define are the kinds of processes that can occur in our systems. We've stated that thermodynamics is a theory for predicting what changes will happen to a material or a system. A key part of making correct predictions about a system is identifying what processes can happen within the system. What is even possible? To do this, we have to first introduce the language for the different types of processes. First, we have the same word—adiabatic—that I used for a type of system. But in this case, I'm referring to an adiabadic process. Here, the meaning is similar; this type of process is one in which no heat is transferred during the process. Second, we have an isothermal process. In this case, the temperature during the process remains a constant. Next, we have an isobaric process, for which, as you might have guessed, the pressure remains constant. and then we have an isochoric process. In this case, the volume remains constant during the process. And last, we can combine these types of processes together. For example, in an isobarothermal process is one in which both temperature and pressure are held constant. This way of thinking about different thermodynamic processes is essential, since these are just the types of variables we know how to control when we manipulate a material.

Let's look at a few examples. Suppose I place water in a pot to boil on the stove to make some pasta. As we know, it will boil faster if I put a cover on the pot. In this case, the system is the saucepan with water in it, so it's a closed system, since at the boundaries, including the lid, make it so that no matter can pass in or out. Note, though, that although mass does not pass through the boundary of my system, heat can and does pass through the metal boundary and into the water.

As for the types of processes that can occur, it depends on the water temperature. If the water is in the process of getting hot, then we would define the type of process that occurs as isobaric and isochoric, since both the pressure and volume are constant. On the other hand, if the water is already boiling, then it's also isothermal, since the temperature of my system—the pot and the water—is constant; note that once the water boils, little bits of matter in the form of steam will leak out of the top, so technically, it's no longer a closed system. If I were to seal the lid, then it stays a closed system, but now the process is no longer isobaric, at least at first, since the pressure of steam will build up inside.

Let's consider another example. What if my system is simply a nice cold glass of water that I set out on the porch on a warm, sunny day? In this case, what type of system do we have? Well, the open glass container allows pretty much anything to pass—heat, mass, you name it. So the system here is open. And what about the types of processes that can occur? The temperature is not held constant, since the water is cold but in a warm environment. The volume could change if it wanted to, since the top of the glass is open. But one thing that will remain fixed here is the pressure; with the open top of our glass, the water is going to be at a constant pressure set by the pressure of the environment; that would be one atmosphere. So the processes that will occur in this case, we can categorize as isobaric processes. And finally, if I poured the glass of water into a perfect thermos, instead of an open glass, we'd have only adiabatic processes that can occur in my system, which would become an isolated system, since both matter and energy can no longer pass across the boundaries of the thermos. So those are systems and types of processes in which thermodynamic variables can either change or be held constant.

But what about the classification of the variables themselves, like temperature, pressure, and volume? The broadest kind of classification is whether a variable is intensive or extensive. And the definitions are quite simple. An intensive variable is one whose magnitude does not vary with the size of the system but can vary from point to point in space. Let's use a box to illustrate.

Suppose this box is at some pressure and temperature, designated by the variables P and T. Now suppose I have two such boxes at the same values of P and T. I bring them together. Once I merge them, I have a box that is twice as big, but the values for P and T did not change; they are simply the original values of each box separately. Thus, pressure and temperature are intensive variables.

The other kind of variable is called extensive. In this case, the magnitude of the variable varies linearly with the size of the system. So, if we go back to those two boxes and consider the volume of each box, then you can see quite easily that when I bring the two boxes together, the volume doubles. This means that volume is an extensive variable. As we'll see later, intensive and extensive variables can form coupled pairs that, when multiplied together, give a form of thermodynamic work, which means, a kind of energy. Pressure and volume are one example of such a pair, which we'll dive into in some detail in the next lecture.

But now, let's move on to two other very important definitions in thermodynamics. First, we need to define the meaning of the state of the system, and second, we'll discuss what it means for a system to be in something called equilibrium.

The state of a system is a unique set of values for the variables that describe a material on the macroscopic level. Remember our discussion in the second lecture regarding the difference between a microscopic and macroscopic picture? A macroscopic variable means something you can measure related to the whole entire system. Examples of macroscopic variables are pressure, temperature, volume, and amount of the material, and all of these variables are also called state variables. It doesn't matter if the variable is intensive or extensive for it to be considered a state variable. What does matter, is whether the variable depends on how it got there. Allow me to explain.

The molecules in materials heat up, react, rearrange, change shape, form and break bonds with one another, and undergo myriad other changes in response to changes in their environment. Remember that we do not keep track of those changes at the molecular scale in classical thermodynamics. Rather, we keep track of the changes in macroscopic properties that occur due to these internal molecular interactions. These are the changes in the state of the material. Now, for a variable to be a state variable, it must not depend on the sequence of changes it went through to get to where it is. So, I can take a system in some state, let's call it state A, and I can apply some kind of thermodynamic force to the system in this state, like pressure or temperature, and the system will go into some new state, call it state B. The state variables for the initial and final states do not depend in any way on the intermediate steps, or the path taken to get from A to B, only on the initial and final values of the variable of the system.

By the way, if this is a little confusing, since it's the first time you're hearing such a thing, don't worry! The concept of state variables and state functions is so important in thermodynamics that we're going to devote our entire next lecture to them! For now, let's keep going with some more definitions.

Here's another really important aspect of thermodynamics we need to know, and it's related to our definition of state and also gets us to our next important concept. That is, thermodynamics only makes predictions for equilibrium states. Thermodynamics does not make predictions for the manner in which the system changes from one state to another. Rather, it takes the approach that eventually it will happen, and it is a science that tells us what the system will be like once it does happen. So, in that example of a system going from state A to state B, thermodynamics will let us predict, given the forces applied, what state B is. But, it will not predict what happens to the system along the way. For that, one needs to study kinetics, which is a different topic that we might mention here and there, but will not be the focus of this course.

So I've used the word equilibrium now a few times, and I've told you that thermodynamics tells us about the behavior of equilibrium states. Let's look more closely at what this means. Equilibrium is a state from which a material has no tendency to undergo a change, as long as external conditions remain unchanged. That's one of those statements that's so important, I'd like to repeat it. Equilibrium is a state from which a material has no tendency to undergo a change, as long as external conditions remain unchanged.

Take a look at this simple plot of the energy of a system versus some parameter; it could be any variable of the system, but let's just call it generically *X*. The energy of the system can vary in different ways as a function of this variable, as shown here. I'll use this to point out how this variation in energy can tell us something about the equilibrium of a system. Now, the lower the energy of the system, the more stable the system will be, and the higher the energy, the less stable. So on this graph, going down means going to more stability, while going up is less stable.

If our variable of interest were position, and the energy represented by only gravitational potential energy, then you could imagine this path as a roller coaster, and our discussion about a ball rolling around on top of it. So you can imagine that, at this point here, the system is in an unstable state, that is, it is actively going to tend toward a new equilibrium state. On the other hand, at this point in the energy landscape, the system is in what is called a metastable equilibrium state, that is, the system can change to a lower total energy state, but it may stay stable in this higher energy state as well. At this other point, the system is in an unstable equilibrium state, which means that any perturbation or small change will cause the material to change to a new equilibrium state. And finally, at least for this variable and energy landscape, at this point for the variable *X*, the system is in a stable equilibrium state; the material does not change its state in response to small perturbations.

In physics, you learn that stable mechanical equilibrium is achieved when the potential energy is at its lowest level; the potential energy is minimized when the ball rolls down to the bottom of the track, or a roller coaster car comes to the bottom of a hill. Similar principles will come into play in reaching internal equilibrium in materials; we may look for the maximum or minimum of a thermodynamic function to identify equilibrium states.

Now that we've talked about what the state of a system is and what equilibrium means, I'd like to turn to two more extremely important thermodynamic definitions, namely, the internal energy of a system, and entropy. Now, in freshman physics you learn that the total energy is equal to sum of two terms. First, there's the potential energy, which we usually write as PE, and which is related to how much energy the system has to potentially be released. Gravity is a good example of this, since, if I lift up a block in the air, it has the potential to fall back down due to gravity. A spring is another good example of potential energy, since if I stretch it out, now it now has the potential to spring back to its starting point.

Second, there's the kinetic energy, which is due to the motion of the system. But the thing is that here, we must add to these two energy terms a third one, namely, the thermodynamic concept of internal energy, such that the total energy is now a sum over all three of these individual energy terms—potential, kinetic, and internal. The additional energy term represented by the internal energy is of central importance. And what does it represent? Well, stated in words, the internal energy of a system is a quantity that measures the capacity of the system to induce a change that would not otherwise occur. As always, it's nice to ground such statements with a concrete example.

Take a look at this comparison between ice and water. Both are sitting on a table, so their potential energy—at least the gravitational potential energy—will be exactly the same, since they are at the same height off of the ground. And we could call this potential energy zero at the table surface. And if the water is perfectly still, neither the ice nor the water is moving, then they both also will have zero kinetic energy as well. Without the concept of internal energy, you might then think that the ice and water both have zero total energy. But this is not true! We certainly know intuitively that these two forms—or phases—of the same material are very different from one another. For example, one is a solid, and the other a liquid; one is cold and the other, well, not as cold; it could even be quite hot.

Internal energy in a material can be thought of as its stored energy. But it's not the same kind of stored energy as in the potential energy. In this case, we're talking about energy storage in the basic building blocks of the material itself, namely, the atoms. Energy is transferred to a material through all of the possible forces that act upon it—pressures, thermal energy, chemical energy, magnetic energy, and many others. And this internal energy is stored within the random thermal motions of the molecules, their bonds, vibrations, rotations, and excitations.

These are the things that lead to the difference between the two states of our same material, the same material being water. And this is quantified by the concept of internal energy. The internal energy of the ice is different than the internal energy of the liquid, and that is the only energy term that captures the difference between these two phases. As we'll see, using the laws of thermodynamics we're going to be able to predict the phase of a material for any given a set thermodynamic state variables, and these predictions will lead to phase diagrams, which are one of the holy grails of thermodynamics. It's like having GPS navigation for materials.

And now last, but certainly not least, let's turn to the one other concept I'd like to introduce to you today, namely entropy. Entropy is a non-intuitive but absolutely critical parameter of materials, along with the more common parameters like volume, pressure, temperature, and number of molecules that I've already mentioned. This concept is different than these other concepts, because it's not as straightforward to grasp, at least at first. For example, when I just discussed the concept of internal energy a minute ago, you may already have had a sense of what it means, or at least could mean, since you may have already heard of the concept of other forms of energy, like kinetic or gravitational potential energy. For entropy, it's likely that we'll be blazing an entirely new trail.

In fact, there has been quite a bit of confusion over what entropy really is, and unfortunately, it gets misrepresented or misunderstood often, even, by the way, among scientists. Entropy has been used synonymously with words such as "disorder," "randomness," "smoothness," "dispersion," and "homogeneity," to name only a few. The great thermodynamicist Gibbs called it "mixed-up-ness." And another brilliant scientist famously said, "Nobody really knows what entropy is anyway." That was von Neumann.

But, actually, we do know what entropy is, and today, we can measure it, calculate it, and understand it. I'm going to be talking a lot about entropy throughout this course, and we will introduce more rigorous thermodynamic definitions for entropy in the coming lectures. We'll be using lots of examples as we go. But for now, I'd like to give you just the first glimpse of what this critical thermodynamic variable is about in simple conceptual terms. By the end of these lectures, I hope that you will have an intuitive grasp of entropy.

Okay, so we just talked about how internal energy is a measure of the stored energy in a material, and we gave an example of the difference between ice and water. Entropy can be thought of as an index or a counter of the number of ways a material can store this type of energy. Let's consider again that molecular view of water. Here's a snapshot of the water molecules at one instant in time. You can see that the water has a number of degrees of freedom; that means that each molecule can rotate, translate, vibrate, and bond to another molecule, and so on. And at any instant in time, if I take a snapshot like this of the water molecules, they would each have their given position, velocity, bonding, and so on. All of these individual molecular details of each water molecule come together and give rise to a macroscopically observed temperature, pressure, and volume.

Now, you may already know this, but a water molecule is pretty small, like only a few angstroms across. An angstrom is a distance of only 1 over 10 to the 10th meters, so, pretty small. So what that means is that there are a whole lot of water molecules in just an ordinary cup of water, around 10^{22}, to be more precise. Now, here's the key point. If I measure, say, the temperature of that cup of water, then the temperature I'm measuring is a collective effect of those degrees of freedom of each water molecule, collected over a 10 with 22 zeroes number of molecules—quite amazing when you think about it!

As you can imagine, with so many molecules and so many possible arrangements and degrees of freedom, more than one snapshot could give the same temperature. In fact, if I hold the temperature of this cup of water constant, then even though at this macroscale I see a constant value, all the way down at the microscale, the scale of the water molecule, things are in continuous flux, and each water molecule is undergoing continuous changes in its position, velocity, bonding, and so on.

For a given set of state variables, like the temperature, pressure, and volume of this cup of water, it is equally likely to be in any of the possible microstates consistent with the macrostate that gives that temperature, pressure, volume. By the way, I also just stated one version of the 2nd law of thermodynamics, but I'm getting a little bit ahead of myself here.

So what is entropy? It's directly related to the number of those microstates that the system can have for a given macrostate. Basically, entropy is a measure of the number of degrees of freedom a system has. Sometimes people confuse this with randomness or things being somehow more spread out and smooth. And you may hear people say that if something is more uniform, or less ordered, then it must have more entropy. But that's not quite right.

As a counter example, take a look at these two pictures showing simply a grid of points. These two pictures represent a situation where 169 squares have been filled in a grid originally containing 1,225 empty spaces. So, which one do you think has the higher entropy; the one on the left or the one on the right? Well, if you went with the one on the right, you'd be going with the one that appears to be more random or uniform. And in fact, that's the one that most people I've asked this question will pick.

But actually, it's the one on the left with the higher entropy, much higher in fact. If you look carefully, you can see, that in the grid on the right, no black squares are next to one another. In fact, for this case, it was an explicit rule in the way the squares were filled in. On the left, on the other hand, no rules were set forth, and so you see some black squares right next to others. The rule that we used for placing the squares on the right limits the number of ways the grid can be filled in. So, it has much fewer degrees of freedom or "possible microstates" if we think back to the water example, and in this case, the state that is shown on the right, therefore, has a much lower entropy than the one on the left.

So I've talked about internal energy and entropy as key thermodynamic variables that describe the state of a system. I'll conclude this lecture with an example that shows how they are connected in real life. Let's use a bowl of soup that needs some salt. As you shake the salt into the soup, what's happening to those little crystals of salt that come out of the container? If we want to have a nice, evenly salted soup, hopefully they dissolve! So I'm talking about dissolving salt in a liquid. What does that have to do with entropy? Well, as it turns out, everything.

Just as it is quite a common experience for us to salt our soup, it is an equally common experience that once dissolved, the salt crystal will never spontaneously re-form. And here's what is actually strange about that. The bonding energy between atoms in the salt crystal is really strong. In fact, the internal energy of the salt crystal is much more favorable when it's a crystal as opposed to when those salt atoms are dissolved in the liquid. From an energetic point of view, it really, really wants to be a solid crystal, even when it's put into the soup. And yet, we all know from experience that salt does not remain a crystal but rather simply dissolves in our soup. So what's going on? As you may have already guessed, it's all in the entropy—or number of degrees of freedom—in the salt. When the salt dissolves, it increases the entropy of the system. The balance between internal energy and entropy is what determines the behavior of our salted soup, and indeed, all real systems.

Let's go back to those microstates of the water, just to make sure we understand this example. You can see by looking at this picture, that if you count the number of possible ways the atoms can arrange to form a solid piece of salt, or a crystal of salt, well, there's only one possible way those atoms can be arranged. Now, if you look over here on the right, you can see that on a molecular level, once the atoms are released from the crystal, their thermal energy will scramble them thoroughly and randomly through the solution. And once the salt atoms are scrambled like this, there are many, many ways in which they can be arranged. The number of degrees of freedom, meaning, the entropy of the system, has skyrocketed compared to the crystal. And it's that higher entropy that drives the salt to dissolve in your soup. Make sure you point that out to your friend next time you have soup together!

Finding the balance between energy and entropy is what defines equilibrium states. And the connection between energy, entropy, and those equilibrium states, well that connection is governed by four thermodynamic laws.

The Ideal Gas Law and a Piston

Lecture 5

This lecture focuses on state functions and the first law that relates them together. This is not the first law of thermodynamics, which will be introduced in the next lecture. Instead, in this lecture, you are going to learn about the ideal gas law, which relates the three very fundamental thermodynamic variables you have been learning about together: pressure, volume, and temperature. These variables are also all known as state variables, so the function that relates them together is known as a state function.

State Functions and State Variables

- A given state function is a thermodynamic variable that depends solely on the other independent thermodynamic variables. This means that we can think of state functions as models of materials. And once we know what the state function is—that is, how the state variables are related to one another—we can plot one state variable against another because it unlocks the secrets of the behavior of a system.

- A state variable is a variable that describes the state of a system. Pressure, volume, temperature, and internal energy are just a few examples. For any given configuration of the system—meaning where things are, like the atoms or molecules that make up the system—we can identify values of the state variables.

- The single most important aspect of a variable being a state variable is that it is path independent, which means that different paths can be used to calculate a change in that state function for any given process.

- So, if you want to know the change in a state variable (for example, let's call it variable U) in going from some point A to some other point B, then it can be written as an integral over what is called the differential, dU, which represents a tiny change in a variable, and the integral is a sum over all of those tiny changes. In this case, the integral from A to B over dU is equal to U evaluated at point B minus U evaluated at point A. This difference is ΔU.

- Point A and point B are certain values for certain state variables of the system, and U represents some other state variable that can be expressed as a function of the ones involved in A and B.

- For example, suppose that A corresponds to a pressure of 1 atmosphere and a temperature of 300 kelvin, and B corresponds to a pressure of 10 atmospheres and a temperature of 400 kelvin. Also suppose that the variable U corresponds to the internal energy of the system (which is, in fact, designated by the letter U).

- The integral over dU from A to B could be performed in different ways. For example, first we could integrate over pressure from $P = 1$ to $P = 10$ atmospheres, holding temperature constant at 300 K, and then we could add to that the integral over temperature from $T = 300$ to $T = 400$, holding pressure constant at 10 atmospheres. That would give us a value for ΔU.

- We could also arrive at point B by integrating first over temperature, from 300 to 400 K, but now at a pressure of 1 atmosphere, and then we could add to that the integral over pressure from 1 to 10 atmospheres, but now at a temperature of 400K.

- Because this is a state variable, the ΔU that we obtain no matter which of these integration paths we take is going to be the same. The path does not matter: The value of a state function only depends on the other state variables, in this case temperature and pressure.

- From this definition, you can see that this also implies that if we were to travel back to point A, then the integral must be zero. The change in U from point A to point A is always zero because U is a state function and only depends on where it is in terms of these other variables, not how it got there. We could mess with the system in all kinds of ways, pushing it all over pressure and temperature space, but if we push it back to the starting place, the internal energy will be the same as when it started. That's path independence.

- In thermodynamics, we often want to measure or know the change in a given variable—such as the internal energy or temperature or pressure or volume—as we do something to the system. The fact that these variables are path independent means that in order to measure a change, we can use the easiest path between two states. For a state variable, we can choose whatever path we want to find the answer, and then that answer is valid for all possible paths.

- There are also variables in thermodynamics that are path dependent. The two most important ones are work and heat. These are both path dependent because they involve a process in which energy is transferred across the boundary of the system. This means that such variables are associated with a change in the state of the system (a transfer of energy), as opposed to simply the state of the system itself.

- Pressure is a state variable that is associated with a force distributed over a surface with a given area: Pressure (P) equals force (F) divided by an area (A). This means that pressure has units of newtons per meter squared N/m^2, which is also known as a pascal, named after the famous French mathematician of the 17th century.

Blaise Pascal (1623–1662) was a French scientist who is known for creating a unit of measure for pressure.

- One pascal is a pretty small amount of pressure—about that exerted by a dollar bill resting flat on a table. At sea level, the pressure of the atmosphere is around 100,000 pascals, also equal to 1 atmosphere, which is just another unit of pressure.

- In terms of what pressure means at the molecular level, we can think of a gas of atoms or molecules moving randomly around in a container. Each particle exerts a force each time it strikes the container wall. Because pressure is force per area, we just need to get from these individual collision forces to an average expression: Pressure is equal to the collision rate (that is, number of collisions per unit time) times the time interval over which the collisions are observed times the average energy imparted by a collision divided by the volume in which the collisions are taking place.

The Ideal Gas Law

- In the case of gases, the ideal gas law is a simple relationship of the state variables pressure (that is, pressure of the gas), volume, and temperature. This relationship makes a few important assumptions, the two most important ones being that the particles that make up the gas do not interact with one another and that they have negligible volume compared to the volume of the container.

- The origins of the ideal gas law date back to the mid-17th and mid-18th centuries, when a number of scientists were fascinated by the relationship between pressure, volume, and temperature.

- Boyle's law, published in 1662, states that at constant temperature, the pressure times the volume of a gas is a constant. Gay-Lussac's law says that at constant volume, the pressure is linearly related to temperature. Avogadro's law relates the volume to the number of moles. Charles's law states that at constant pressure, volume is linearly related to temperature.

- If you have a container of gas particles at some fixed temperature, then the pressure exerted by the gas on the walls of the container is set by how often they collide with the walls and also how much kinetic energy they have when they do.

- If you decrease the volume of the container at constant temperature, the collisions will occur more frequently, resulting in higher pressure. If you increase the volume but hold the temperature constant, the collision frequency goes down, lowering the average pressure.

- On the other hand, if you keep the volume fixed but increase the temperature, then both the collision frequency as well as the amount of force per collision will increase, resulting in an increase in pressure.

- All of these different relationships can be combined together to form the ideal gas law, which says that the pressure (P) times the volume (V) equals the number of moles (n) times what is known as the universal gas constant (R) times the temperature (T): $PV = nRT$.

- One mole of any substance is equal to 6.022×10^{23} of the number of entities that make up the substance. For example, a mole of water is equal to that number—also called Avogadro's number—of water molecules.

- Because an ideal gas has no interactions, the only kind of energy going on inside the container is due to the kinetic energy of the gas particles. That means that the internal energy of the gas is only related to the average kinetic energy, which is directly related to the temperature of the system.

- So, the internal energy of an ideal gas must be proportional to the temperature of the gas. Specifically, the internal energy of an ideal gas is equal to 3/2 times the number of moles of the gas times a constant called the gas constant times the temperature.

- Keep in mind that any time you use temperature in a thermodynamic relationship or equation, you must use the units of Kelvin.

Suggested Reading

DeHoff, *Thermodynamics in Materials Science*, Chaps. 2.2.1, 4.2.4, and 9.3.

Moran, Shapiro, Boettner, and Bailey, *Fundamentals of Engineering Thermodynamics*, Chap. 3.

Serway and Jewett, *Physics for Scientists and Engineers*, Chaps. 19 and 21.

Questions to Consider

1. What does it mean for a system to reach equilibrium?

2. Consider the task of inflating a balloon. What thermodynamic variables are you actually changing? How are you changing them? Are these variables intensive or extensive?

The Ideal Gas Law and a Piston

Lecture 5—Transcript

Welcome to our lecture on state functions and our first law that relates them together. No, this isn't the first law, that would be the first law of thermodynamics, which we will get to in the next lecture. Instead, today we're going to talk about the ideal gas law, which relates the three very fundamental thermodynamic variables we've been talking about together—pressure, volume, and temperature. These variables are also all known as state variables, so the function that relates them together is known as a state function.

A given state function itself is a thermodynamic variable that depends solely on the other independent thermodynamic variables. This means that we can think of state functions as models of materials. And once we know what the state function is, that is, how the state variables are related to one another, it's a whole lot of fun to plot one state variable against another, since it unlocks the secrets of the behavior of a system. So we'll be doing that a bunch today for pressure and volume.

At its core, a state variable is named so because of exactly that; it is a variable that describes the state of a system. Pressure, volume, temperature, and internal energy are just a few examples. For any given configuration of the system, we can identify values of the state variables. And by configuration, I just mean where things are, like the atoms or molecules that make up the system. The single, most important aspect of a variable being a state variable is that it is path independent. I said this briefly in the last lecture, but here, we're going to really make sure we feel comfortable with the meaning of this definition. The fact that a state variable is path independent means that different paths can be used to calculate a change in that state function for any given process.

In order to illustrate this concept, first, I'll show you the math, and then, I'll show a mountain. So, if I want to know the change in a state variable, let's call it variable U, in going from some point A to some other point B, then it can be written as an integral over what is called the differential, dU. The differential represents a tiny change in a variable, and the integral is a sum over all those tiny changes. So, in this case, the integral from A to B over dU is equal to U evaluated at point B minus U evaluated at point A. We write this difference as ΔU. And when I say "point A" and "point B," I mean certain values for certain state variables of the system. And U represents some other state variable that can be expressed as a function of the ones involved in A and B.

So, for example, suppose A corresponds to a pressure of 1 atmosphere and a temperature of 300 Kelvin, and B corresponds to a pressure of 10 atmospheres and a temperature of 400 Kelvin, and we'll suppose that the variable U corresponds to the internal energy of the system, which is, in fact, designated by the letter U. The integral over dU from A to B could be performed in different ways. For example, first I could integrate over pressure from $P = 1$ to $P = 10$ atmospheres, holding temperature constant at 300 Kelvin, and then add to that the integral over temperature from $T = 300$ to $T = 400$, holding pressure constant at 10 atmospheres. That would give me a value for ΔU.

I could also arrive at point B by integrating first over temperature, from 300 to 400 K, but now at a pressure of 1 atmosphere, and then add to that the integral over pressure from 1 to 10 atmospheres, but now at a temperature of 400 Kelvin. The point is that, since this is a state variable, the ΔU that I obtain, no matter which of these integration paths I take, is going to be the same. The path does not matter; the value of a state function only depends on the other state variables, in this case, temperature and pressure.

From this definition, you can see that this also implies that if I were to travel back to point A, then the integral must be zero. The change in U from point A to point A is always zero, since U is a state function and only depends on where it is in terms of these other variables, not how it got there. I could mess around with the system in all kinds of ways, pushing it all over pressure and temperature space, but if I push it back to the starting place, the internal energy will be the same as when it started. Now that's path independence.

Okay, that's a bit of math. Now, how about that mountain? Suppose one day I feel like hiking up to the top of Mount Everest. That's a height of 8,848 meters above sea level. Now there are two main routes to getting to the top.

Notice that each of these routes is a different distance. Some require less total hiking distance, while others represent a much longer trek. But the key here is that regardless of which path you take to get to the top, once you're at the top, you are always the same exact distance from sea level. The height of the mountain is a state function; it is not at all dependent on the path taken to get there. As long as your configuration in space is the same, in this case, that would be the location of the peak, then your height variable will always have the same value.

So why am I making such a big deal out of all this path independence stuff? Here's the thing; in thermodynamics we often want to measure or know the change in a given variable, like the internal energy, or temperature, or pressure, or volume, as we do something to the system. The fact that these variables are path independent means that in order to measure a change, I can use the easiest path between two states. In the case of my PV diagram from before, this path here might be a whole lot easier to measure in an experiment than this one here. For a state variable, I can choose whatever path I want to find the answer, and then that answer is valid for all possible paths.

Now, since I'm making the distinction of path independence so important, you may have guessed by now that there are also variables in thermodynamics that are path dependent. The two most important ones, and the only ones we will care about in this course, are work and heat. Work and heat are both path-dependent variables, since they involve a process in which energy is transferred across the boundary of the system. This means that such variables are associated with a change in the state of the system, a transfer of energy, as opposed to simply the state of the system itself.

I'll come back to these in a moment, but first, I want to stick with the key state variables of this lecture, namely pressure, volume, temperature, and the number of atoms or molecules in the system. Temperature, we already spent a whole lecture on, and you probably have a good sense of volume and number of particles. But pressure may not be as intuitive, so I want to make sure I define that clearly first. Pressure is a state variable that is associated with a force distributed over a surface with a given area. So pressure equals force, divided by an area.

This means that pressure has units of Newtons per meter squared, which also is known as a Pascal, named after the famous French mathematician of the 17th century. One Pascal is a pretty small amount of pressure—about that exerted by a dollar bill resting flat on a table. At sea level, the pressure of the atmosphere is around a hundred thousand Pascals, also equal to 1 atmosphere which is just another unit of pressure.

In terms of what pressure means at the molecular level, we can think of a gas of atoms or molecules moving randomly around as we saw in the Brownian motion that I discussed in the temperature lecture. Each particle exerts a force each time it strikes the container wall. Since pressure is force per area, we just need to get from these individual collision forces to an average expression, which we do by writing that pressure is equal to the collision rate, that is, the number of collisions per unit time, times the time interval over which the collisions are observed, times the average energy imparted by a collision, divided by the volume in which the collisions are taking place.

From basic physics, we learn that energy equals force times distance, so, if you look at this expression for pressure, it makes sense; we've taken our force-per-area definition and put energy in place of force and added a distance term on the bottom, since E equals F times d. This shows that pressure can also be viewed as energy per unit volume. So, now that we've defined what it is, let's play around with the pressure in our demo to make sure we have a good intuition for the variable, and in particular, how it's connected to another state variable, volume.

In this demo, we're going to explore the relationship between volume and pressure. These are state variables that are related to one another via the ideal gas law. What that law lets us do, is to make predictions about the effects of changing one of those variables, the effect that that has on other variables. So, how am I going to do what? Well, here's a glass jar, and it's hooked up to a pump, and what this pump does is it sucks air out of the jar, and that's what we call being under vacuum. Now, it's not a perfect vacuum, since there's still going to be air inside there, just a lot less of the air than we have outside.

So, inside this jar, the pressure is going to be different than the pressure outside. Now, roughly speaking, given the strength of this pump, there are going to be about $^{1}/_{20}{}^{th}$ the number of air molecules inside this jar, compared to outside. So, that means that the pressure inside is about .05 atmospheres when I turn the pump on, compared to 1 atmosphere for sort of normal, standard sea-level pressures. So, let's now play around with this and see what the effects of changing the pressure are on various items. I have different things that I'm going to put into this jar and lower the pressure, and we're going to take a look at what happens.

First, I'm going to start with the balloon. Now, this balloon is something I hope many of us have experienced; it has pressure inside, from the air inside. And combined with the tension of the balloon and the pressure of 1 atmosphere outside, it's in stable equilibrium; it's not changing—it's not expanding or contracting. But, if I change the pressure outside, which is the force acting on it, what do you think will happen? Let's find out. I'm going to put the balloon inside, and, I'm going to turn on the pump.

So what you can see, is that by lowering the pressure outside—that's the force that's being imposed on the outside of the balloon—by lowering that force, the balloon was able to find a new equilibrium. It was able to expand more. Now, if I take that lower pressure away by undoing this hose here, and you see, it goes right back to the initial place where it started. Okay, so that gives us a little bit of a sense of what happens when we play with pressure for a balloon.

What about a can of potato chips. What do you think would happen if I put a can of potato chips inside of this vacuum chamber? Well, let's take a look.

So what you saw there is that I was actually able to open the can of potato chips just by lowering the pressure outside. What's happening is that the air molecules outside the can, outside the top of the can, are a lot less in terms of the force they're putting down on top, than the air molecules inside of the can, because the ones inside of the can are still at one atmosphere. So I've got a big pressure differential across the top of the can. And it's so much, that it's enough to burst the top open. That's a pretty fun way to open a can of potato chips.

There's no better way to shave than with shaving cream that's been put under pressure. So what I'm going to do now is fill this container up with some shaving cream. And let's see, what do you think is going to happen when I put this under pressure? Okay, so I'll turn on my pump again.

So as I lower the pressure outside of the shaving cream bubbles, those bubbles wanted to expand, just like the balloon, and find a new equilibrium given the lower pressure outside. Pressure is related to volume. And we know that from the ideal gas law, because, well, because $PV = nRT$. And since T and n, and R were all constant in this experiment, if I lower the pressure, volume must change in reaction. So, this shows you in a pretty intuitive way, I hope, how we have this inter dependence between two of the most important state variables in thermodynamics—pressure and volume.

So there we were able to see first hand the close relationship between volume and pressure. In order to dive deeper into this connection, let's go back now to the piston we discussed in Lecture 2. With a piston, we will be able to really connect three of these state variables—pressure, volume, and temperature—and their changes together. The setup here is pretty simple; it's a cylinder filled with some gas, and at the top we have fitted a movable piston.

Now, at equilibrium, the gas occupies some volume V, and it exerts a uniform pressure P on the cylinder's walls and on the piston. Remember that equilibrium means that the macroscopic variables of the system do not change with time. At a given temperature, the gas particles, which could be either atoms or molecules, will have a certain average kinetic energy. With that energy, the gas particles will be ricocheting off of the container walls.

If the piston has a cross-sectional area A, then the force exerted by the gas on the piston is Force = pressure times area. But now, let's suppose that I have a bunch of tiny pebbles handy, and I start to put them on top of the piston one by one. Just by their weight, these pebbles will exert a force down on the piston, inward into the container, resulting in a compression of the gas inside. Now, remember that I just recalled from intro physics that energy equals force times distance? Well, as I add pebbles here to the top of the piston, we certainly have a force times distance. The force is external from the pebbles, and the distance is just how far down the piston gets displaced. This type of energy is called work, which is a path-dependent thermodynamic variable. It involves energy flow across the boundary of the system, measuring change of the system and not the state of the system itself.

Now, once the piston has had a chance to adjust to the force that was applied by the pebble and come to a new equilibrium position, we can set the magnitude of the external force exactly equal to the pressure of the gas inside the container times the area of the piston. The two forces cancel one another when the piston is at rest. So, each time we place a pebble on top, we can think of a little bit of work being done by the piston on the gas. A little bit of force times distance happens.

This is a good time to put up a plot of the pressure vs. volume of the gas inside the container. Say we start at some point here on the lower right part of this graph, with some pressure and volume. Now, we add a pebble to the top. What happens in this plot? Well, we go down a little bit in volume, so we move over to the left, and the pressure increases, so we move up a little. As I add more pebbles, the volume of the gas continues to decrease while the pressure increases. And at each step, work is being done on the gas equal to the area under this curve. I can sum up these pieces area along the way to see how much total work is being done.

Some of you who have had some calculus will notice that what I'm doing here is taking an integral. Remember, a sum over many small changes in one or more variables. If I move from this pressure and volume point over to this one by adding all those pebbles, then the work that I have done to the gas is equal to the total area under this curve. That's also equal to the integral over the pressure times the change in volume, or the integral over $P\,\Delta V$.

In order to evaluate this integral, we have to know how the pressure of the gas varies with volume during the process. In general, the pressure is not constant during the process, but rather, the pressure depends on the volume and temperature. But the most important thing here is the following; notice the big difference between this integral and the one we did before for a state function. In this case, even though I'm going from one PV point to another, the way in which I go there is crucial. The integral will be completely different depending on which path I take. The work is a path-dependent variable. And heat falls into the same category as work—path dependent—in the sense that it is energy that can be put in or taken out of a system during a process, and not a state of the system. Because heat and work depend on the path, neither quantity is determined solely by the end points of a thermodynamic process. Okay, we're going to come back to this piston, so please don't forget about it.

But before I go any further, I need to turn back to the idea of a state function in order to tell you about a very nice and simple relationship of state variables, that is, pressure, and volume, and temperature. In the case of gases, we have the ideal gas law. This relationship makes a few important assumptions; the two most important ones being that the particles that make up the gas do not interact with one another and that they have negligible volume compared to the volume of the container. We're going to see how this relationship helps us to better understand the work of our piston here.

The origins of the ideal gas law date back to the mid-17th and 18th centuries, when a number of scientists were fascinated by the relationship between pressure, volume, and temperature. For example, Boyle's law, published in 1662, states that at constant temperature, the pressure times volume of a gas is a constant. Gay-Lussac's law says, at constant volume, the pressure is linearly related to temperature, Avogadro's law relates the volume to the number of moles, and Charles's law states that at constant pressure volume is linearly related to temperature. Clearly, for that period in history, playing with gases was all the rage.

If I have a container of gas particles at some fixed temperature, then the pressure exerted by the gas on the walls of the container is set by how often they collide with the walls, and also, how much kinetic energy they have when they do. If I were to decrease the volume of the container at constant temperature, you can see that the collisions will occur more frequently, resulting in higher pressure. If I increase the volume but hold the temperature constant, the collision frequency goes down, lowering the average pressure.

On the other hand, if I keep the volume fixed but increase the temperature, well, then both the collision frequency, as well as amount of force per collision, will increase, resulting in an increase in pressure. All of these different relationships can be combined together to form the ideal gas law, which says that the pressure times volume equals the number of moles times what is known as the universal gas constant times the temperature, or, $PV = nRT$. As a reminder, one mole of any substance is equal to 6.022 times 10, raised to the 23rd power of the number of entities that make up the substance. For example, a mole of water is equal to that number—also called Avogadro's number—of H_2O molecules.

Okay, now that we're armed with our state function that relates these variables to one another, let's return to our piston example, as promised. Here's the same piston as before, with some pebbles placed on top exerting a force down, which in equilibrium, is exactly canceled by the pressure the gas exerts up. By the way, just to be sure there's no confusion on this point, the pressure of the gas is the same on all sides of the container. It's the average force per unit area of the gas, and it has no particular directional preference; the gas fills any container it's placed in, and its pressure is the force exerted on any boundary of the container.

Now, remember that I said that if I take pebbles off of the top, the pressure from the gas pushes the piston up to a new equilibrium position that is higher, since there's less force pushing down, so less pressure is needed to cancel that force and be in equilibrium again. We can use our intuitive picture to see how the volume expansion of the container is what allows the pressure to decrease. More volume for the same number of gas molecules to roam around in means fewer collisions, on average, with the container walls—bigger volume, smaller pressure.

And when I do take pebbles off, the change in volume means that the system is doing work, and as we saw, the work done is the integral over P times dV which is the area under the PV curve on this diagram. So I could get from, say, state 1 with $P1$ and $V1$; to state 2 with $P2$ and $V2$ by any path and each path will involve a different amount of work that the gas does. The way I've drawn it here, $P1$ and $V1$ have higher pressure and lower volume, so going from state 1 to state 2 implies removing pebbles from the top, while going back to state 1 would mean putting pebbles back on.

From state 1 to state 2, the system is doing work, although, in this case, the volume is increasing, instead of decreasing. Because of that, it has to have the opposite sign as the work done when the volume decreased. For this course, we use a sign convention where a volume decrease is positive work while volume expansion is negative work. It's still the area under this curve, but we put a minus sign in front, since it is work done by the system on the piston—the system being the gas. We'll discuss in more detail the sign conventions for both work and heat in the next lecture.

So, in going from state 2 to state 1 we are decreasing the volume and the integral over PdV will be the area under that curve, which is the work done by that process. Our sign convention tells us that, in this case, it is positive work, since we are doing work to the system. So, if I go along this path here to get from state 1 to state 2, and then let's say this other path here to get from state 2 back to state 1, then the total work done over this full loop is going to be the difference in the areas under the two curves. That happens to be the area inside of this loop region.

So for this process, in which work is done by the gas on the piston as it expands, and then work is done by the piston on the gas to push it back to its original starting point, the total net work that is the area in here is going to be work done by the system, since overall, it has a negative sign in front of it. That's because we started in state 1.

Let me emphasize again that I'll be talking a whole lot more about work as well as heat in the next lecture. But here, it's such an important part of the pressure-volume-temperature state functions that I can't possibly cover the ideal gas without talking about work. It just wouldn't feel right. And another thing that might not feel right to you at this time is that I still haven't brought temperature to the piston. Let's do that now.

The first point I want to make about temperature is to remind you that since an ideal gas has no interactions, the only kind of energy going on inside that container is due to the kinetic energy of the gas particles. That means that the internal energy of the gas is only related to the average kinetic energy, which we also happen to know is directly related to the temperature of the system. So, the internal energy of an ideal gas must be proportional to the temperature of the gas. Specifically, the internal energy of an ideal gas is equal to $^3/_2$ times the number of moles of the gas times a constant—called the gas constant—times the temperature.

The next thing I want to tell you about temperature is to please remember that any time you use temperature in a thermodynamic relationship or equation, you must use units of Kelvin.

And finally, the third thing I'd like to do with temperature is to include it now in our piston example. We'll consider two distinct cases. First, suppose that the container walls of the piston are adiabatic, which as you may recall from last lecture, means the gas inside cannot exchange heat with its surroundings. Now, suppose again, I start the piston in state 1, with a bunch of pebbles sitting on top, and I begin removing the pebbles one by one. We already know that P goes down and V goes up, but what happens to the temperature? We can figure that out for this case by realizing that the gas does work on the piston when the volume expands, as I described just a few moments ago. But how does the gas do that work?

Remember, work means energy has flowed across the system boundary. In this case, the system gives energy in the form of mechanical work to the piston. But where does that energy come from in the first place? You may have already guessed; it has to come from the kinetic energy of the gas. That must go down in this situation. Since no thermal energy can enter or leave the system, the only place for energy to come from to do that mechanical work is the temperature. The kinetic energy of the gas is traded for mechanical work of the piston. Therefore, in this case, the temperature goes down.

Okay, how about a different scenario. What if we were to hold the temperature constant? The way we do this is now to assume the boundaries are closed, as opposed to adiabatic, so heat can now pass across them. And we then envision the bottom of the container, on the side opposite the piston, as being in contact with an enormous reservoir. This reservoir is so incredibly big that we can think of it as an infinite either source or sink for heat to flow. But since its so massive compared to the container, the reservoir itself is unaffected by any energy transfer. And since it's in contact with the container, such a reservoir will hold the temperature of the gas inside the container fixed to whatever the temperature of the reservoir is.

Now, if I do the same thing, starting with a heap of pebbles on top of the piston in state 1 and slowly removing them, the behavior of the system is different. In this case, I can use the ideal gas law to find an exact relation between pressure and volume. For a constant temperature, PV equals a constant, since the number of moles of the gas isn't changing either, and R is just a constant. This means that P must be inversely proportional to V for a constant temperature process, and we get a curve like this in going from state 1 to state 2.

This type of PV curve is called an isotherm, since it's what you get with constant temperature. If I had a different temperature, I'd get a different, though similarly shaped curve, say, down here below this one. And notice that for this particular path, one at constant temperature, I can use the ideal gas law to calculate the exact amount of work done by the piston. In going from state 1 to state 2, the work done is equal to the integral PdV over that curve. Now before, when we didn't know the path, we also didn't know what to write for the function P. But now, armed with our state function for the ideal gas, we can write the pressure as nRT / V. And since T is a constant we can pull that out of the integral, along with n which is also a constant since no gas is escaping or entering, and R which is a constant by definition. That leaves us an integral from V_1 to V_2 of 1 over V, which we know is a natural logarithm.

The work done by this isothermal process, or the area under the curve, is then nRT times log of $V2$ over $V1$.

In this constant temperature case, the kinetic energy of the gas is still what did the work on the piston, but since the gas particles are in contact with an infinite reservoir of constant-temperature, they get replenished so as to maintain their overall average temperature. You can think of the energy that went into moving the piston as ultimately coming from the heat transfer from the reservoir.

I won't go much into heat at this point, but we'll come back to this piston later in the course, when we talk about an engine. But when I talk about engines, I want to be able to talk about different kinds of engines based on all sorts of different thermodynamic forces, from heat to magnetism to phase change to entropy itself. So please don't forget about these beautiful P-V-T processes that we can understand using our ideal gas state function. Before we return to them, though, we're going to get a whole bunch more thermo under our belt!

Energy Transferred and Conserved

Lecture 6

While all four of the laws of thermodynamics are important, it's the first and second laws that stand out because they set the stage for the two most important concepts: energy and entropy. In this lecture, the key concepts that you will learn about are the first law of thermodynamics, different types of energy and work, and energy transfer. The first law describes how energy is conserved but changes and flows from one form and process to another.

Energy Transfer

- In general, energy transfer from one system to another happens routinely in our daily lives, and one form of energy gets converted to another form in many of these processes. At the most general level, there are two types of energy transfer: Energy can be transferred in the form of heat (Q) and work (W).

- In the case of work, there is a thermodynamic force plus a corresponding thermodynamic motion. In basic physics, energy is equivalent to force times displacement, but in thermodynamics, the concept is much broader: Work is any thermodynamic force times its corresponding displacement.

- Each type of work—including chemical work, electrical work, and work done by entropy—can be described by a thermodynamic force term times a displacement term, and in each case, the work is part of the overall picture of energy transfer in the system.

- Heat is the workless transfer of energy. In other words, it's energy that transferred without any work terms. Importantly, heat is energy transferred during a process; it is not a property of the system. This is true for work as well: Heat and work only refer to processes of energy transfer.

- There are many different ways—or paths—that can get you from the initial state of the system to the final state. And heat and work are path-dependent quantities. You could use a combination of heat and work in many ways to achieve the same end result, and they will therefore have different values depending on the process that is chosen.

- Something that only depends on initial and final states is precisely what a state function is. Pressure and volume are state functions, while heat and work are not state functions, since unlike pressure and volume, which are path independent, heat and work are path dependent.

- The simple combination of heat plus work ($Q + W$) for each of the paths from the initial to the final state of the system is a constant—it remains the same for all of the processes. In addition, the quantity $Q + W$ depends only on the two states of the system: its initial and final states. This means that $Q + W$ must represent a change in an intrinsic property of the system. Because it only depends on the initial and final states, it is a state property.

- It turns out that this quantity, this change in a system's intrinsic property, is defined as the internal energy of the system (U), which is a quantity that measures the capacity to induce a change that would not otherwise occur. It is associated with atomic and molecular motion and interactions.

- In basic physics, we learn that total energy equals potential plus kinetic energies. Now we must add this thermodynamic internal energy term U, which we can think of as a kind of microscopic potential energy term.

- Internal energy allows us to describe the changes that occur in a system for any type of energy transfer, whether from heat or work. And those energy transfers are exactly what the first law of thermodynamics keeps track of.

The First Law of Thermodynamics

- The first law of thermodynamics states the energy conservation principle: Energy can neither be created nor destroyed, but only converted from one form to another. Energy lost from a system is not destroyed; it is passed to its surroundings. The first law of thermodynamics is simply a statement of this conservation.

- Stated in terms of our variables, the first law reads as follows: The change in the internal energy of a system that has undergone some process or combination of processes is the change in U (or ΔU). So, you could think of it as the final value for U minus the initial value for U over some process that a system undergoes; ΔU is equal to the sum of the heat and work energies transferred over those same processes. The equation for the first law is $\Delta U = Q + W$.

- Stated in simple language, the first law says that a change in internal energy is exactly accounted for by summing the contribution due to heat transferred (into or out of the system) and the work performed (on or by the system).

- The sign conventions for this course are as follows: When heat is transferred *into* a system, Q is positive; when it transfers *out of* a system, Q is negative. Similarly, when work is done *on* a system, Q is positive; when work is done *by* a system, Q is negative.

- Suppose that heat transfers into the system, but no work is done during that process. In that case, the sign of the heat transfer is positive, and the change in internal energy of our system is $Q + W$, so ΔU will also be positive because W is zero. On the other hand, if work is done on the system, the change in the internal energy of the system would be positive, and Q is zero because there is no heat being put in or taken out that does not involve a work term.

- The first law of thermodynamics itself can be used to define heat. Because the change in the internal energy of a system during some process is equal to the sum of heat and work energy transfers, this means that heat can be defined in the following way: Heat is the workless transfer of energy stored in the internal degrees of freedom of the material. $Q = \Delta U - W$.

- Degrees of freedom are the ways in which the atoms and molecules that make up a system can move around. The internal energy is the energy that is distributed over these different movements.

- The first law of thermodynamics can be thought of as a statement of energy conservation, but it is also a statement of the equivalence of work and heat. The first law says that you could change the internal energy of a system by some amount in two entirely different ways.

- Suppose that the internal energy of a system, U, is increased by 5 joules during some process. So, ΔU = 5 joules. This could be accomplished by performing 5 joules of work on the system with no heat transfer at the boundaries of the system. In this case, $Q = 0$, and $\Delta U = Q + W = 0 + 5 = 5$ joules. (This type of process—where no heat is allowed to enter or leave the system, so $Q = 0$—is referred to as an adiabatic process.)

- On the other hand, you could transfer 5 joules of heat into the system while performing no mechanical work, or any other work for that matter. In this case, the process is non-adiabatic, and $W = 0$. In addition, $\Delta U = Q + W = 5 + 0 = 5$ joules.

- This simple example shows how both work and heat can be used to cause the same changes to the internal degrees of freedom of a system. Combining knowledge of the types of processes occurring with the first law allows you to calculate changes in internal energy directly from measurable quantities like heat and work.

The Work of James Joule

- You might be intuitively comfortable with the idea that mechanical work can convert to heat through processes such as friction, but it was James Joule who was the first to rigorously test this theory.

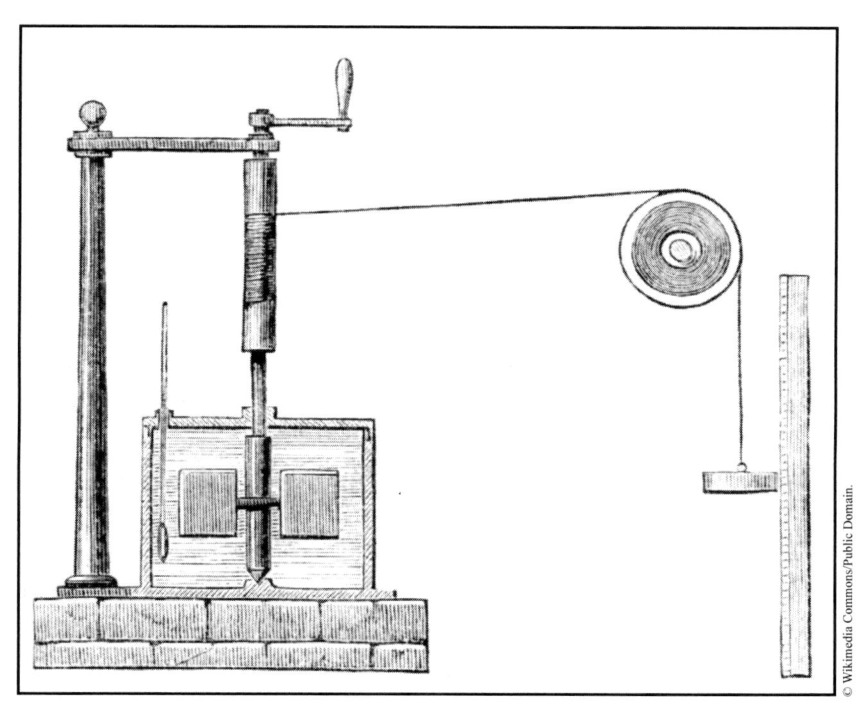

Through his experiments, English physicist James Joule was able to measure the mechanical equivalent of heat.

- Joule probed deeply the question of how much work could be extracted from a given source, which in turn led to one of the most important questions of the day—namely, how energy is converted from one form to another.

- Before Joule, the caloric theory dominated the scientific consensus on the theory of heat. The caloric theory, which dated back to the time of Carnot in the late 1700s, held that heat could neither be created nor destroyed. The idea was that there existed a kind of weightless fluid or gas—referred to as caloric—that flowed from hotter bodies to colder bodies.

- Caloric was thought of as a substance that could pass in and out of pores in solids and liquids. Caloric was thought to be the substance of heat and a constant throughout the universe that could neither be created nor destroyed.

- Joule's experiments in the mid-1800s on the mechanical equivalent of heat challenged the caloric theory. In what was perhaps his most famous demonstration, he used a falling weight, in which gravity did mechanical work, to spin a paddle wheel in an insulated barrel of water. When the wheel spun, mechanical work was done on the system, some of which went into the form of heat and increased the temperature of the water.

- From such experiments, Joule was able to measure quite precisely the mechanical equivalent of heat, which he placed just over 800 foot-pounds per British thermal unit (Btu). There are about 550 foot-pounds in one horsepower, and a Btu is defined as the amount of heat required to raise the temperature of a pound of water by one degree.

- These old-fashioned units are sometimes still used today, but they were the units that Joule and others at the time thought in. A horsepower was roughly the power of one horse, and, just as for temperature, units of energy were first measured in terms of water as a good reference point (like raising a certain amount of it by one degree). In modern units, a horsepower is roughly 746 watts, or joules per second, and a Btu is a bit over 1,000 joules.

- In both cases, the current way we think about energy is in the context of a joule. This illustrates how important James Joule's experiments were; they set the stage for the first law of thermodynamics to be developed. Following Joule, many other scientists confirmed that all forms of work—not just mechanical work—can be converted to heat.

Suggested Reading

Anderson, Thermodynamics of Natural Systems, Chap. 3.

DeHoff, Thermodynamics in Materials Science, Chap. 3.1.

Halliday, Resnick, and Walker, *Fundamentals of Physics*, Chaps. 18.7–18.12.

Moran, Shapiro, Boettner, and Bailey, *Fundamentals of Engineering Thermodynamics*, Chap. 2.

Serway and Jewett, Physics for Scientists and Engineers, Chap. 20.5.

Questions to Consider

1. What is the difference between the two forms of energy work and heat?

2. If heat and work are not properties of a system, what is?

3. Pick any technology you'd like: What are all of the types of work being displayed in this technology? Chart the energy pathways.

Energy Transferred and Conserved

Lecture 6—Transcript

Hi, and welcome to our lecture on the First Law of Thermodynamics! You may remember that we discussed the zeroth law earlier when we talked about temperature, so even though it's called the first law, this is actually the second law we're seeing so far in the course. Now, while I absolutely love all four of the thermodynamic laws we'll be discussing throughout the class, it's the first and second laws that kind of stand out, since they set the stage for the two most important concepts of all—energy and entropy. So I'm quite excited to be talking about the first law in this lecture.

The key concepts that we'll discuss are the first law itself, different types of energy and work, as well as energy transfer. We'll use a simple motor as an example of the first law to describe how energy is conserved but changes and flows from one form and process to another. Let's dive right in!

In general, energy transfer from one system to another happens routinely in our daily lives. And one form of energy gets converted to another form in many of these processes. For example, when you drive a car, the energy stored in gasoline, which is in the form of chemical energy, is converted to heat energy by burning the gasoline. This heat energy is then used to drive a piston, which, in effect, has turned the heat energy into mechanical energy. At the most general level, we need to define two types of energy transfer. First, energy can be transferred in the form of heat, and we use the variable Q in thermodynamics to represent energy in the form of heat. And second, energy can be transferred in the form of work, and here, we use the variable W to represent such a type of energy transfer.

How do we distinguish between work and heat? Let's take a very simple example. Suppose I hold my hands above a campfire to warm them up on a cold day. In this case, my hands will feel warmer from the heat radiating from the fire, some of which is absorbed by my hands. Now, let's suppose there's no fire, but I rub my hands quickly together. In this case, I produced the same warm feeling in my hands, but in a completely different way. With the campfire, heat was the form of energy transfer. When I rubbed my hands together, the mechanical work I performed was the form of energy transfer.

So in the case of work—here's a really key point—there was a thermodynamic force plus a corresponding thermodynamic motion. In this case, the force was created by friction between my hands, which opposes the direction of the rubbing. The motion in this case is simply the movement of the hands back and forth. We can think of this type of work as similar to what we learned in basic physics, namely, that energy is equivalent to force times displacement.

But here in thermodynamics, the concept is broader. We refer to work as any thermodynamic force times its corresponding displacement. This can be a friction force times a distance displacement, as in the example of rubbing our hands, or, as in the classic example used in introductory physics, a mass attached to a spring. But this work term in thermo has many other forms, and in fact, as we'll see, in thermo we account for all of them, or at least, all of the ones that matter for the problem at hand. And we'll need to go far beyond a mass on a spring to describe most realistic scenarios.

Let's illustrate this last point with a simple example. Suppose the only work term we know about is that of force times distance. In that case, would we be able to describe the energy processes going on for this weight lifter? When this person holds the weight above their head like this and stays perfectly still, are they doing any work? Is energy being expended? Of course, the answer is yes, since, for any of you who have tried it, you know how hard it is. But if the only work term we know about is force times distance, then, since the weights are held still, no work is being performed. The motion part of our force times motion is zero.

But as we know, even though no work is done with respect to the weights, the body is doing a whole lot of work. Each muscle cell's contraction is generated by billions of molecular machines, which take turns supporting the tension caused by the weights. When a particular molecule goes on or off duty, it moves, and since it moves while exerting a force, it's doing work. So work is done by one molecule in a muscle cell on another. There is a transformation of energy, but it is taking place entirely within your own muscles, which are converting chemical energy into heat.

Throughout the course, we'll be adding these different thermodynamic work terms to our repertoire, ranging from chemical work, to electrical work, to work done by entropy. As we'll learn, each type of work can be described by a thermodynamic force term times a displacement term, and in each case, the work is part of the overall picture of energy transfer in the system.

Okay, so, that's our initial discussion about work, but what about the other type of energy transfer I mentioned, namely heat? I mentioned in the hand-warming example that heat would be transferred from the campfire to the hands. Let's get a little more specific. In a way, the first law of thermodynamics itself can define heat for us, but I'll come back to that in a moment. For now, let me state that heat is the workless transfer of energy. In other words, it's energy that transferred without any of those work terms I just discussed. A really important thing we need to know about heat is that it is energy transferred during a process. Heat is not a property of the system. This is true for work as well. Heat and work only refer to processes of energy transfer. Because that's so important, I'd like to make it my repeat sentence for this lecture—Heat and work only refer to processes of energy transfer.

This means that we cannot, or at least, should not, make statements such as the following: "This system has 100 joules of heat." That's an incorrect statement, since while heat is a form of energy, it only refers to energy transfer into or out of a system, not the energy content of the system itself.

This is a good time to take a look at our demo for this lecture. Even though it's perhaps the simplest electric motor one can make, it's a system that helps bring to life a bunch of different forms of energy transfer.

So in this lecture, we're talking about the first law of thermodynamics, which is a statement of energy conservation. What I'd like to do in this demo is talk about the different forms of energy and how they can get transformed from one of those forms into another. So, let's start with this. I think you all know what this is; this is a battery. But, what can we do with this? Well, you also probably all know we can power just about anything. We can power lights; we can power motors; fans; computers; cell phones; and so forth.

But where did this energy come from in the first place? And how is it stored? Well, the way it's stored is in terms of chemical energy that's inside of this battery. And what happens to that chemical energy when I hook it up to, say, a light, is that it gets transformed into electricity. And then that electricity gets formed into light energy.

When the battery is empty, if it's rechargeable, I can fill it up again with more chemical energy. And just like I can do a lot of things with the stored chemical energy inside the battery, like power the lights, motors, and so forth, well, there are a lot of ways I could fill the energy inside of this battery back up. Like, I could plug it into a wall. But I could also use mechanical energy, turning a crank to provide electricity to fill the chemical energy back up.

So, there are a lot of different pathways for energy to flow, and that's kind of what I want to show you today. So, the first example that I'd like to show is one where we have this device right here, which is a piezoelectric. And in this one, what we can do, is we can use mechanical motion to convert directly into electrical energy. So, as I strike this piece of piezoelectric, what happens is I get a little current is these wires. And what you can see is that here I've hooked it up to a light, and so when I hit it with my finger, the light goes on.

Now, what I'd like to do is, so that you can see this more clearly, is I'd like to lower the lights here in the studio just so that you can see that we get a little, tiny amount of energy when I hit this piezoelectric.

So, in that case, what was the energy pathway that we had? Well, we had the mechanical motion of my finger, which in turn, converted into a force on the piezoelectric, so it converted into a stress in the material, which in turn created a current, so that's electricity, which in turn, made a light go on. We can make these kinds of energy maps for any process that involves energy transfer, and that's, in fact, something that we do all the time in thermodynamics.

By the way, as an aside, when I hit this piezoelectric, what is the real source of energy? Well, if you think about that, it's actually the food that I had to eat to give me the strength to hit the piezoelectric. And that food? Well, that was grown by the Sun. So, actually, this light that I just powered by tapping on this material, it's powered by the sun.

So here's another example of converting energy from one form into another. It's about the simplest way that you can make a motor. What I have is I have a piece of wire; I have some magnets; and I have a battery. I've taped some washers to the top to give the wire a little bit more stability. And, what I'm going to do is, I'm going to put the battery on top of the magnets, like that, and then, I'm going to put this wire on top of the battery. And you can see that it starts spinning, so I've actually made a motor. And what I'm doing here is I'm converting chemical energy into electrical energy, which goes through the wire and makes magnetic energy, which makes the wire spin, which is kinetic energy. So those are the energy pathways that give me this motors.

And the important point is that all of the energy pathways can be accounted for. That's the first law of thermodynamics. The first law tells us about the interplay between work and heat in any and all types of processes.

Now, in the demo you just saw, energy stored in a battery system is transferred to a metal loop that experiences a torque, that is, a turning force, and starts rotating. Here, we are converting one form of work, that is, electrochemical, to another form of work, that is, mechanical. As I mentioned, some of the energy is also rejected in the form of heating the wire. But in addition to these energy transfers, there's now a very important point I'd like to make related to heat and work, namely, that there are many different ways—or paths—that one can take from the initial state of the system to the final state. And as I mentioned before, heat and work are path-dependent quantities.

In our demo example, let's consider the state of the wire loop. In its initial state, the loop is stationary. In its final state, it's rotating. One of the ways we used to get the loop from its initial state to the final state was by using electrochemical work from the energy stored in the battery.

But another way to achieve the same result is by manually using my own muscular force. In other words, I could just spin the loop with my finger.

Here I'm converting mechanical work of my finger to mechanical work of the loop. That's a very different path, but it results in the same initial and final states; that is, the loop not spinning to the loop spinning.

And yet, another possibility would be to heat up a gas in a container connected to a piston that then rotates the loop. In that case, I'd be using heat to do mechanical work. And again, same initial and final states would occur—the loop not spinning to the loop spinning.

Basically, there are infinitely many ways to go from the initial state to the final state. Heat, remember we call it Q, may or may not be involved in many of these processes. So, one could use a combination of heat (Q) and work (W) in many different ways to achieve the same end result, and Q and W will, therefore, have different values depending on the process I choose. This is the reason why Q and W are called path-dependent quantities. They do not depend on the loop's initial and final states. If they did, they'd have to be the same for any process you chose.

And speaking of something that only depends on initial and final states, remember from our last lecture on the ideal gas law that this is precisely what a state function is. Two of the variables we discussed that are state functions are pressure and volume. Here, we can conclude that heat and work are most certainly not state functions, since, unlike pressure and volume, which are path independent, heat and work are path-dependent. Okay, so now we're at a very exciting point in this lecture, since we are equipped to develop the first law of thermodynamics.

Now that we know energy transfer occurs in the forms of heat and work, or Q and W, the question we ask is, can we connect these variables quantitatively? Let's look at the three possible paths I described just a moment ago to reach the final rotating state of the loop, starting from its initial stationary state. One case involved the battery's energy, one my finger, and the other heat to get the loop to spin. As I mentioned, in each of these cases very different amounts of Q and W are used along the path.

But, is there some combination of Q and W that is a constant for any process chosen? It turns out the answer is yes. If we were to experimentally measure the simple combination of heat plus work, or $Q + W$ for each of these paths, it would be a constant. It remains the same for all of the processes. That's pretty cool. And it tells us that the quantity $Q + W$ depends only on the two states of the system—its initial and final states. This means that $Q + W$ must represent a change in an intrinsic property of the system. And because it only depends on the initial and final states, it is a state property.

It turns out that this quantity, this change in a system's intrinsic property, is something we have discussed before, back in our lecture on basic thermodynamic concepts. It's defined as the internal energy of the system, or the variable U. As a reminder, internal energy is a quantity that measures the capacity to induce a change that would not otherwise occur. It's associated with atomic and molecular motion and interactions. In basic physics we learn that total energy equals potential plus kinetic energies. Now, we must add this thermodynamic internal energy term U, which if we want, we can think of as a kind of microscopic potential energy term.

Remember our example of a glass of water and a glass of ice sitting on a table. Both systems are at rest and at the same height, so the gravitational potential energy and kinetic energy of both systems would be zero. But the systems are, of course, very different, since one is liquid and the other solid. It is the internal energy that captures this difference. And you can see why this concept of internal energy is so important when we bring heat into the picture. If I apply heat to the ice, it will melt, changing its phase and completely changing its properties. But during that process of heating, the kinetic and potential energies of the system would still be zero.

Internal energy allows us to describe the changes that occur in a system for any type of energy transfer, whether from heat or work. And those energy transfers are exactly what the first law of thermodynamics keeps track of. The first law of thermodynamics states the energy conservation principle: "Energy can neither be created nor be destroyed, but only converted from one form to another." Energy lost from a system is not destroyed; it is passed to its surroundings. The first law of thermodynamics is simply a statement of this conservation.

Stated in terms of our variables, the first law reads as follows: The change in the internal energy of a system that has undergone some process or combination of processes, we write as ΔU. The delta here just means the change in the variable U. So, you could think of it as the final value for U minus the initial value for U over some process that a system undergoes. That ΔU is equal to the sum of the heat and work energies transferred over those same processes. Written out, the first law is: ΔU equals Q plus W.

Stated in simple language, the first law says that a change in internal energy is exactly accounted for by summing the contribution due to heat transferred into or out of the system and the work performed on or by the system.

Now, in order to use this equation to solve problems, we need to define a sign convention. Unfortunately, different thermodynamics textbooks actually use different conventions. There's no agreed-upon standard. But that's okay, since we just need to know how we're defining it here, and then we'll be consistent with that definition throughout the course.

So the sign conventions I'll adopt for this class are as follows. When heat is transferred into a system, we say that Q is positive; and when it transfers out of a system Q is negative. Similarly, when work is done on a system, it will be positive; and when work is done by a system our convention will be that it is negative. In terms of our equation, we can see, then, what this means for the internal energy of the system. Suppose heat transfers into the system but no work is done during that process. Going back to our hand-warming example, this would be the case if the system is our hands and the energy-transfer process occurring is that the system is being warmed over the campfire. In that case, the sign of the heat transfer is positive, and the change in internal energy of our system is $Q + W$, so ΔU will also be positive, since W is zero.

On the other hand (no pun intended), if I rub my hands together, then I am doing work on the system, so again, the change in the internal energy of the system would be positive. In that case, Q is zero, since there is no heat being put in or taken out that does not involve a work term.

Now remember, I mentioned that the first law itself can be used to define heat. We can see that pretty clearly simply by looking at the equation for the first law. Since the change in the internal energy of a system during some process is equal to the sum of heat and work energy transfers, this means that heat can be defined in the following way. Heat is the workless transfer of energy stored in the internal degrees of freedom of the material. Q equals ΔU minus W.

By degrees of freedom, what I mean are the ways in which the atoms and molecules that make up a system can move around. The internal energy is the energy that is distributed over these different movements. For the example of ice and water sitting on a table, there is less energy spread over those atomic-scale movements for solid ice, as compared to liquid water. That's why the ice is a solid; those internal motions, which are captured by the internal energy, are small enough for the water molecules to become locked into place and form a crystal.

The first law can be thought of as a statement of energy conservation, as I defined it. But it is also a statement of the equivalence of work and heat. The first law says that I could change the internal energy of a system by some amount in two entirely different ways. Suppose the internal energy of a system, U, is increased by 5 Joules during some process. So ΔU equals 5 joules. This could be accomplished by performing 5 Joules of work on the system with no heat transfer at the boundaries of the system. In this case, Q equals zero, and ΔU equals Q plus W, which equals zero plus five, which equals 5 Joules.

By the way, as a reminder of one of our important definitions from Lecture 4, this type of process, where no heat is allowed to enter or leave the system, so $Q = 0$, is referred to as an adiabatic process.

On the other hand, I could transfer 5 Joules of heat into the system while performing no mechanical work, or any other work for that matter. In this case, the process is non-adiabatic, and W equals zero; ΔU is equal to Q plus W, which equals 5 Joules plus 0, which equals 5 joules.

Now, I know this is a pretty simple example, but I want you to really see how both work and heat can be used to cause the very same changes to the internal degrees of freedom of a system, and also, I want you to start to get comfortable with the terms in the first law equation. In the next lecture, we'll dive deeper into this equivalence of work and heat in thermodynamics. But for now, the key point is that combining knowledge of the types of processes occurring with the first law allows one to calculate changes in internal energy directly from measurable quantities, like heat and work.

By the way, you may be intuitively comfortable with the idea that mechanical work can convert to heat through processes such as friction. That's what happens in the example of rubbing our hands together to warm them up. But did you know that it was James Joule himself who was the first to rigorously test this theory? Joule probed deeply the question of how much work could be extracted from a given source, which in turn, lead to one of the most important questions of the day, namely, how energy is converted from one form to another.

Before Joule, the caloric theory dominated the scientific consensus on the theory of heat. The caloric theory, which dated back to the time of Carnot and Lavoisier in the late 1700s, held that heat could neither be created nor destroyed. The idea was that there existed of a kind of weightless fluid or gas referred to as caloric, which flowed from hotter bodies to colder bodies. Caloric was thought of as a substance that could pass in and out of pores in solids and liquids. Caloric was thought to be the substance of heat and a constant throughout the universe that could neither be created nor destroyed.

Joule's experiments in the mid-1800's on the mechanical equivalent of heat challenged the caloric theory. In what was perhaps his most famous demonstration, he used a falling weight, in which gravity did mechanical work to spin a paddle wheel in an insulated barrel of water. When the wheel spins, mechanical work is done on this system, some of which goes into the form of heat and increases the temperature of the water.

From such experiments he was able to measure quite precisely the mechanical equivalence of heat, which he placed just over 800 foot-pounds per Btu. In case you're wondering, there are about 550 foot-pounds in one horsepower, and a Btu is defined as the amount of heat required to raise the temperature of a pound of water by one degree. Now, these old-fashioned units are sometimes still used today; just think of the power ratings given to air conditioners, but I wanted to explicitly use them, since they were the units that Joule and others at the time thought in. A horsepower was just that, roughly the power of one horse. And just as for temperature, units of energy were first measured in terms of water as a good reference point, like raising a certain amount of it by one degree. In modern units, a horsepower is roughly 746 Watts, or Joules per second, and a Btu is a bit over a thousand Joules.

Notice that in both cases the current way we think about energy is in the context of a Joule. This illustrates how important James Joule's experiments were; they did no less than set the stage for the first law of thermodynamics to be developed. Following Joule, many other scientists confirmed that all forms of work, not just mechanical work, can be converted to heat.

So now I'd like to conclude this lecture by returning to the first law and the example I showed you of the electric motor. As I described, chemical energy converts into electrical energy, which flows through a wire that allows a magnet to exert a force on it, doing magnetic work, which is then converted into the mechanical motion of the loop, or mechanical work. In each of these processes, some heat is generated. So W of one form converts to W of another form plus some Q. As long as we account for each type of energy transfer, the first law states that the internal energy of the system equals the addition of all Q and W terms.

And that brings me to the following question. For this motor, if I just let it go, what is the final state? In the end, where does all of the energy and work go? Well, let's take a look. If I let this motor keep on going, we know that the loop will keep on spinning in addition to heat being generated until the chemical energy in the battery is exhausted. At that point, it will simply stop spinning, and also, no more heat will be generated, since no more work is being done. So we started from energy stored in the form of chemical energy inside the battery, and after all of those various energy flows, where did it all wind up?

Well, some of it went into mechanical work to spin the loop, which in turn, did mechanical work on its surroundings by moving air molecules nearby. So you could say that some of it went to increasing the temperature of the air. Notice that this would be a good example of work being done on the surroundings, so the W term in the first law would be negative according to our sign conventions.

The heat that was generated during the various energy transfer processes also flows from the system into its surrounding environment. So that energy also went into heating the air. So it's also going to be a negative value for the process, since heat transferred out of the system, giving a negative Q in the first-law equation. Taken together, this means that after everything is said and done, we have ultimately converted chemical energy into an ever-so-slight temperature change of the environment.

Now, we did do something useful along the way, namely, we made something spin around. But one of the basic questions we can answer with thermodynamics is the following. What is the usefulness of energy? In other words, to what extent can we take energy from one form and make it perform tasks for us by converting it into another form? As we'll learn, it turns out that while we can convert work into heat with 100 percent efficiency, we cannot do the reverse. We can never convert heat back into work with 100 percent efficiency. That statement, by the way, is the reason why a perpetual motion machine can never work. This seemingly one-sided nature of work and heat flow cannot be understood from the first law of thermodynamics alone; for that, we'll need to know the second law, which don't worry, we'll get to in just a few lectures from now.

Work-Heat Equivalence

Lecture 7

This lecture digs deeper into the equivalence of work and heat. By the end of this lecture, you should have a solid understanding of the first law of thermodynamics and the three terms that go into it: work, heat, and internal energy. In addition to these concepts, which have been introduced in previous lectures, you will learn some important new concepts—specifically, some mathematical relationships that are used in thermodynamics to describe the interdependence of these quantities.

Work Terms

- A property is a quantity we can measure and observe for a given system. It's something that differs depending on the thermodynamic state of the system.
- An intensive variable does not get bigger or smaller with the size of the system, while an extensive variable does. A given intensive property has what is called a conjugate pair that is an extensive property. We use the term conjugate pair to mean that the variables are connected or coupled.
- The product of the intensive property and the differential change in the conjugate-paired extensive property gives us a work term. More precisely, it's called differential configurational work, which is just a fancy way of saying that there is a change in the configuration, or structure, of the system. All work involves such a change.
- When we write a change in work, we don't use d to write dW as we would for a change in, for example, volume or pressure. Instead, we write it as δW. We do this to signify that we're talking about a change in a variable that is path dependent.

- There are many of these conjugate pairs of thermodynamic variables. Each pair when multiplied gives a thermodynamic force times a thermodynamic displacement, to produce a certain type of configurational work.

- For example, pressure is an intensive property (it doesn't scale with the system size) that, when multiplied by the change in the extensive property volume (which does scale with the system size), gives us a configurational work term. This is a term representing mechanical work.

- Surface tension is an intensive property (it doesn't scale with size) that, when multiplied by the change in the extensive property area (which will get bigger as the system gets bigger), gives us the surface energy configurational work term.

- The chemical potential is an intensive property that, when multiplied by the extensive property of the change in the number of particles, gives us the configurational work term known as chemical work.

- In all cases, the units of the work term are energy. In all cases, we have [thermo force] × [thermo displacement] = energy.

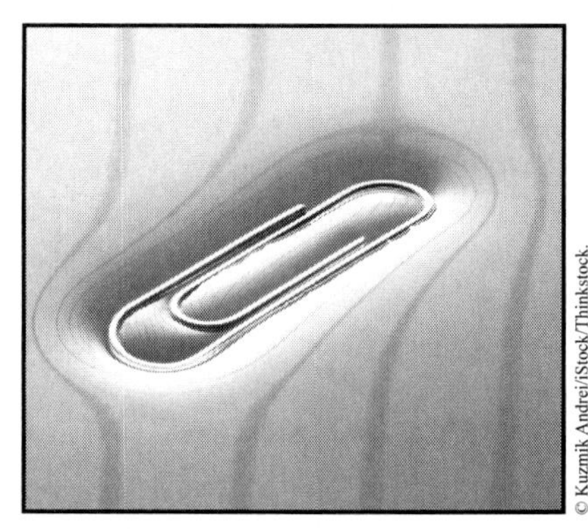

Surface tension is a property of a liquid surface that makes it seem like the liquid is being stretched.

- These pairs of variables are called conjugate pairs, which simply means that the variables are connected or coupled. In other words, not just any two variables will give a thermodynamic work term but, rather, only these special conjugate pairs.

- The word "configurational" is used to define the work terms. This means that it is a type of work that changes the configuration, or structure, of the system. We could have used the word "work" instead, but it is common in thermodynamics to use the term "configurational work."

Reversibility and Irreversibility

- The terms "reversibility" and "irreversibility" are used to describe the type of process occurring, which depends on the type of work being done during that process. Irreversible work arises from the fact that some types of work are always done on the system regardless; in this case, reversing the force or displacement direction does not affect the sign of the work. This is also known as dissipative work.

- A reversible process is not real: It is an idealized process for which the following conditions all hold throughout the entire process. First, the system is always in equilibrium. Second, there are no dissipative processes occurring (in other words, no energy is lost to the surroundings—there's no friction, viscosity, electrical resistance, and so on). Finally, a reversible process can occur both forward and backward with absolutely no change in the surroundings of the system.

- Irreversible processes represent all natural processes. For example, if you put a drop of food coloring into a glass of water, you can observe the food color diffuse slowly out into the liquid. The irreversible process going on is diffusion, and of course, once it is spread out through the liquid, you never see the food coloring spontaneously diffuse back into a single droplet.

- It is quite intuitive in the example that the process described will never reverse. In fact, the second law of thermodynamics serves as a rule to account for the one-way nature of this process.

- The word "spontaneous" is a very important one in thermodynamics because thermodynamics tells us about equilibrium states of a system, and in order to find those places of equilibrium, we need to be able to predict which processes are spontaneous and which ones are not. It is the second law of thermodynamics that provides a theoretical criterion to let us determine which direction processes will occur, even when the answer is not intuitively obvious, as it was in the example of food coloring in water.

- Real processes are never reversible. They may approach the conditions of reversibility, but they will never adhere to them fully. These unreal, highly theoretical reversible processes can be used to perform calculations for very real irreversible processes.

- The power of using the concept of reversible processes in thermodynamic analysis is that the variables used to analyze thermodynamic systems are state variables, which means that if the final and initial states of a system are known, then calculations on those systems can be done with any arbitrary path because the final state is independent of the way the system came to that state.

- This means that you could choose to take a theoretical reversible path in your calculation of a final state property, and the property you calculate will hold true for the real irreversible path.

- In addition to work, another way—that is just as important—to transfer energy into a system is through heat. It is the addition of these two transfers that gives us the energy-conservation principle in the first law.

- Similar to the case for work, a differential form of heat (δQ) can be found in terms of an intensive property multiplied by the differential of an extensive property. In this case, the intensive property is going to be temperature (T), and the extensive property is entropy itself (S).

- Entropy has already been defined as a counting of the number of possible microstates a system can possess. The following is another way to understand what entropy means: Just like volume change is the conjugate pair for pressure to give us a work term for mechanical energy, entropy is the conjugate pair for temperature to give us heat energy.

- This means that heat itself can be defined as the energy transferred into a system in terms of thermal interactions with a transfer of entropy. Entropy provides the link between heat and temperature.

Differentials

- Another name for a delta, in general, is a differential. So, the differential volume would be dV, and as we learn in introductory calculus, we can integrate over a differential variable to get the total value of that variable. If we integrate over the differential dV from, for example, V_1 to V_2, we would get the change in volume, or $V_2 - V_1$.

- That type of differential—when d is used—is called an exact differential. It means that the integral over it does not depend on the path taken. We know this intuitively from the calculus: We integrate over the dV to get a "V" and then plug in the values of the integrand (V_1 and V_2) to get the actual value of the integral. We didn't worry about how the volume of some system went from V_1 to V_2.

- In fact, it's called a state variable when it doesn't depend on the path taken. We describe changes in state variables by exact differentials. That brings us to the two energy transfer terms of the first law: work and heat, or W and Q.

- In this case, we have variables that are not state variables, but rather, they are path dependent. That's why we use a different differential to describe a change in such a variable. The δ is also called an inexact differential because if we integrate over it, we must specify a path for the integral.

- That makes sense if you think about it: We cannot integrate over a path-dependent variable without knowing its path. The δ is just our mathematical notation to make sure that we know what type of variable we're describing.

- The first law states that for a given process, any change in U, the internal energy of the system, is equal to the sum of the energy transfer terms Q and W. But now let's write the first law in its differential form: The differential dU equals the differential change in the internal energy of the system, which equals $\delta Q + \delta W$.

- Notice that for U, we use the exact differential (d)—that's because internal energy is a state function and does not depend on the path taken, only on the end points of the integral itself (or, in thermodynamic terminology, the initial and final states of the system).

- One of the nice things about a differential being exact is that it means there are a number of mathematical consequences that we can take advantage of in thermodynamics. For example, when a differential is exact, we can use certain relationships and rules that hold rigorously for the variable being differentiated.

- One class of relationships, known as Maxwell relations, are extremely important because they allow us to transform one thermodynamic equation into another one, sometimes making it much more useful for the problem at hand.

- The mathematical rules associated with exact differentials also allow us to equate one thermodynamic variable—such as temperature, entropy, volume, pressure, or internal energy—with derivatives of other variables.

Suggested Reading

Callen, Thermodynamics and an Introduction to Thermostatistics, Chap. 7.

DeHoff, Thermodynamics in Materials Science, Chaps. 3–4.

Serway and Jewett, Physics for Scientists and Engineers, Chap. 20.4.

Questions to Consider

1. If you compress and expand a gas, it is possible to recover all of the energy you put in. Can you do the same thing for rubbing your hands together?

2. What is the difference between reversible and irreversible work?

3. Under what conditions is enthalpy equivalent to thermal energy?

4. Can you think of other ways to raise temperature instead of using mechanical work and heat?

Work-Heat Equivalence
Lecture 7—Transcript

In this lecture I want to dig a bit deeper into the equivalence of work and heat. This is something we touched on in the last lecture, when I introduced the first law of thermodynamics. But the idea that work and heat can lead to the same change in internal energy is so powerful, that I want to devote an entire lecture to it here in order to do the concept justice.

And speaking of concepts, the ones I'd like you to be sure you know from this lecture are a solid grounding of the first law of thermodynamics, and the three terms that go into it, namely, work, heat, and internal energy. These are not new concepts, since I've introduced them all in previous lectures, but today, I want you to really feel your oneness with them. We'll also learn some important new concepts. And to complement the feeling side of things, in this lecture, we're also going to bring in some important mathematical relationships that we use in thermodynamics to describe the interdependence of these quantities.

But before we get into all that, I'm going to begin with a smashing spark—literally! In the demo for this lecture I show a couple of examples of how work can get converted into heat. I'll start by lighting a piece of cotton on fire with a hammer. Let's take a look.

The first law of thermodynamics shows the equivalence of using heat or work to raise the internal energy of a system. What that means is that, for example, if I have boiling water, I can't tell you whether I put mechanical work into making it boil, or thermal energy. That's what the first law says, is that those two are equivalent in terms of raising the internal energy of that water. So I want to put this into action today, and the way I want to do that, to start with, is by lighting something on fire, which is always a fun thing to do in a thermal course.

So, let's take this piece of cotton. My question is, how could I light this cotton on fire with thermodynamics and a hammer? Well, one way you might think is that if I just put the cotton down on the table, and I hit it really hard, maybe the kinetic energy from this hammer will spark and will lead to the cotton lighting on fire. Let's give that a shot. So there's the piece of cotton; I'm going to hit it with a hammer.

Now, what you can see is that, while I've certainly smooshed the cotton a little bit, but it hasn't lit on fire. So using that transformation of energy doesn't work. But, we in this course, are armed with more. We know about the ideal gas law. And the question now is, can I use the ideal gas law to light this cotton on fire using a hammer. And in fact, the answer is yes. Let's take a look.

In here, what we have is a chamber, and what I've done is I've put another little piece of cotton inside, as you can see, okay. And what I'm going to do is, I have here a little piston that fits perfectly into the chamber, okay, like that. And what I'm going to do is I'm going to screw the top on tight so that we have a closed system. And then, what I'm going to do is take a hammer, and I'm going to hit the top of this piston. And what that's going to do is it's going to compress the gas inside and raise the temperature. Okay? Let's see if we can get it hot enough. What I'm going to do, though, is lower the lights here for you so we can see really clearly if this piece of cotton catches on fire.

So there you saw that I was able to light the cotton on fire with the same initial amount of work from my arm and the hammer. The way that worked is because we got the pressure up so high that it increased the temperature to above the auto-ignition temperature of the cotton, which is a toasty 407°C, by the way.

So let's now look at another example of the equivalence of work and heat. In this case, I'm going to use soup. And what I've done here is I've poured the same soup into two different containers. In the first, it's in a bowl, sitting on what is, perhaps, a kind of stove that many of us have seen, right. It's just a normal stove. But in the other case, I'm going to heat the soup up using a blender. So, if we were to look at the stove, and here's a thermometer, this is an IR thermometer, which some of you may remember from our lecture on temperature, Lecture 3, I can look at the soup, and I can see that it's getting warmer. It's about 30°C right now, and getting slowly warmer, because the burner is on, okay.

But, a different way to heat that soup up would be with a blender, and that's not usually how we use the blender, but that's how I want to use it today. So, let's take an initial read of the temperature of the soup before we turn the blender on. So you can see it's about 21°C. But now, if I turn the blender on, what's going to happen? So I'm going to turn it on.

And now you can see we have nice, blended soup. But what about the temperature? Let's take a look. Okay, so you can see it's gone up by a couple of degrees Celsius. Well, that's good, but it's not quite hot enough to serve, since I like to serve my soup piping hot. So let's turn the blender back on and see how hot we can get it.

Okay, let's take another look and see how hot we got it. So you can see that now, with just a little bit more mechanical work, I've got myself some nice, tasty, piping-hot soup. And the point is that the first law tells us that the internal energy has changed. But it didn't tell me how it changed. Work and heat are equivalent in this sense.

Now, what we just witnessed were examples of energy flowing from one type to another—from mechanical to heat just like in the original experiments of James Joule. What if we were to track all of the places the energy comes from, or goes to, even beyond our system as the blender and its contents? If we really track it down, the initial energy to power the blender started from sunlight that struck the earth 50 million years ago, give or take. That sunlight energy created biomass in the form of plants and animals. And then, under intense amounts of pressure and a whole lot of time, the biomass was converted, underground, into coal, natural gas, or oil.

This fossil fuel was then dug out of the ground and lit on fire. The heat from burning that ancient sunlight was then used to boil water, which generated steam, which at high enough pressure can be used to turn a massive turbine. And the motion of the turbine was then used to generate electricity, which was brought through wires to this building and ultimately into the blender.

At all points along this total energy pathway, different forms of work and heat are involved, and at all points, the first law of thermodynamics holds true. After the blender has mixed and heated its contents, we turn it off. Where has the energy gone at that point? Well, some went into mixing the contents, and as we showed in the demo, some went into raising its temperature. If we wait a while longer, the blender contents will equilibrate with the room temperature by giving off heat to its surroundings. So, ultimately, in that experiment, what we did is the following; we took million-year-old sunlight energy and brought it to the air in this room in the form of an imperceptibly tiny bit of heat. Along the way, we—hopefully—did something useful with that energy. In most cases involving a blender, that doesn't mean heat, but rather, work. In particular, our usefulness in this case is the work of mixing.

As I mentioned in the last lecture, one of the crucial questions thermodynamics can answer involves the usefulness of energy in general. After we introduce the second law in the next lecture, we'll be able to actually quantify this usefulness. Meanwhile, let's go back to thinking about all of those different work terms. Remember that a property is a quantity we can measure and observe for a given system. It's something that differs depending on the thermodynamic state of the system. Remember that in Lecture 4, we defined intensive and extensive variables? Put concisely, an intensive variable does not get bigger or smaller with the size of the system, while an extensive variable does.

Now, I defined these two types of variables before, but what I didn't tell you is how they are related. It turns out that a given intensive property has what is called a conjugate pair that is an extensive property. We use the term conjugate pair to mean that the variables are connected or coupled. And here's the key point; the product of the intensive property and the differential change in the conjugate paired extensive property gives us a work term. More precisely, it's called differential configurational work, which is just a fancy way of saying that there is a change in the configuration, or structure, of the system. All work involves such a change.

Now, when we write a change in work, we don't use just a plain old "d" to write dW as we would have for a change in, say, volume or pressure. Instead, we write it as δ. We do this squiggly thing to signify that we're talking about a change in a variable that is path dependent; I'll come back to this point in detail in a few moments. I want to highlight that there are many of these so-called conjugate pairs of thermodynamic variables. Each pair, when multiplied, gives a thermodynamic force times a thermodynamic displacement to produce a certain type of configurational work. For example, pressure is an intensive property, so, it doesn't scale with the system size, which when multiplied by the change in the extensive property volume, which does scale with the system size, gives us a configurational work term. This is a term representing mechanical work. Surface tension is an intensive property, so it doesn't scale with size, which when multiplied by the change in the extensive property area, which will get bigger as the system gets bigger, gives the surface energy configurational work term.

The chemical potential, and oh, this is a really important one, is an intensive property, which when multiplied by the extensive property, the change in the number of particles, gives us the configurational term known as chemical work. And so on, and so on. Notice that in all cases, the units of the work term are energy. In all cases, we have a thermo force times a thermo displacement equals energy.

Now, remember our sign convention. We are assuming a convention where positive work is work done on a system and negative work is done by a system. This means that in the case of the PdV work term, we'll need to put a minus sign in front, so that δW equals $-P$ times dV. This ensures that if mechanical work from a volume change is done on the system, it will be positive. Just think about squeezing a piston with a gas inside; in that case, I am applying a pressure to do work on the system. The volume change is negative since the piston compresses, so that means that the total work done will be positive, which follows our sign convention correctly.

Now, you may have noticed that I slipped in a couple of new terms there. First of all, I'm calling these pairs of variables conjugate pairs. That one is easy; it simply means that the variables are connected, or coupled, which I hope is now pretty clear. In other words, not just any two variables will give a thermodynamic work term, but rather, only these special conjugate pairs. The other word I used was configurational to define the work terms. This means that it is a type of work that changes the configuration, or structure, of the system. We could have just used plain old "work" instead, but it is common in thermodynamics to use the term configurational work, so I wanted to introduce that terminology to you here.

And now we need to turn to another important concept that is crucial in thermodynamics, namely, the concept of reversibility, or irreversibility, as the case may be. And the way we use these terms is to describe the type of process occurring, which depends on the type of work being done during that process. Irreversible work arises from the fact that some types of work are always done on the system regardless. In this case, reversing the force or displacement direction does not affect the sign of the work. This is also known as dissipative work, and I'll come back to this in a moment.

First, let's talk about reversible processes. A reversible process is not real; it's an idealized process for which the following conditions all hold throughout the entire process. First, the system is always in equilibrium; second, there are no dissipative processes occurring, in other words, no energy is lost to the surroundings, there's no friction, viscosity, electrical resistance, and so on; and third, a reversible process can occur both forward and backward with absolutely no change in the surroundings of the system.

Now, irreversible processes, they represent, well, they basically represent nothing short of all natural processes. We are quite familiar with the concept of irreversibility. For example, if I put a drop of food coloring into a glass of water, we can observe the food color diffuse slowly out into the liquid. The irreversible process going on here is diffusion, and of course, once it is spread out through the liquid, we never see the food coloring spontaneously diffuse back into a single droplet. That's what I mean by irreversibility. Other examples include the expansion of gas into vacuum; if you allow gas to see a vacuum chamber, it will quickly flow into it, but you would never see gas doing the opposite, namely, spontaneously leaving a chamber to form a vacuum. For that, we need to put energy into the system using a pump.

And here's one last example. Suppose I place a cup of hot coffee on my desk. The coffee will start to cool off, because it will give off heat to the surroundings. But, if I place a cup of room-temperature coffee on my desk, it would never spontaneously heat up and become a nice, warm cup of coffee again. For that, I'd need again to input energy by, say, nuking it for a few minutes. Now, nuking cold coffee, there's a particular work term that I'm guessing a lot of us are quite familiar with. I know I am.

It's quite intuitive in the examples I just gave that the process described will never reverse. Later, we'll see that the second law of thermodynamics serves as a rule to account for the one-way nature of these processes. And notice that I used, quite purposefully, the word "spontaneous" in each example. That's a very important word in thermodynamics, because remember, thermo tells us about equilibrium states of a system. In order to find those places of equilibrium, we need to be able to predict which processes are spontaneous and which ones are not. It's the second law that provides a theoretical criterion to let us determine which direction processes will occur, even when the answer is not intuitively obvious as it was in those examples.

Now, that's irreversibility, and as I mentioned, it describes all processes in nature. But back to reversibility for a moment. I mentioned before that real processes are never reversible. They may approach the conditions of reversibility that I outlined a few moments ago, but they will never adhere to them fully. So, why do we even discuss or define such a type of process? As we'll see, these unreal, highly theoretical reversible processes can be used to perform calculations for very real irreversible processes. The power of using the concept of reversible processes in thermodynamic analysis is that the variables used to analyze thermodynamic systems are state variables. Remember, this means that if the final and initial states of a system are known, then calculations on those systems can be done with any arbitrary path, since the final state is independent of the way the system came to that state. This means I could choose to take a theoretical reversible path in my calculation of a final state property, and the property I calculate will hold true for the real irreversible path. Don't worry if that sounds confusing right now; it will become clearer as we work through some specific examples.

Now, with all this talk of work, I don't want you to forget that another way, just as important, to transfer energy into a system is through heat. Remember that it is the addition of these two transfers that give us the energy-conservation principle in the first law. And here's something very exciting I want to show you. Similar to the case for work, a differential form for heat, that would be δQ, can be found in terms of an intensive property multiplied by the differential of an extensive property. In this case, the intensive property is going to be temperature T, and the extensive property is none other than entropy itself, or S.

We'll be going all entropy crazy in the next lecture, where I introduce the second law, but I couldn't help but bring up this other crucial conjugate pair. For one thing, because of the fact that it completes our first-law equation, so that you have an expression for δQ, but for another, it's the sheer coolness factor of being able to view entropy in this way.

Remember that I've already defined entropy as a counting of the number of possible microstates a system can possess; remember that salt dissolution example from Lecture 4, and how counting the allowed microstates explains why salt dissolves in your soup. But remember, also, how I promised you that we'd be revisiting entropy in a number of different ways throughout the class to really try to feel it intuitively. Well, here's another way to see what entropy means. Just like volume change is the conjugate pair for pressure to give us a work term for mechanical energy, entropy is the conjugate pair for temperature to give us heat energy. This means that heat itself can be defined as the energy transferred into a system in terms of thermal interactions with a transfer of entropy. Entropy provides the link between heat and temperature. And that there is my repeat for the day—Entropy provides the link between heat and temperature.

The last topic I want to discuss in this lecture, as promised, involves digging in a bit into the math that's important in thermodynamics. Specifically, we need to go into more detail on the nature of that δ I've introduced earlier. Another name for a delta, in general, is a differential. So the differential volume would be dV, and as we learn in introductory calculus, I can integrate over a differential variable to get the total value of that variable. If I integrate over the differential dV from, say $V1$ to $V2$, I would get the change in volume, or $V2$ minus $V1$.

Now, that type of differential, when I write it as a d like that, is called an exact differential. It means that the integral over it does not depend on the path taken. We know this intuitively from the calculus; we integrate over the dV to get a V and then plug in the values of the integrand—$V1$ and $V2$—to get the actual value of the integral. We didn't worry about how the volume of some system went from $V1$ to $V2$.

And you may be thinking, hey, I know another, more thermodynamic way to call that; it's a state variable when it doesn't depend on the path taken. And that's absolutely right; we describe changes in state variables by exact differentials. And that brings us to the two energy transfer terms of the first law, work and heat, or W and Q. Here, as I've mentioned a number of times, we have variables that are not state variables, but rather, they are path dependent. That's why we use a different differential to describe a change in such a variable. The δ is also called an inexact differential, since, if I integrate over it, I must specify a path for the integral. That makes some sense if you think about it. We cannot integrate over a path-dependent variable without knowing its path. The squiggly is just our mathematical notation to make sure we know what type of variable we're describing.

Let's return to the first law itself. Remember that for a given process, any change in U, the internal energy of the system, is equal to the sum of the energy transfer terms Q and W. But now, let's write the first law in its differential form. The differential dU equals the differential change in the internal energy of the system, which equals δQ plus δW. Notice that for U, I used the exact differential, so, no squiggly, and that's because internal energy is a state function and does not depend on the path taken, only on the end points of the integral itself, or, in our thermodynamic terminology, the initial and final states of the system.

One of the nice things about a differential being exact is that it means there are a number of mathematical consequences that we can take advantage of in thermodynamics. For example, when a differential is exact, we can use certain relationships and rules that hold rigorously for the variable being differentiated. One class of relationships, known as Maxwell relations, are extremely important, since they allow us to transform one thermodynamic equation into another one, sometimes making it much more useful for the problem at hand.

The mathematical rules associated with exact differentials also allow us to equate one thermodynamic variable, such as temperature, entropy, volume, pressure, or internal energy, with derivatives of other variables. So that covers all of the topics I wanted to go over in this lecture. But I'd like to finish the lecture with a calculation. We don't do this all the time, but with all this talk of variables and math, I'm kind of in the mood to get quantitative!

In the demo, you saw that with a hammer, I could create enough pressure to ignite a piece of cotton. Let's use thermodynamics to answer the following question. What is the minimum velocity of that hammer that is required to ignite the cotton? Apart from our thermo, one piece of information we'll need to know is the temperature at which the cotton auto ignites. It turns out to be a nice and toasty 407°C. So, first we need to think about what it means to know the velocity of the hammer. What type of energy is that? Well, the velocity is directly related to the kinetic energy of the hammer, since that energy is one half, times the mass of the hammer, times its velocity squared. So our goal in this problem is to find the kinetic energy of the hammer necessary to make the cotton reach 407°C.

Let's assume the gas inside of the cylinder is ideal, and I hope that you may remember from Lecture 5 what that means, since now we'll treat the behavior of the gas using the ideal gas law. That is, PV equals nRT, or pressure times volume equals the number of moles of the gas times the gas constant R times the temperature. Also, for an ideal gas, remember that the internal energy is equal to $^3/_2$ times n times R times T.

Since for an ideal gas the internal energy only depends on the temperature, at least when the number of moles of the gas is constant, then, we can just use changes in the internal energy to get the changes in temperature of the gas. If all of the kinetic energy of the hammer goes into the gas, then it goes directly into changing the internal energy of the gas. So, ΔU of the gas, which equals $^3/_2$ times n times R times ΔT of the gas, must equal the kinetic energy of the hammer, or ½ mass of the hammer times velocity of the hammer squared. The change in temperature of the gas is equal to the auto-ignition temperature of cotton minus room temperature, which is 407°C minus 25°C.

The cylinder in the experiment I performed, while open to air, has a volume of 8.5 cm cubed at atmospheric pressure and room temperature. From the ideal gas law, the total number of moles of gas in the cylinder is given as follows: n equals PV divided by RT. So plugging in the values for P, V, R and T, we get that the number of gas molecules is roughly 3.5 times 10 to the minus 4th moles. And we can assume that this value does not change. So when the rod is placed in the cylinder, the number of gas molecules remains a constant since the cylinder is now closed.

So now we have *n*, and we have the change in temperature, which means we can compute the change in internal energy of the gas during the process. Plugging everything in, we get a value of around 1.7 joules. And from that, we get our velocity of the hammer, since we can now set the kinetic energy of the hammer equal to this value. It turns out that for a hammer that weighs bout a kilogram, this minimum velocity needed to ignite the cotton is around two meters per second. Also, since we have the ideal gas law relationship, if we wanted, to we could use that to compute the final pressure at the point of auto-ignition.

So there you have it, a nice example of how to calculate the transformation of work. In this case, from the motion of the hammer into heat. By the way, if there had been heat dissipation, our first law covers that too. We'd just have had to know what it was and include that in the calculation. One way to know if heat dissipated in that process would be to measure temperature during the process. Another way would be to know something about the entropy change during the process, and that, will be the topic of our next lecture.

Entropy—The Arrow of Time

Lecture 8

In this lecture, you will learn about the second law of thermodynamics. The concept of entropy has been introduced in previous lectures, but in this lecture, you are going to learn about the thermodynamic law that deals directly with entropy and the important consequences of this law. The key concepts you should feel comfortable with by the end of this lecture include entropy, ordered and disordered states, the idea of irreversibility, and the second law and what it means.

Entropy and the Second Law of Thermodynamics

- We can think about entropy (S) as the property that measures the amount of irreversibility of a process. If an irreversible process occurs in an isolated system, the entropy of the system always increases. For no process in an isolated system does the entropy ever decrease.

- In an isolated system—which is a system that has boundaries through which nothing can pass, including energy, heat, and matter—we know that energy will be conserved. But we just stated that entropy may not be conserved in such a system, and it can only stay the same or increase.

- Recall that reversibility refers to a process for which the system is always in equilibrium. In addition, there are no dissipative processes occurring (in other words, no energy is lost to the surroundings—there's no friction, viscosity, electrical resistance, and so on). Furthermore, a reversible process can occur both forward and backward with absolutely no change in the surroundings of the system.

- No reversible process exists—it's only a theoretical concept because all real processes dissipate some amount of energy. But reversibility is an extremely useful theoretical concept.

- If a process is fully reversible, then we can define a change in entropy during that process as the following: $dS = \delta Q/T$. So, the change in entropy is the change in heat flow over the temperature. In this definition, the units of entropy are energy divided by temperature. So, in standard units, this would be joules per kelvin.

- We use δ to remind us that the variable under consideration is path dependent. Also, standard notation is δ_{rev}, which denotes that this relationship holds only if the path is reversible.

- So, if we want to know the change in entropy for a whole process, then we simply integrate both sides of this equation. The integral of the left side gives the final entropy minus the initial entropy, or ΔS.

- When the integral over the differential of a thermodynamic variable is just equal to taking the difference between the final and initial values of that variable, it means that the variable is a state function. Part of what the second law does is to confirm the very existence of entropy as a state function.

- Heat is not a state function because it's path dependent, but entropy is, effectively, a way to change heat into a quantity such that it is a state function and, therefore, no longer path dependent.

- In the late 1800s, Rudolf Clausius showed that when an amount of heat flows from a higher-temperature object to a lower-temperature object, the entropy gained by the cooler object during the heat transfer is greater than the entropy lost by the warmer one.

German mathematical physicist Rudolf Clausius (1822–1888) is credited with making thermodynamics a science.

- This result allowed him to state that what drives all natural thermodynamic processes is the fact that any heat transfer results in a net increase in the combined entropy of the two objects. We know that heat flows this way simply from our own experience, but with entropy, we have a rigorous variable that establishes quantitatively the direction that natural processes proceed.

- All natural processes occur in such a way that the total entropy of the universe increases. The only heat transfer that could occur and leave the entropy of the universe unchanged is one that occurs between two objects that are at the same temperature. However, we know that is not possible, because no heat would transfer in such a case.

- All processes—that is, all real processes—have the effect of increasing the entropy of the universe, and that is the second law of thermodynamics. So, we have that a change in entropy for a given process, ΔS (which equals S final minus S initial ($S_f - S_i$)), is equal to the integral from some initial state to the final state of δQ divided by temperature, for a path that is reversible.

- This equation allows us to write the first law of thermodynamics in a new way. The first law states that the change in internal energy during a process is equal to the heat flow plus all work done during that process: $dU = \delta Q + \delta W$.

- Now, we are in a position to make a substitution, because the second law gives us a relationship for δQ. For a reversible process, it's equal to TdS. Let's say that the only type of work being done during this process is mechanical work, so pressure-volume work. The work term for that is $-PdV$.

- In this case, we can now write out the first law as follows: $dU = TdS - PdV$. Remember that this holds for a process that is reversible. If no work was being done at all, then we would have the direct relationship between entropy and internal energy: $dU = T \times dS$.

- One of the important consequences of entropy being a state property is that because it's path independent, a change in its value can be obtained by artificially constructing any path.

Ways of Stating the Second Law

- In essence, the second law of thermodynamics makes predictions about what processes will occur. There are actually a number of different ways of stating the second law, but the most common and also simplest form of the second law is as follows. We consider a system and call the rest "surroundings." We could then say that the universe consists of the system plus the surroundings, because we said that the surroundings were everything else.

- We can write the entropy of the universe as the entropy of the system plus the entropy of the surroundings. The second law of thermodynamics states that the entropy of the universe (that is, system plus surroundings) always increases in an irreversible process and remains constant in a reversible process.

- Thus, ΔS universe is greater than or equal to zero for all processes. When the change in entropy is exactly equal to zero, then the system must be in equilibrium. When it's greater than zero, it tells us the directionality of the process. That's the second law.

- The Clausius definition of entropy, which says that $dS = \delta Q/T$ for a reversible process, is based on the macroscopic variables temperature and heat. There is also a very important microscopic definition of entropy.

- Temperature and pressure are macroscopic variables that average over a very wide range of microscopic behaviors. For temperature, this is an average over the kinetic energy of the particles, while for pressure, it's an average over their collision frequency and speed.

- For the case of entropy, which is also a macroscopic thermodynamic variable, the microscopic quantity that matters is the number of ways in which these particles can be arranged, or the number of microstates of the system.

- The definition for entropy that connects S to this microscopic property comes from statistical mechanics, and it states that the entropy of a system is equal to a constant times the natural logarithm of the number of available states the system has. The constant, written as k_B, is known as the Boltzmann constant, and it has this name because this definition of entropy is due to him.

Counting the Microstates for a Material

- How do we count the microstates for a material? It has to do with the number of different ways in which the atoms or molecules can move around and be arranged. It has to do with the atomic-scale degrees of freedom.

- Take the example of a perfect crystal, in which case the atoms are all locked into rigid positions in a lattice, so the number of ways they can move around is quite limited. In a liquid, the options expand considerably, and in a gas, the atoms can take on many more configurations. This is why the entropy of a gas is higher than a liquid, which is in turn higher than a solid.

- A very common example in thermodynamics of the power of thinking of entropy as a counting of states is the expansion of an ideal gas into a vacuum. Suppose that we have a chamber that has an adiabatic wall around it, which means that no heat can pass across the boundaries. There is also a little membrane down the middle, and we start with the gas all on just one side of that membrane. On the other side, there's nothing, so it's a vacuum.

- When we remove the membrane, the gas will expand into the vacuum. But why does this happen? The temperature of the gas does not increase. That's because when we remove the membrane, the gas particles can travel over to the other side, but they're not going to travel any faster or slower on average, and because the temperature is related to the average kinetic energy of the gas, it will not change. And no heat can transfer in or out of the system because it's an adiabatic wall.

- Is any work done during this expansion? During the expansion, the gas has zero pressure, because it's expanding into a vacuum. So, there cannot be any work done during the process. There is no work done, no heat transfer, and no change in temperature or internal energy, so what is it that governs the behavior?

- Suppose that we slice up the container into a bunch of boxes; for example, there are M boxes on the left side, which is where the gas starts. If we have N particles, then the number of ways we can arrange these particles—that is, the number of possible microstates for them to be in—would be equal to M raised to the Nth power (M^N). However, once we remove the membrane wall, there are $2M$ boxes, and the number of possible ways to arrange N particles over those $2M$ boxes is $(2M)^N$, which is a whole lot more.

- It's only because of that reason that the gas expands into the vacuum; there are more ways in which the gas can be. Whenever there is a way for a system to have more ways to be arranged, it tends toward that state.

- The first law of thermodynamics does not distinguish between heat transfer and work; it considers them as equivalent. The distinction between heat transfer and work comes about by the second law of thermodynamics.

- One more important consequence of entropy is that it tells us something about the usefulness of energy. The quantity of energy is always preserved during an actual process (the first law), but the quality is likely to decrease (the second law). This decrease in quality is always accompanied by an increase in entropy.

Suggested Reading

Anderson, Thermodynamics of Natural Systems, Chap. 4.

DeHoff, Thermodynamics in Materials Science, Chaps. 3.2–3.5 and 6.1.

Halliday, Resnick, and Walker, *Fundamentals of Physics*, Chap. 20.

Lambert, "Shuffled Cards, Messy Desks, and Disorderly Dorm Rooms."

Moran, Shapiro, Boettner, and Bailey, *Fundamentals of Engineering Thermodynamics*, Chap. 5.

Serway and Jewett, *Physics for Scientists and Engineers*, Chaps. 22.6–22.8.

http://www.job-stiftung.de/pdf/skripte/Physical_Chemistry/chapter_2.pdf.

http://www.felderbooks.com/papers/entropy.html

Questions to Consider

1. How would you measure the difference in entropy before and after an egg drop? If you crack the egg into a bowl, then fry it in a pan to make an omelet, is there any way to put the egg back to its starting point? If you could, would any laws of physics be broken?

2. Think about your favorite movie, and imagine that you are watching it played in reverse. How could you tell it was being played backward?

3. If energy favors salt to stay in a crystalline form, what causes it to dissolve in your soup?

4. Is it possible for the entropy of a system to decrease?

Entropy—The Arrow of Time
Lecture 8—Transcript

Hello and welcome! Today we're going to talk about the second law of thermodynamics. We've already introduced the concept of entropy several times in previous lectures, but today, we're going to learn the thermodynamic law that deals directly with entropy, and we'll discuss the important consequences of this law. The key concepts I want you to feel comfortable with by the end of this lecture include entropy itself, ordered and disordered states, the idea of irreversibility, which is something we touched on the last lecture, and of course, the second law and what it means.

We begin this lecture with the very nature of the word "begin" itself, that is, the nature of time. What is the connection of time to various processes in nature? I think that most of us would agree; time is unidirectional. But why is this? Take the following examples; an egg drops to the floor accidentally and breaks apart. Ice is placed into a coffee to make iced coffee; the ice melts while the coffee gets colder. A piece of paper catches fire and burns, turning into ashes. What is common to all of these processes? Do we ever observe these processes happening in reverse? What we know intuitively is that in all of these examples, the reverse never happens. The broken egg does not suddenly reassemble into an unbroken shell; the coffee would never become warm while ice cubes form spontaneously in the glass, and the burnt ashes would never un-burn to turn into paper. But why is this the case?

So time is unidirectional and intuitively related to all of these processes that occur naturally, but the question we are really asking is, can we connect the unidirectionality of time to some property of these processes that occur? The answer lies in the concept of irreversibility. As I mentioned in the last lecture, all natural processes are to some extent irreversible due to the presence of dissipative forces in nature. No process is perfectly ideal or reversible. But take the egg example; in that case, would the process occurring in reverse—that is, the shattered egg reassembling—would that violate any laws of physics? Would it violate the first law of thermodynamics?

The answer is no, the reverse process does not violate the laws of physics, and the reverse process does obey the energy conservation of the first law.

In fact, as far as we currently know, all physical laws are time reversible, meaning that when viewed in reverse, no laws of physics are broken. From our day-to-day experience, though, time definitely flows only in one direction.

As we'll learn today, the reverse processes do not happen because they are statistically improbable. We will need to use entropy to understand such statistics. Entropy is our only measure for time moving forward; entropy serves as nothing less than the arrow of time and explains the unidirectionality of all of these processes. We can, in fact, think about the property called entropy, which remember, we write with the letter S in thermodynamics, as the property that measures the amount of irreversibility of a process. And here is an important postulate about entropy. If an irreversible process occurs in an isolated system, the entropy of the system always increases. For no process in an isolated system does the entropy ever decrease— ever.

Let's think about the difference here with energy. In an isolated system—remember, that would be a system that has boundaries through which nothing can pass, not energy, nor heat, nor matter—we know that energy will be conserved. But we just stated that entropy may not be conserved in such a system, and it can only stay the same or increase. This is a postulate for now, which I'm going to turn to more rigorously into a law shortly.

But first, let's look at a mathematical definition of entropy, which I briefly alluded to in our last lecture. Now, I gave those examples of irreversibility in my discussion of the natural direction of time a moment ago. As a quick reminder, the concept of reversibility refers to a process for which the system is always in equilibrium, there are no dissipative processes occurring; in other words, no energy is lost to the surroundings, there's no friction, no viscosity, electrical resistance, and so on. And a reversible process can occur both forward and backward with absolutely no change in the surroundings of the system.

And remember that when I talked about reversible processes, I also mentioned that no such process exists. It's only a theoretical concept, since all real processes dissipate some amount of energy. But, reversibility is an extremely useful theoretical concept as we're about to see. That's because, if a process is fully reversible, then I can define a change in entropy during that process as the following: dS equals δQ divided by T. So the change in entropy is the change in heat flow over the temperature.

Now we can see from this definition that the units of entropy are energy divided by temperature. So, in standard units, this would be Joules per Kelvin. Remember to always be aware of the units for any variable you're working with. As we said in the last lecture, we use the δ to remind us that the variable under consideration is path dependent. Also, standard notation is that we put the letters R-E-V as subscript on the δ, to denote that this relationship holds only if the path is reversible.

So, if we want to know the change in entropy for a whole process, then we simply integrate both sides of this equation. The integral of the left side gives the final entropy minus the initial entropy, or we can just write it as ΔS. Here, I use the capital Greek letter for delta, signifying to our mathematically inclined friends that this is not a differential, but rather, a full difference between some initial and final state.

Now, remember what it means when the integral over the differential of a thermodynamic variable is just equal to taking the difference between the final and initial values of that variable? You got it. It means the variable is a state function. Part of what the second law does is to confirm the very existence of entropy as a state function. Remember that heat is not a state function since it's path dependent. Well, effectively, entropy is a way to change heat into a quantity such that it is a state function, and therefore, no longer path dependent.

The origins of entropy go back to Rudolf Clausius. Back in the day, that would be the late 1800's, scientists were trying to figure out a way to understand and combine the ideas of Kelvin, Joule, and Carnot, that is, they were trying to figure out why energy is conserved in thermodynamic processes and why it is that heat always flows downhill in temperature, that is, heat always flows from a hot body to a colder one. Clausius defined the idea of entropy as I just described above, namely, the change in entropy for a process is the ratio of the heat exchanged in the process with the absolute temperature at which that heat is exchanged. Now, what this does is give a mathematical framework to the idea that heat always flows from higher to lower temperature.

By his new concept and definition of entropy, Clausius showed when an amount of heat Q flows from a higher temperature object to a lower temperature object, the entropy gained by the cooler object during the heat transfer is greater than the entropy lost by the warmer one. You can see this from the math, since Q / T cold is greater than Q / T_hot, since T hot is larger than T cold. This was a really powerful and important result. It allowed him to state that what drives all natural thermodynamic processes is the fact that any heat transfer results in a net increase in the combined entropy of the two objects. We know that heat flows this way simply from our own experience, but, with entropy, we have a rigorous variable that establishes quantitatively the direction that natural processes proceed.

Before I move on, let me give a brief analogy to help visualize the role of dividing by temperature in our definition of entropy. Imagine that out of the blue you decide to scream at the top of your lungs; let's say you scream "HEY YOU!" Now, if you did that in a quiet library, then every single person, including perhaps a quite irritated librarian, would turn and look at you. But, if you did that in, say, a football stadium right when a touchdown is scored by the home team, in that case, no one would even notice. Tying this back to our thermo, the change in heat is our scream, and the temperature is related to whether we're in a library or a football game, or, somewhere in between. Dividing by temperature gives a kind of relative impact that the heat has on the system.

Okay, back to our natural processes. All natural processes occur in such a way that the total entropy of the universe increases. The only heat transfer that could occur and leave the entropy of the universe unchanged is one that occurs between two objects which are at the same temperature. But we know that is not possible, since no heat would transfer in such a case. All processes, that is, all real processes, have the effect of increasing the entropy of the universe. And that, is the second law of thermodynamics.

So, we have that a change in entropy for a given process, call it ΔS which, equals S-final minus S-initial, is equal to the integral from some initial state to the final state of δQ divided by temperature, for a path that is reversible. Now, this equation allows us to write the first law of thermodynamics in a new way. The first law states that the change in internal energy during a process is equal to the heat flow plus all work done during that process. Written in differential form, this amounts to dU equals δQ plus δW.

Now, as you can see, we are in a position to make a substitution here, since the second law gives us a relationship for δQ. For a reversible process, it's equal to TdS. And let's say the only type of work being done during this process is mechanical work, so pressure-volume work. Well, as we've learned, the work term for that is minus PdV.

In this case, we can now write out the first law as the following: dU equals TdS minus PdV. And then again, remember, this holds for a process that is reversible. If no work was being done at all, then we'd have the direct relationship between entropy and internal energy: dU equals T times dS. As a side note, check the units here to see that they work out properly. You should get energy on both sides of the equality.

And one of the important consequences of entropy being a state property is that since it's path independent, a change in its value can be obtained by artificially constructing any path. Let's dig into this a bit more, since I know all this talk of reversibility and path dependence can be confusing the first time or two you hear it. Don't worry, my students at MIT are just as confused. It usually takes thinking about this at least two or three times and seeing it used in specific examples before it all starts to makes sense.

In essence, the second law of thermodynamics is about predictions. It is a law that predicts what processes will occur. Let me repeat that crucial fact: The second law of thermodynamics makes predictions about what processes will occur. There are actually a number of different ways of stating the second law. I'll let you know when I come to other ones, but for starters, let's go with the most common and also simplest form of the second law, which goes as follows. We consider a system and call the rest surroundings. We could then say that the universe consists of the system plus the surroundings, since we said the surroundings was everything else. Now, we can write the entropy of the Universe as the entropy of the system plus the entropy of the surroundings. The second law of thermodynamics states that the entropy of the universe, that is the system plus surroundings, always increases in an irreversible process and remains constant in a reversible process. That means that ΔS for the universe is greater than or equal to zero for all processes. When the change in entropy is exactly equal to zero, then the system must be in equilibrium. When it's greater than zero, it tells us the directionality of the process, that concept we discussed in the beginning of this lecture. That's the second law.

Let's put this law into action for a very simple system to determine equilibrium. Suppose I have a system of two blocks, call them block A and block B, which are in contact with one another but isolated from everything else, and held at constant volume. Now suppose that block A initially has a temperature *T*_A, and block B initially has a different temperature *T*_B. When will heat transfer stop between these two blocks? When is the system at equilibrium? Of course, we know intuitively that heat transfer stops when the temperatures of the blocks are equal. But how can we prove this from thermodynamics? In order to do that, the question we ask is as follows: what is the condition of blocks A and B when the point of maximum entropy is reached? According to the second law, this will be the point at which the system is no longer changing its entropy, so, the point where the derivative is zero.

Let's start with the fundamental equation for the change in entropy for a reversible process. Remember, this comes from the first law of thermodynamics assuming only work of expansion. To get this, I simply take the equation I just showed you for *dU* and divide through by temperature. This gives us *dS* equals *dU* divided by *T* plus *P* times *dV* divided by *T*. We can write this equation for each block, so we get an expression for the change in entropy of block A and a similar one for block B. Now, I mentioned these blocks are not going to change their volume, so we can eliminate the *dV* terms; that's nice. And that leaves us with *dS* for block A equals *dU*_A over *T*_A, and *dS* for block B equals *dU*_B over *T*_B.

And from the second law when the change in entropy of the universe is equal to zero, the system will be in equilibrium. Since the system is isolated, we know there cannot be any change in entropy of the surroundings, and we're left with dS universe equals dS system. So, dS universe equals dS for block A plus dS for block B, which, from the equation above equals dU of block A over T_A plus dU of block B over T_B.

And again, we take advantage of the knowledge that the system is isolated to apply a constraint, namely, that the total internal energy must equal a constant for any process, so the change in total internal energy must equal zero, meaning that dU of block A equals minus dU of block B. Plugging that in, we get that the change in entropy of the universe for this system is equal to dU for block A times the quantity 1 over T_A minus 1 over T_B.

And there we have it! This quantity can only be zero if T_A equals T_B. The second law of thermodynamics told us exactly how to find the point of equilibrium. Now, I know this seems a bit trivial, because again, it's kind of intuitively obvious what will happen for this problem. By the way, it's also intuitively obvious for the examples I gave in the beginning of the lecture, like the egg smashing apart, the burning paper turning to ash, or the ice melting in hot coffee. Those examples and this one were by design intuitive because I want you to feel the intuitive nature we all have for the flow of time and the irreversibility of processes.

However, I also want to point out that the reason we need a thermodynamics-based approach for determining rigorously which direction a process flows is that in many, many cases the answer is not at all intuitive. As you can imagine, the more complex a process is, say, with many different thermodynamic driving forces acting at once, the more difficult it is to predict whether a given process will occur by intuition alone. And even when only one driving force is present, such as the chemical potential and how it directs chemical reactions, we often have very little natural insight into what will happen. The second law of thermodynamics provides that guidance.

Before we move on, I want to clarify an important distinction. Some of you may be remembering our zeroth law of thermodynamics and thinking, wait a minute, doesn't that kind of explain this problem of two blocks of different temperatures next to each other? In fact, didn't I use this example of temperature blocks to explain the zeroth law itself? If that's what you're thinking, then great memory; that's exactly right! But, there's an important distinction here. You see, the zeroth law of thermodynamics in this case would tell us the following statement: If heat stops flowing between the two blocks, then the temperature of one block must equal that of the other. The second law, however, tells us that this is also the point of equilibrium.

And the second law is extremely general. You may be thinking, but that example was kind of a special case, wasn't it? I mean, it was for an isolated system, where the transfer of heat, work, or molecules at the boundaries are not allowed. In such constant internal energy systems, it's straightforward to directly apply the fundamental equation for the entropy and the second law to calculate equilibrium properties of the system. However, controlling heat and work flow at the boundaries of a real system is often not a simple task, and in fact, it is unnecessary for most experimental procedures one would be interested in performing. The second law can be applied to define the conditions for equilibrium for any arbitrary system. And as we'll see in a later lecture, the Gibbs free energy is another state function that is extremely useful for determining equilibrium in the most common type of experimental system, namely, one where the pressure and temperature are maintained at a constant value.

But for now, back to entropy. So the Clausius definition of entropy, which says that dS equals δQ over T for a reversible process, is based on the macroscopic variables temperature and heat. There is also a very important microscopic definition of entropy, which I'd like to explain now. Remember how we talked about the fact that temperature and pressure are macroscopic variables that average over a very wide range of microscopic behaviors? For temperature, this is an average over the kinetic energy of the particles, while for pressure, it's an average over their collision frequency and speed. For the case of entropy, which is also a macroscopic thermodynamic variable, the microscopic quantity that matters is the number of ways in which these particles can be arranged, or, the number of microstates of the system.

The definition for entropy that connects S to this microscopic property comes from statistical mechanics, and it states that the entropy of a system is equal to a constant times the natural logarithm of the number of available states the system has. The constant, written as k_B is known as the Boltzmann constant, and, as you may have guessed, it has this name because this definition of entropy is due to him.

Now, what exactly do I mean by the number of available microstates in a system? Let's take dice as an analogy, since it lets us do some simple counting. Suppose I roll three dice together; then, how many ways could I roll a 3? Well, only 1; the only way to get a three is by rolling three 1s. But how many ways are there to roll a 12? Well, quite a number of ways; I could roll a 4, 4, 4, or a 3, 3, 6, or a 4, 5, 3, and so on. So the number of states available to rolling a 12 is much higher than the number available to rolling a 3. So, let's now think about the microstates for a material. How do we count those?

Well, as you might imagine, it has to do with the number of different ways in which the atoms or molecules can move around and be arranged. It has to do with the atomic-scale degrees of freedom. Take the example of a perfect crystal. In that case, the atoms are all locked into rigid positions in a lattice, so the number of ways they can move around is quite limited. In a liquid, the options expand considerably, and in a gas, the atoms can take on many more configurations. This is why the entropy of a gas is higher than a liquid, which in turn, is higher than a solid.

Just to give you a sense of how many actual ways I'm talking about here, take one mole of water. In that amount of water at room temperature, the number of possible states we have is about ten raised to the number 2 with 24 zeros after it; that's an almost unimaginable number of microstates! Let me give one more example to illustrate the power of thinking of entropy as a counting of states. And I'll use a very common example in thermodynamics, that is, the expansion of an ideal gas into a vacuum.

Suppose I have this chamber, and just to keep it simple, I'll make it have an adiabatic wall around it; that means that no heat can pass across the boundaries. But I'll also put a little membrane down the middle, and I start with the gas all on just one side of that membrane. On the other side, there's nothing, so, it's a vacuum. What happens when I remove this membrane? Well, intuitively we all know what will happen, right? The gas will expand into the vacuum. But hang on. Why does this happen? If you think about it, the temperature of the gas does not increase. That's because when I remove the membrane, now the gas particles can travel over to this side over here where the vacuum was, but they're not going to travel any faster or slower on average, and since the temperature is related to the average kinetic energy of the gas, it won't change, and no heat can transfer in or out of the system since it's an adiabatic wall.

What about work? Is any work done during this expansion? That's sometimes tricky, since most students who see this at first will assume that work is done, because the volume of the gas changes, and we know that work is equal to minus pressure times the change in volume. But hang on a minute. During the expansion, the gas has zero pressure, since it's expanding into a vacuum. So actually, there cannot be any work done during the process. So we have no work done, no heat transferred, no change in temperature or internal energy; seems kind of like not much of anything is going on. What is it that governs the behavior? It all comes back to counting.

Suppose I slice up the container into a bunch of boxes; let's say m boxes on the left where the gas starts, and if I have n particles, then the number of ways I can arrange these particles, that is, the number of possible microstates for them to be in, would be equal to m raised to the mth power. But once I remove the membrane wall, there are now $2m$ boxes, and the number of possible ways to arrange n particles over those $2m$ boxes is going to be $2m$ raised to the nth power, so, a whole lot more. It's only because of that, only because of that reason that the gas expands into the vacuum. There are more ways in which the gas can be, just like rolling a 12 instead of a 1 with the three dice. Whenever there is a way for a system to have more ways to be arranged, it tends towards that state.

Again, I know that some of this stuff is a bit confusing the first time you see it. Don't worry; you're not alone. If nothing else, as you think about the concepts I've discussed in this lecture, start from this very simple statement. The first law of thermodynamics does not distinguish between heat transfer and work. As we learned, it considers them as equivalent. The distinction between heat transfer and work comes about by the second law of thermodynamics.

Now, before I conclude, I want to bring up one more important consequence of entropy. That is, it tells us something about the usefulness of energy. Suppose I have an axle that can rotate, and it has a weight attached to one end by a string. Now, when I put energy into rotating the axle, which in turn lifts the weight, all the molecules of the rod are moving the same direction; this helps to keep the energy organized, in a sense, so that I can perform the useful work of lifting up the weight.

And here's an important observation. If I assume that there is no dissipation, then there would also be no entropy transfer associated with the work the rod is doing on the weight. Of course, I always do have some degree of dissipation, like from friction, but I could try to get pretty close to the ideal case, where no additional disorder is produced by moving the weight up in this way. Any process that does not produce net entropy is reversible, and therefore, the process described here can be reversed by lowering the weight. The energy is not degraded during this process, and the potential to do work is maintained.

But now, instead of using this axle to raise and lower a weight, let's do something different; let's put the axle into a box filled with gas molecules, and instead of a weight on a string, I'll put a fan on the end of it. Now, I can put the same amount of energy into turning the rod, and the fan will rotate, which in turn will heat up the gas. The hotter the gas gets, the more disorder it has and the higher its entropy will be. Notice how different this process is compared to raising a weight: We're converting this organized form of energy into a highly disorganized form of energy, that is, rotation of the rod is converted into higher internal energy of the gas.

But I cannot convert this energy back to the rod as the rotational kinetic energy. It cannot be reversed. No matter how much energy those gas molecules have, they're not going to spin the fan (or in other words, they will not do work on the fan). This is because the energy of this gas is in no way organized. It's hard to extract useful work directly from disorganized energy. Maybe if I wanted I could use the heat in a heat engine to make the rod turn, but as we'll see in a later lecture on Carnot, I can never do this with 100 percent efficiency.

So, energy is degraded during this process, the ability to do work is reduced, disorder is produced, and associated with all this is an increase in entropy. The quantity of energy is always preserved during an actual process (that's the first law), but the quality is likely to decrease (that's the second law). This decrease in quality is always accompanied by an increase in entropy.

The Chemical Potential
Lecture 9

In this lecture, you will learn about the concept of molar and partial molar quantities. It is the concept of molar, or partial molar, quantities that allows us to describe the thermodynamic variables associated with different components of a material when we have more than one component in the first place, as is very often the case in reality. It is also this concept that leads to one of the single most important variables in all of thermodynamics: the chemical potential.

Molar Value

- Perhaps somewhat counterintuitively, we'll define mixture to mean a heterogeneous system with multiple components that are not mixed on the molecular scale. A solution, on the other hand, is defined as a homogeneous multicomponent system with the components mixed on the molecular scale.

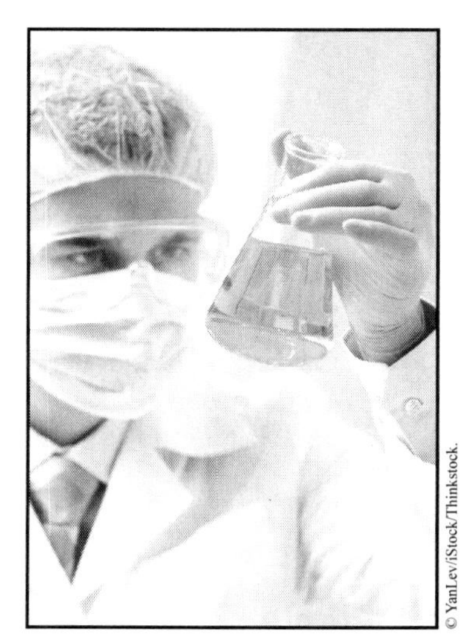

The components of a mixture are not mixed on the molecular scale, whereas the components of a solution are.

- The mixture of oil and water isn't a solution, and it only even mixes when we shake it up, forming large droplets of each type of liquid interspersed in the other. The mixture of ethanol and water, on the other hand, forms a solution because they can mix together all the way down to their molecular core. Salt and pepper are two solids that form a mixture when in their usual state.

- When we have more than one component in a system, we call it a multicomponent system. And for an open multicomponent system, in addition to the state variables such as pressure, temperature, and volume, each additional component is treated as another variable of the system.

- This variable is represented in terms of the number of moles of each component. So, for a system with two components, component A and component B, we would have the variables n_A and n_B which correspond to the number of moles of each of those components, respectively.

- The molar quantity of a given extensive variable is written as that variable with a bar on top. So, if we have a property Y, then its molar quantity is equal to $\bar{Y}$, which equals the variable divided by the number of moles of the component we have.

- For a single-component system, this simply gives the property per mole as opposed to the total amount. That can sometimes be a convenient way of writing a variable for a single-component system. However, it's when we have more than one component where the idea of molar quantities—and partial molar quantities—really kicks in some value.

- For the system with two components, A and B, if we want to know the molar quantities of a given property for each of these components, we would divide the variable for that property by the number of moles of the component.

- So, for that same variable Y, we would have $\bar{Y}_A$, which is the extensive variable Y for the total system (that is, the value of that variable for the full mixture of A and B), divided by the number of moles of A, which is n_A. Similarly, we can write the molar quantity for component B.

- Note that the fraction of the material that has component A is simply n_A divided by the total number of moles, which is $n_A + n_B$. The total fraction of the material that has component B is n_B divided by the same total number of moles n. We like to use the variable X to represent fractions of components in materials. Note that by definition, X_A plus X_B has to equal 1.

Molar Volume

- Recall that if a property is extensive, it means that it scales with the system size. The molar volume, on the other hand, will no longer be an extensive property. Now, we're dividing the property by the number of moles of the material we have.

- So, writing a variable like volume in terms of its molar quantity changes it from being an extensive variable into an intensive one. If we double the system size, then the total volume doubles, but the molar volume remains the same because the number of moles also doubles. Because it does not change with the size of the system, molar volume is an intensive variable.

- The formal mathematical definition of a partial molar quantity is as follows: For a given component, call it component k, and a given property, call it property Y, the partial molar value of Y is equal to the partial derivative of Y with respect to the number of moles of component k—holding all other variables fixed while the derivative is taken.

- In terms of notation, when we write the partial molar quantity for a variable, we put a bar over the variable. The subscript of the variable with a bar on top is the component that the partial molar value is for.

- With this definition, we can also define the total value for the variable, obtaining its original extensive value. Namely, it would simply be a sum over all components in the system of the partial molar variable for that component times the number of moles of that component.

- The partial molar volume does not remain constant as the composition of the mixture changes. And the behavior is not a simple linear or quadratic function but, rather, varies in a complex manner as the materials are mixed. The mixing can indeed be predicted using sophisticated models, ones that account for the ways in which the materials interact at the molecular level.

- Because the partial molar volume varies with concentration, the total volume equation should really be written as an integral from zero to the number of moles of one component of the partial molar volume for that component times the differential number of moles of that component. The total volume would be the sum of such an integral for each component of the system. However, keep in mind that because volume is a state function, the final volume will be the same and this equation will be valid no matter how the solution was prepared.

- Another point about the partial molar volume is that while the total volume is always a positive value, the partial molar volume does not necessarily have to be. In some cases, the volume goes up when you add more of a given material, while in other cases, the volume goes down, corresponding to either a positive or negative partial molar volume.

The Chemical Potential

- We can look at the partial derivative—and therefore partial molar value—of any thermodynamic variable. Recall that the internal energy is the variable that forms the basis for the first law of thermodynamics and equals heat plus work ($Q + W$).

- So, the partial derivative of the internal energy with respect to a given component of a system while holding everything else fixed is such an important quantity that we have a special name for it: the chemical potential, which is represented by mu (μ).

- Physically, the chemical potential represents the portioning of potential energy attributable to the presence of a particular component in a system. For a material with more than one component, there is a chemical potential associated with each component.

- So, we can think of the chemical potential as a thermodynamic force resisting the addition or removal of molecules to a system; chemical work is performed to move molecules against this force. And this chemical work occurs because the internal energy of the system changes in response to changes in the composition of a system.

- The chemical potential is the partial derivative of U with respect to a given component, holding all other variables such as entropy, volume, and the number of moles of all other components fixed.

- Put simply, the chemical potential is a measure of the general tendency of matter to transform. This transformation could be a reaction, or a change of phase, or a redistribution in space. Either way, it is the chemical potential that expresses the likelihood of transformation.

- If we think about a given transformation, then we can think about the system before the transformation, in some initial state, and after the transformation, in some final state. There is a chemical potential for each component in the system, which means that the chemical potential for the system is a sum over the chemical potentials of each component in the system.

- Did the total chemical potential increase or decrease as a result of a given transformation? If the chemical potential is higher on one side than the other, then the transformation will occur. And the transformation will be driven to occur in the direction of high chemical potential to low chemical potential. In other words, it will be favorable if the chemical potential is lower in the final state than in the initial state.

- Think of a seesaw as an analogy. The weight on each side is the only thing that determines which way the seesaw tips. More weight will tip one side down, and equilibrium (the seesaw perfectly horizontal) can be reached only when the weights on each side are the same. We can think of the chemical potential as the weight, tipping the balance of a given transformation to either happen or not. More chemical potential on one side of a transformation means that this side will "win."

- It's important to consider that barriers may exist between transformations. The chemical potential only tells us about the tendency of a given transformation to occur, but it doesn't always mean that it will indeed occur.

- Think about the seesaw example again: What if there were a big rock stuck under one side? Then, even if there were more weight on that side, the seesaw would not go down that way. So, a drop in the chemical potential is a necessary but not sufficient condition for a transformation to occur. Water wants to spill out of a cup but will not because the walls of the cup prevent it from doing so.

Suggested Reading

Anderson, *Thermodynamics of Natural Systems*, Chap. 2.

DeHoff, *Thermodynamics in Materials Science*, Chaps. 8 and 11.

Questions to Consider

1. If you fill a jar with different-size objects—such as steel balls, sand, and then water—does the order you fill the jar matter in terms of what it looks like when it's full? From a purely thermodynamics perspective, *should* the order matter?

2. Why doesn't a candle self-ignite? Think about the answer to this question in terms of the partial molar derivative of the free energy (otherwise known as the chemical potential).

The Chemical Potential

Lecture 9—Transcript

In this lecture, we'll learn about the concept of molar and partial molar quantities. These concepts are indispensible for describing what happens when we have more than one material together, as is very often the case in reality. For example, steel is not just made of iron, but rather, it is mostly iron with a little bit of carbon mixed in. The silicon in your computer chips has oxygen mixed in with it at the surface. And Teflon has carbon and fluorine mixed in polymer chains. Cement is a mixture of calcium, silicon, and oxygen, among other stuff, and even a plain old glass of water isn't just H_2O; rather, it has small amounts of calcium, sulfur, iron, among many other elements mixed in.

It is the concept of molar, or partial molar quantities that allows us to describe the thermodynamic variables associated with different components of a material when we have more than one component in the first place. It is also a concept that leads to one of the single most important variables in all of thermodynamics, something called the chemical potential. But first, as in a number of other lectures, let's start off with a few important definitions.

Since we'll be talking about what happens when mixing materials together, the first word we should define is the word mixture itself. Perhaps somewhat counter intuitively, we'll define mixture to mean a heterogeneous system with multiple components that are not mixed on the molecular scale. A solution, on the other hand, is defined as a homogeneous multicomponent system with the components mixed on the molecular scale.

So, oil and water isn't a solution, and it only even mixes when I shake it up, forming large droplets of each type of liquid interspersed in the other. Ethanol and water, on the other hand, forms a solution, since, as we'll see in our demo today, they can mix together all the way down to their molecular core. Salt and pepper are two solids that form a mixture when in their usual state.

Now, when we have more than one component in a system, we call it a multicomponent system. And for an open multicomponent system, in addition to the state variables, such as pressure, temperature, and volume, each additional component is treated as another variable of the system. This variable is represented in terms of the number of moles of each component. So, for a system with two components, call them component A and component B, we'd have the variables n sub A and n sub B, which correspond to the number of moles of each of those components, respectively.

The molar quantity of a given extensive variable is written as that variable with a bar on top, like this. So if I have a property, call it Y, then its molar quantity is equal to $\bar{Y}$, which equals the variable divided by the number of moles of the component we have. Now, for a single component system, this simply gives the property per mole, as opposed to the total amount. That can sometimes be a convenient way of writing a variable for a single component system. However, as I mentioned, it's when we have more than one component where the idea of molar quantities—and as we'll see, partial molar quantities—really kicks in some value.

For the system with two components, A and B, if I want to know the molar quantities of a given property for each of these components, I would divide the variable for that property by the number of moles of the component. So, for that same variable Y, I would have $\bar{Y}_A$, which is the extensive variable Y for the total system, that is, the value of that variable for the full mixture of A and B, divided by the number of moles of A, n_A. Similarly, I can write the molar quantity for component B.

Note that the fraction of the material that has component A is simply n_A divided by the total number of moles, which is just n sub A plus n sub B. The total fraction of the material that has component B is n_B divided by the same total number of moles n. We like to use the variable X to represent fractions of components in materials. Note that by definition, X_A plus X_B has to equal 1.

Let's use a simple example to bring these variables and definitions to life. We'll work with the extensive property volume, since it's a nice and easy one to visualize. By the way, remember that when I say a property is extensive, it means that it scales with the system size. The molar volume, on the other hand, will no longer be an extensive property. Now, I'm dividing the property by the number of moles of the material I have. So, writing a variable like volume in terms of its molar quantity changes it from being an extensive variable into an intensive one. If I double the system size, then the total volume doubles, but the molar volume remains the same, since the number of moles also doubles. Since it does not change with the size of the system, molar volume is an intensive variable.

The molar volume of, say, water, is 18 cm cubed per mole. That means that 1 mole of water, that is, about 6.022 times 10^{23} water molecules, that much water takes up a volume of 18 cm cubed. So, if I add a mole of water to a beaker filled with water, then the volume of the liquid in the beaker will increase by 18 cm cubed. But, as I mentioned, this is not where things get interesting. You see, it turns out that the volume change would have been different if I added the same amount of water to something else. For example, if the beaker had been filled with ethanol instead of water, the volume change upon adding the same 1 mole of water would have been only 14 cm cubed instead of 18, and that's pretty strange! The change in volume from adding something depends on the things I'm adding it to. That's where the concept of partial molar quantities comes in. And this is the perfect time to turn to our demo, where we'll be able to see and hold this concept in our hands.

In this demo we're going to talk about the concept of partial molar quantities, and we're going to bring them to life and show how they work in action with a particular variable, namely, volume. So, in this jar, we see that we have filled up to about a third to half the way, it's filled with steel balls. But the question is, is it really filled up to that point? Well, it seems so, but, as you can see, if I put smaller balls into there, then they find their way down, and they can go in between the bigger balls. And you can see that it's still filled up to the same point as it was before.

So then the question is, well, what about now? Is it full now? The answer is no, because now I can add even smaller steel balls, and they'll find their way to the spaces in between and trickle down into the open volume that you see. So, is it full yet? Is it filled up to that same starting point? No. Not yet, because now, I can add something even smaller. Here I have sand. How much could I put in?

Well, let's pour some in. And you can see that the level of the steel balls that I started with is not changing, but I'm adding something to it. I'm still adding mass and material to the container. Okay? So I'm adding sand, adding sand, and the volume does not change. Let's shake it up a little bit and let it settle, and you can see that the height of the steel balls is still exactly the same.

So what about now? Can I put anything else into it? Well, the answer is yes. The sand is a small, little, tiny particle, but it's still pretty big compared to, say, molecules. And it's also a solid. If I go to a liquid, well then I can really start filling up the remaining space in this container. And you see that if I pour liquid in, then it can settle and go down into the sand and find it's way even further into those voids that are still remaining in between the little sand particles.

And then we can ask, well, is it full now? The answer, still not quite yet. I can fit another liquid into the water, namely, ethanol. Let's take a look. Over here, I have exactly 50 mL of water, and 50 mL of ethanol measured out. Ethanol is on the left, and water is on the right. If I mix them together, what happens? Well, let's take a look.

First, I'll pour the water into this bigger beaker, and what you can see is that that comes up to a measurement of exactly 50 mL, since that's how much I started with. But now, when I add 50 mL of ethanol, what you see is that the level that this mixture comes up to, and I'll pour it all in, is not 100 mL. I only get 97 mL. That's based on the same, exact thing you just saw with the steel balls and the sand, but it's happening at a molecular scale.

What's happening is that the ethanol molecules are able to fit inside of some of the open spaces in between the water molecules, and that's why the two volumes, when you add them together, are not the same as if you just doubled them each independently. And that is the concept of partial molar volume.

So now you have a good sense of the reason we need to be able to quantify the amount of a given property per mole of a component in the system, in the context of that component being mixed with others. Pouring sand into a bowl would change the volume by a certain amount with each sand grain, but if I've already filled the bowl with marbles, that amount of change is different; it's effectively zero until we cover all the marbles.

The same thing goes at the molecular level, which is why, if I add water to ethanol, the volume change is different than if I add water to water. Now the word I'd like to focus on there is change. As you may have noticed, the most important quantity we want here is the change in volume per mole. And when we want a change in a variable, what we're talking about is a derivate. So that brings us to the formal mathematical definition of a partial molar quantity. For a given component, call it component k, and a given property, call it, again, property Y, the partial molar value of Y is equal to the partial derivative of Y with respect to the number of moles of component k, holding all other variables fixed while the derivative is taken.

In terms of notation, that's written like this, where this type of d in the derivative just means "partial," and the list of variables in the subscript are the ones that are held fixed when the derivative is taken. In terms of notation, when we write the partial molar quantity for a variable, we like to put a bar over the variable. The subscript of the variable with a bar on top is the component that this partial molar value is for. Notice also, that with this definition, we can also define the total value for the variable, obtaining its original extensive value. Namely, it would simply be a sum over all components in the system of the partial molar variable for that component times the number of moles of that component.

Just to be sure, we can connect this definition to what we just saw in the demo, let's work out the problem of the volume of the ethanol-water mixture. Specifically, let's calculate the final volume of the mixture of 50.0 mL of ethanol and 50.0 mL of water. We know the original volume of each component, so, what we need to do is to find the partial molar volumes. Assuming constant temperature and pressure and that the liquids are fully miscible, that is, they don't separate like oil and water, then we can write down the total volume as the sum over the partial molar volume of water times the number of moles of water, plus the partial molar volume of ethanol times the number of moles of ethanol.

So, we'll need to know the partial molar volume for the water and for the ethanol, and we'll need to know how many moles we have of each. First, how do we get the partial molar volumes? Well, it's actually quite easy; we simply look it up. Now, I'm not just saying that lightly. In fact, this is quite an important aspect of thermodynamics, namely, the fact that very often, important quantities are tabulated, and the way we access them is simply to look them up in a table. For the case of partial molar volumes, thousands and thousands of them have been carefully measured and put into tables and graphs. So, when we mix ethanol and water together all we need to know is how much of each one we have and the table will guide us to the values for partial molar volume.

And that gets us to the heart of the problem, that is, figuring out how much stuff we actually have of each component in the mixture. I could try to do this by telling you the volume, but that's not good, since the whole point here is that the volume a substance has can change when it's mixed with something else, and it depends on the fraction it has within the mixture. Aha, and the key here is that word—fraction. What we need to know in order to know how much stuff we have, is the mole fraction of each component.

Unlike volume or other variables that can change depending on the mixture, the mole fraction is a direct measure of the mass of each component. It's based on the molecule count, since a mole of anything is always the same number of molecules of that thing. The mole fraction of a component in a mixture is so important in thermodynamics that we're going to be seeing it over and over and over again. It's the basis for how we examine the behavior of mixtures.

So back to our water and ethanol. How many moles of each liquid do we have before mixing them together? 50 mL of water times the density of water, which is 1 g per mL, gives us the number of grams, that would be, 50 g. And for water, we have 18.02 g for every mole, so the number of moles in 50 mL would be equal to 2.77, as you can see here. For ethanol, the density is 0.79 g per mL, and so, that's 46.08 g per mole, so 50 mL of ethanol corresponds to 0.86 moles of ethanol, as you can see from the math shown here. Now I pointed out that this was a calculation for the number of moles of each liquid before we mixed them. You may be wondering, what about after they're mixed? Well that's just the thing; it's the same. No chemical reactions occurred to transform one component into something else, so the number of moles of each cannot change when they're mixed.

And now that we know how many of moles of each one we have, we can compute the mole fraction of each. For ethanol, the mole fraction, for which we like to use the variable X, is equal to the number of moles of ethanol divided by the total number of moles for both ethanol and water, or 0.86 divided by 2.77 plus 0.86, which gives a number of 0.24. Note that it's a fraction, so this number has no units. And so this means that when I mix 50 mL of water with 50 mL of ethanol, in terms of a count of the number of molecules, the ethanol stands at around 24%. And now, we're in a position to look up the partial molar volume, since we know the mole fraction.

From the tables for the water-ethanol mixture at a mole fraction of 24% ethanol and 76% water, the partial molar volume of water is 17.5 mL per mole, and that of ethanol is 55.6 mL per mole. The total volume of the mixture is each of these times their respective number of moles, so that gives the summation you see here, which amounts to about 96 mL; in fact, exactly what we measured in the demo.

But, let's return to this idea of a look-up table. I know that some of you may be thinking, well that's sort of cheating isn't it? I mean, you solved the problem by kind of looking up the answer. And that's true to a certain extent, although, most importantly, what we have is a general framework for dealing with properties of systems with multiple components. In this case, we looked up the property in a table, but as we'll see in other cases, we may rely on a formula or a relationship derived from a model or some physical insight.

And you may also wonder how much we really need such tables or equations? What that question really asks is how much does a property really change upon mixing? The answer to this question is that it depends on what is mixed with what, and in particular, on the nature of the interactions between components. As we'll learn, the two major factors in mixing are the strength of the interactions and the changes in entropy of the system.

For now, to give you a sense of why tables are so useful, take a look at the change in partial molar volume on the mole fraction. The x-axis in this graph corresponds to the mole fraction of ethanol, and since it's the mole fraction, it goes from 0 to 1. The y-axis is the partial molar volume of water, with its scale shown on the left. Notice how this quantity changes quite a bit as a function of the mole fraction! When the ethanol mole fraction is zero, that means we just have pure water, and the partial molar volume of water is around 18 cm cubed per mole. This, remember, is what we said the volume of a mole of water is when it's just by itself. As it mixes with the ethanol, this number actually first increases, with a peak at just under 10% ethanol, after which it decreases nonlinearly down to the value of 14, the volume change that occurs when I add water to pure ethanol.

So as you can see, the partial molar volume does not remain constant as the composition of the mixture changes. And the behavior is not so straightforward; it's not a simple linear or quadratic function, but rather, varies in a complex manner as the materials are mixed together. The mixing can, indeed, be predicted using sophisticated models, ones that account for the ways in which water and ethanol interact at the molecular level. But, a look-up table is also extremely useful for such complicated dependencies.

Now, one last point on the math; since the partial molar volume varies with concentration, our total volume equation should really be written as an integral from zero to the number of moles of one component of the partial molar volume for that component, times the differential number of moles of that component, like this. The total volume would be the sum of such an integral for each component of the system. However, keep in mind that, since volume is a state function, the final volume will be the same and this equation valid no matter how the solution was prepared.

Okay, so before I move on to discuss another partial molar variable, I'd like to highlight one more point about the partial molar volume. Notice that while the total volume is always a positive value, the partial molar volume does not necessarily have to be. Let's take an example. The compound magnesium sulfate is an inorganic salt found naturally and used in many applications. Some of you may be more familiar with it as Epsom salts, used by a lot of folks to take a nice, relaxing bath. But, did you know that when you add those bath salts to the tub filled with water, the level of that water actually goes down? This is because the partial molar volume of magnesium sulfate, when mixed with water at low concentrations, is negative, negative 1.4 centimeters cubed per mole, to be specific.

We can understand this behavior from the fact that when we add salt to water, the salt dissolves into ions, that is, it separates into charged species. The simpler salt we all love to put on our food, sodium chloride, separates into a sodium atom with a positive charge and a chlorine atom with a negative charge. This slightly more complicated salt separates into a magnesium atom with positive charge of plus 2, and a sulfate group, SO_4, with a net negative charge of minus 2. The point is that these charged species really like water molecules, and water molecules really like them; so much so, in fact, that the water tends to cluster around them closely and not want to let go. The water likes the salt ions more than it likes itself! And that's why the volume goes down, since the clustering to the salts is closer than the normal spacing among water molecules.

So, let's go back to our original definition of a partial molar quantity and take a look at it graphically. Here's a plot of the volume of a mixture as a function of the number of moles of some component, while holding all other variables in the system fixed. Our definition of the partial molar volume is simply the slope of this graph. And now you can see that in some cases, the volume goes up when you add more of a given material, while, in other cases, the volume goes down, corresponding to either a positive or a negative partial molar volume.

And now, I'd like to spend the rest of this lecture on a different partial molar quantity. I used volume as an example in order to walk through what a partial molar quantity is, but, as I mentioned, we can look at the partial derivative, and therefore, partial molar value, of any thermodynamic variable. And remember that variable called the internal energy? The one that forms the basis for the first law of thermodynamics, which I talked about in Lecture 6, and the one that equals Q plus W, or heat plus work? That's the one.

So the partial derivative of the internal energy with respect to a given component of a system while holding everything else fixed, that's such an important quantity that we have a special name for it. We don't just call that the partial molar internal energy. No, no, this is so important it gets its very own name; it's called, the chemical potential.

Physically, the chemical potential, for which we use the symbol μ, represents the portioning of potential energy attributable to the presence of a particular component in a system. For a material with more than one component, there is a chemical potential associated with each component. So we can think of the chemical potential as a thermodynamic force resisting the addition or removal of molecules to a system. Chemical work is performed to move molecules against this force. And this chemical work occurs because the internal energy of the system changes in response to changes in the composition of a system.

So, mathematically, the chemical potential is written as seen here, the partial derivative of U with respect to a given component, holding all other variables such as entropy, volume and the number of moles of all other components fixed. Put simply, the chemical potential is a measure of the general tendency of matter to transform. This transformation could be a reaction, or a change in phase, or redistribution in space. Either way, it is the chemical potential that expresses the likelihood of transformation.

If we think about a given transformation, then we can think about the system before the transformation, in some initial state, and after the transformation, in some final state. Remember that I mentioned there is a chemical potential for each component in the system. That means that the chemical potential for the system is a sum over the chemical potentials of each component in the system. The question we can ask, then, about a given transformation is the following; did the total chemical potential increase or decrease as a result of that transformation? If the chemical potential is higher on one side than the other, then the transformation will occur.

And here's the key; the transformation will be driven to occur in the direction of high chemical potential to low chemical potential. That's worth repeating—A given transformation will be favorable if the chemical potential is lower in a final state than in the initial state. Think of a seesaw as an analogy. Here, the weight on each side is the only thing that determines which way the seesaw tips. More weight will tip one side down, and equilibrium, that is, when the seesaw perfectly horizontal, can be reached only when the weights on each side are the same. We can think of the chemical potential as the weight, tipping the balance of a given transformation to either happen or not. More chemical potential on one side of a transformation means that this side will win.

Let's use an example of a reaction as our transformation to help illustrate this point. Take a candle, which is made out of paraffin wax. This material consists primarily of long hydrocarbon chains, a chemical based on the basic unit CH_2. So, why does this material burn when I light it on fire? Well, because of the change in chemical potential of course! When I combine paraffin wax with oxygen, those represent the reactants, or, starting materials for the reaction. The products, or ending materials, that I get by combusting these starting materials are carbon dioxide and water. Written out as a balanced reaction, we have the following:

$$3(O_2) + 2((CH_2) \leftrightarrow 2(CO_2) + 2(H_2O).$$

So, I can look up these chemical potentials, add them up for each side, and I get a value of 8 on the left and −1246 on the right, a pretty big difference. That means that the left side wins, since it's larger, which means that that side will do the reacting so that the system can lower its chemical potential by going to the right-hand side.

Now, I just did a big no-no for this course. I threw some numbers out there without even specifying their units! I hope you caught that and got very concerned. Let's remedy this situation now. As you can see from the definition of the chemical potential, its units are going to be energy per number of moles. The most common units are Joules per mole. But we need to know more than just the units. We also need to know how the chemical potential is referenced. That value of 8 for the reactants doesn't mean anything by itself unless, just like for temperature, we have some standard way to reference and measure it. And of course, we do. Standard conditions for the chemical potential are at a temperature of 298 Kelvin, and a pressure of 1 atmosphere. For the pure elements, we define the chemical potential to be zero.

Here is a table of measured chemical potentials for different materials at standard conditions. Notice that most of the chemical potential values are negative, meaning that they can be produced spontaneously from the elements. This also means that the reverse is true, Most materials do not tend to decompose into their elements, but rather tend to be produced from them. That's a good thing, since it means the materials we deal with in the world, including our own selves, are stable; they don't tend to decompose. On the other hand, if the chemical potential is positive, the material will tend to decompose into its elements, and the material is unstable.

But, there's a really important point we need to consider here, namely, the fact that barriers may exist between transformations. The chemical potential only tells us about the tendency of a given transformation to occur, but it doesn't always mean it will indeed occur. Think about the seesaw example again. What if there was a big rock stuck under one side? Then, even if there were more weight on that side, the seesaw would not go down that way. And it's the same with our candle example; if the chemical potential is lower for the carbon dioxide and water than for the paraffin wax and oxygen, then why doesn't the candle simply self combust? It's because, even though it has a tendency to want to do so, to want to go through that very transformation, a barrier inhibits it from occurring, that is, until we supply a little bit of heat by lighting the candle on fire.

So, a drop in the chemical potential is a necessary but not sufficient condition for a transformation to occur. An apple wants to fall from the tree, but it does not when the stem holds it to the branch. Water wants to spill out of a cup, but will not, because the walls of the cup prevent it from doing so. And so on.

I hope you've gotten a glimpse into the importance of this partial molar quantity. The chemical potential is at the core of understanding so many different thermodynamic processes, from chemical reactions, to the effects of temperature and pressure on a solid or liquid, to the interaction of light with matter, to the way in which materials mix or not, to how mass transports through materials and membranes; that last one being the topic of our next lecture.

Enthalpy, Free Energy, and Equilibrium

Lecture 10

At this point in the course, you have mastered all the tools necessary to tackle the crucial concept of chemical equilibrium. The notion of equilibrium has been discussed before, but in this lecture, you will examine it in greater detail and learn about how to determine equilibrium under realistic laboratory conditions. As the basis for this discussion, you will need to learn about a few new thermodynamic variables, one of which is named after Josiah Willard Gibbs.

The Gibbs Free Energy

- Let's consider a single-component system—that is, a system with only one type of material—such as water. Think about ice cubes in a beaker closed at the top. What happens when we raise the temperature by heating the beaker? At zero degrees Celsius, the ice begins to melt to form liquid water, and if we continue heating to above 100 degrees Celsius, then the water will boil to form a vapor phase.

- But why does ice become water above zero degrees and vapor above 100 degrees? To put this in thermodynamic terms, the question we need to answer is the following: What quantity governs the equilibrium of this material at different temperatures? And, in particular, what quantity can we use to determine equilibrium that is actually easy to measure?

- This is where a new quantity comes into play—a new thermodynamic variable called the Gibbs free energy (G). The definition of G can be written as an equation in terms of other thermodynamic variables. In particular, G equals the internal energy (U) minus the temperature times the entropy of the system (T times S) plus the pressure times volume (P times V): $G = U - TS + PV$.

- G is important if we have a closed system, which means that the system has fixed mass and composition because no matter can pass through the boundaries. For such a system, G tells us when the system is in stable equilibrium; this occurs when it has the lowest possible value of the Gibbs free energy.

Josiah Willard Gibbs (1839–1903) was an American theoretical physicist and chemist who applied his knowledge of thermodynamics to physical chemistry.

- If a system is always trying to minimize its G, because this is what it means for it to be in equilibrium, then for different temperatures, the system tries to find the best compromise between a low value for the combined internal energy plus PV term and a high value for the entropy (related to the number of degrees of freedom for the system). This balance is simply due to the plus and minus signs in the definition.

- At higher temperatures, the effects of increasing entropy will be more and more important because a high T will make S have a bigger effect. When the temperature is very high, the entropy contribution dominates over the contribution from internal energy and PV.

- The derivative is dG, which is equal to dU plus P times dV plus V times dP, minus T times dS minus S times dT. The first law gives us a definition for dU for a reversible process: $dU = TdS - PdV$, assuming that only mechanical work is being done. If we substitute this expression for dU from the first law into the equation for dG, dG simplifies to V times dP minus S times dT—again, for a system where the only work term is from mechanical work.

- Suppose that we were to perform experiments in which the pressure is held constant. In that case, we could look at the dependence of G on temperature and arrive at the following relationship: The partial derivative of G with respect to T, holding P fixed, is simply equal to $-S$. That tells us that in a constant-pressure experiment, the free energy depends linearly on temperature, and the slope of that dependence is the entropy of the system.

- The definition for dG can be equally useful for a system held at a constant temperature. In this case, we would have that the partial derivative of G with respect to pressure, with temperature held constant, is simply equal to V, the volume of the system. With this information, a similar curve for G versus P can be constructed.

The Second Law and Equilibrium

- The second law of thermodynamics can be applied to define the conditions for equilibrium for any arbitrary system. But the way that it was stated a few lectures ago says that the change in entropy of the universe is always increasing—that is, for any spontaneous process (one that will occur under a given set of conditions without any additional external forces acting on the system), it will increase—or, if the system has reached equilibrium, then the change in entropy is zero.

- In isolated systems, the transfer of heat, work, or molecules at the boundaries of the system is not allowed. In constant internal-energy systems, it is straightforward to directly apply the fundamental equation for the entropy and the second law to calculate equilibrium properties of the system.

- However, we rarely have an isolated system, and we are almost never able to even measure the internal energy. For that matter, it is extremely difficult to really keep a system at constant volume.

Controlling heat and work flow at the boundaries of a real system is often not a simple task, and it's also completely unnecessary for most experimental procedures you would be interested in performing.

- And that's where the Gibbs free energy comes in. In most experiments, what we would like to control at the boundaries of our system are things like temperature and pressure.

- The Gibbs free energy is defined as the internal energy plus pressure times volume minus temperature times entropy. With that definition, this equation becomes simply that dG for constant temperature and pressure must be less than or equal to zero for all possible processes. This is because when you take the derivative of G to get dG at constant temperature and pressure, you end up with exactly the expression we derived from our substitutions for entropy.

- This equation is, for practical calculations, perhaps the most important one in all of thermodynamics for predicting the behavior of materials. The second law demands that the Gibbs free energy must be lowered by the change happening to a system, in order for that change to happen spontaneously at constant temperature and pressure. It also dictates that the Gibbs free energy at constant temperature and pressure is a minimum at equilibrium.

- Basically, if you look at the definition for G, what we've done is taken the total internal energy and subtracted off the thermal energy and the compressive energy—what remains is the energy arising from other kinds of work terms. So far, we've mostly focused on two sources of internal energy: compressive work from an applied pressure and thermal energy due to heat transfer.

- However, there are many, many different types of work, including the work of polarization, chemical work, and magnetic work. In many practical problems in science and engineering, it is these other kinds of work in which we are most interested. So, the Gibbs free energy is the amount of energy that is free to do these other types of work. That's why it's called a free energy.

Enthalpy

- Another thermodynamic variable is called the enthalpy (H), and it's equal to the internal energy (U) plus the pressure (P) times volume (V): $H = U + PV$. This is a new energy state variable (because it only depends on other state variables). As with the free energy, enthalpy is not a new independent variable, but rather it is a function of other state variables—in this case, the internal energy, pressure, and volume.

- The enthalpy is useful when discussing constant-pressure processes, because in that case, the change in enthalpy (dH) is related to the heat of the system if only mechanical work is involved. It turns out that $dH - dPV = \delta Q$.

- If the process is a constant-pressure process, then the dP term is zero, and dH simply equals δQ. So, for a constant-pressure process, the change in enthalpy is exactly equal to the heat transfer involved in the process. And it's a state variable, which means that unlike the general variable Q, which is path dependent, the enthalpy can mean the same thing but is path independent, as long as pressure is held fixed.

- Unlike the internal energy, it's difficult to give a simple, intuitive physical description of enthalpy. A change in enthalpy is the heat gain or loss for a constant-pressure process, and it turns the variable for heat energy (Q) into a state variable, because enthalpy is a state function. This is why enthalpy is also referred to as the "heat content."

- And it is pretty common for pressure to be held fixed. As with the Gibbs free energy, the enthalpy is useful because of the fact that volume is not the most convenient variable to hold constant in an experiment; it's often much easier to control the pressure, especially if the system is a solid or a liquid.

- The thermodynamics of constant-pressure environments is extremely important, and enthalpy is a very useful variable. You can think of it as being a thermodynamic variable that contains all the information that is contained in the internal energy but can be controlled by controlling the pressure.

- And notice the connection to the Gibbs free energy. The expression for G can be rewritten as $H - TS$. The Gibbs free energy is a state function useful for determining equilibrium in the most common type of experimental system: one where the pressure and temperature are maintained at a constant value. And a system always tries to make its G as low as possible.

- You can see from this equation that there will always be a compromise between trying to have a low enthalpy and high entropy, with temperature playing an important role in determining which of these is more dominant.

Suggested Reading

Anderson, *Thermodynamics of Natural Systems*, Chaps. 3–4.

Callen, *Thermodynamics and an Introduction to Thermostatistics*, Chap. 2.

DeHoff, *Thermodynamics in Materials Science*, Chap. 5.

Serway and Jewett, *Physics for Scientists and Engineers*, Chap. 20.3.

Questions to Consider

1. What governs the equilibrium of materials?

2. What are typical laboratory conditions, and what thermodynamic variable do we need to understand in order to determine equilibrium under such conditions?

3. What is "free" in the Gibbs free energy?

Enthalpy, Free Energy, and Equilibrium
Lecture 10—Transcript

Hi and welcome to our lecture on chemical equilibrium. At this stage in the course, we've mastered all the tools necessary to tackle this crucial concept. The notion of equilibrium has been discussed before, but in this lecture, we'll look at it in greater detail and talk about how to determine equilibrium under realistic laboratory conditions. And as the basis for this discussion, we'll need to learn about a few new thermodynamic variables, one of which is named after none other than Gibbs himself. I know, it's an emotional moment!

Let's begin by considering a single-component system, that is, a system with only one type of material. One of my favorite materials, which you may have already guessed, is water, so let's consider that as the single-component system. Now, think about ice cubes in a beaker closed at the top. What happens when I start raising the temperature by heating the beaker? Of course, as we know, at zero degrees Celsius the ice begins to melt to form liquid water, and, if I continue heating to above 100 degrees Celsius, then the water will boil to form a vapor phase. But why does ice become water above 0 degrees and vapor above 100 degrees? To put this in thermodynamic-y terms, the question we need to answer is the following; what quantity governs the equilibrium of this material at different temperatures? And, in particular, what quantity can we use to determine equilibrium that is actually easy to measure?

This is where a new quantity comes into play, and, as you may have guessed, it's the one I just took an emotional moment for; it is a new thermodynamic variable called the Gibbs free energy, or G. The definition of G can be written as an equation in terms of other thermodynamic variables. In particular, G equals the internal energy U, minus the temperature, times the entropy of the system, T times S, plus the pressure times volume, P times V. So, written out, it looks something like this: $G = U - (TS) + (PV)$.

So that's the mathematical definition, but, what is it good for? Well, let me first skip to the punch line, and then we'll unpack it further, and I'll explain why we needed yet another thermodynamic variable in the first place. So, here's where G is important. It's important if we have a closed system. Remember, this means the system has fixed mass and composition, since no matter can pass through the boundaries. For such a system, then G tells us when the system is in stable equilibrium, namely, this occurs when it has the lowest possible value of the Gibbs free energy.

Let's look at this statement in terms of the equation we have for G. If a system is always trying to minimize its G, since this is what it means for it to be in equilibrium, then for different temperatures, the system tries to find the best compromise between a low value for the combined internal energy plus PV term, and high value for the entropy, remember, related to the number of the degrees of freedom for the system. This balance is simply due to the plus and minus signs in the definition. You can see from the equation that at higher temperatures the effects of increasing entropy will be more and more important since a high T will make S have a bigger effect. When the temperature is very high, the entropy contribution dominates over the contribution from internal energy and PV.

Now, what I'd like to do is a little bit of math to illustrate how G behaves under different conditions. First, let's look at the derivative, dG. It's equal to dU, plus P times dV, plus V times dP, minus T times dS, minus S times dT. But remember that wonderful first law? Well we can use it here; it gives us a definition for dU for a reversible process as $dU = TdS$ minus PdV, remember, assuming only mechanical work is being done. As we've discussed, there are many other types of work that could be important depending on the processes involved, but for now, as we're just beginning to explore what G is all about, let's keep it simple and include only this one work term. If we substitute this expression for dU from the first law into the equation for dG, we see that dG simplifies to V times dP minus S times dT, again, for a system where the only work term is from mechanical work. And from this equation we can really start to see why G is such a useful variable.

Suppose I were to perform experiments in which the pressure is held constant. In that case, we could look at the dependence of G on temperature, and arrive at the following relationship; the partial derivative of G with respect to T, holding P fixed, is simply equal to negative S. That tells us that in a constant pressure experiment, the free energy depends linearly on temperature, and the slope of that dependence is the entropy of the system.

We can take this analysis back to our water example. Given what we know about entropy, we know that it will be different for the three different phases of water. In the solid phase, it's going to have the lowest value, since the number of degrees of freedom is the lowest, while in the liquid phase, it will be higher, and in the vapor phase, it will be the highest. If we construct a plot of the dependence of G on temperature for water, we'd have three lines corresponding to these three phases, and it would look something like this.

You can see that all the lines have a negative slope, which we know must be the case, since from the equation, G goes down with temperature. Also, note that for the solid phase G has the smallest negative slope, while for the vapor phase, it has the largest negative slope, because of the differences in entropy we just discussed.

And finally, you can see that the lines intersect at important points. In particular, the solid line intersects with the liquid line at zero degrees Celsius. Notice that for T less than zero, the solid line is lower, but for T greater than zero the liquid line is lower. Given what we said about G, namely that it will always be the lowest in equilibrium, we can see how this makes sense. The equilibrium phase of water is ice below zero degrees, so that curve shows the lowest value. We still have curves for the other phases, they're just higher in energy, and therefore, not representative of the equilibrium phases.

We'll see in a later lecture when I cool water below its freezing point but still manage to keep it as a liquid, that it's possible to not be in the equilibrium phase. But, as we've discussed, thermodynamics is about equilibrium, and you can see that if we have a graph like this, we're able to predict the equilibrium phase of the material. So just to finish with this graph, you can see that the curve for the vapor phase is above both other curves until the temperature is greater than 100 degrees Celsius, above which point, it represents the dominant phase, and therefore, must have the lowest G values among the various phases.

And if we go back to the definition for dG, we can see that it can be equally useful for a system held at a constant temperature. In this case, we would have that the partial derivative of G with respect to pressure, with temperature held constant, is simply equal to V, the volume of the system. With this information, we can construct a similar curve for G vs. P, and in this case for any given pressure, it is the curve with lowest G that corresponds to the equilibrium phase of the material.

Here's a plot of the free energy vs. pressure for two phases of carbon you're probably familiar with—graphite and diamond. Notice that the two curves cross at a pressure of around 1 gigapascal. In case you're used to thinking in units of atmospheres, where the air around us represents about 1 atmosphere, 1 gigapascal is equivalent to about 10,000 atmospheres, so it's a pretty high pressure.

We can see that above this pressure, diamond is the equilibrium phase, while below this pressure, it's graphite that is more stable. Now, the diamond some of us have in our jewelry you may have noticed is not spontaneously converting into graphite, even though we're about 10,000 times lower pressure than crossover point in this graph. But actually, at 1 atmosphere of pressure, diamond will, in fact, convert, spontaneously, into graphite. It may take 100,000 years or so, but for these two phases, from this graph, we are able to say which one will be most stable—eventually. In this case, we have barriers that slow the transformation between phases, making it hard to go from one phase to the other. The Gibbs free energy plots only tell us about equilibrium, not about such barriers. But remember that in thermodynamics, this is our primary concern, namely, the equilibrium properties of materials as opposed to their kinetic properties.

Now, why is it that G is so important? I already had a whole bunch of other thermodynamic variables that fit nicely into the thermodynamic laws. Why did I need another one? Let's go back to the second law of thermodynamics to see why. As you may remember, the second law can be applied to define the conditions for equilibrium for any arbitrary system. That's pretty powerful stuff. But the way that we stated it a few lectures ago, it says that the change in entropy of the universe is always increasing, that is, for any spontaneous process it will increase. Or, if the system has reached equilibrium, then the change in entropy must be zero.

By the way, just to be sure we all remember, a spontaneous process is one that will occur under a given set of conditions without any additional external forces acting on the system. One of the examples we looked at when we learned the second law was how to predict when equilibrium is reached when two boxes are placed into contact with one another. We assumed that the system was isolated and at constant volume, and by doing so, we were able to write the change in entropy of the universe in terms of the change in internal energy of the boxes. Remember that in isolated systems the transfer of heat, work, or molecules at the boundaries of the system is not allowed. In such constant internal energy systems, it's pretty straightforward to directly apply the fundamental equation for the entropy and the second law to calculate equilibrium properties of the system.

But hang on a second. How often do we really have an isolated system? And how often are we really able to even measure the internal energy? And for that matter, how hard is it to really keep a system at constant volume? It turns out the answers to these questions are rarely, almost never, and pretty darned hard. Controlling heat and work flow at the boundaries of a real system is often not a simple task, and it's also completely unnecessary for most experimental procedures one would be interested in performing. And that's where the Gibbs free energy comes in.

In most experiments, what we would like to control at the boundaries of our system are things like the temperature and pressure. Let me repeat that statement, since it's kind of crucial. In most experiments, what we would like to control at the boundaries of our system are things like the temperature and pressure.

Imagine we have a solution in a test tube, sitting in a beaker of water. We leave the top of the test tube open, as shown here. If we keep the water in the beaker at a constant temperature, either by heating or cooling the beaker, then we can see that whatever is inside the test tube will also maintain a constant temperature. And since the test tube is open at the top, it will also have a constant pressure. This is a very simple and common laboratory condition, perhaps one that many of you have seen in a high school chemistry class. And instead of our thermodynamics telling us what happens when the internal energy of the system is kept constant, these kinds of experiments require us to apply thermodynamics while keeping the temperature and pressure constant and to predict how the system changes when various thermodynamic driving forces are applied. For the case of our test tube, an example of a driving force would be if we wanted to carry out a reaction on the material in the solution.

So we have two new complications here. If we have our test tube in a heating bath, as shown, then internal energy in the form of heat is being transferred into and out of the system to maintain a constant temperature. This means that we can no longer apply the fundamental equation for the entropy to the system alone to determine the equilibrium state, since the second law only dictates the behavior of the total entropy of the universe. So how do we get around this problem? We'd much rather apply thermodynamic equations to the system alone, rather than having to understand the thermodynamic behavior of the system and its surroundings. The solution is to define a fundamental equation designed for the conditions at hand.

The second law defines equilibrium by the change in entropy of the entire universe. But, as I just mentioned, that's an incredible pain when we're only really interested in what's happening in our test tube. What we can do is to define a new state function that will allow us to apply the second law by looking only at the changes occurring in our system.

So, how do we do that? Well, we need to stop having to worry about whole universe, for starters. And the way we do that is to make the following assumption; at some point, away from my system, I can draw a boundary that makes the system, plus it's surroundings inside, completely isolated. That's a pretty good approximation, as long as I'm far enough away.

Take the test tube in the heat bath, right up close to the test tube, we may feel heat radiating from the test tube to the heat bath, or vice versa. But, if I go far enough away within the liquid heat bath, I'm no longer interacting with the test tube or the surroundings of the test tube. Given that I draw my box far enough away, then I can make it an isolated boundary, meaning that anything that goes on inside the box cannot possibly change the entropy outside.

So, as you can see here, inside this box we have the system of interest, in this case, the test tube, plus it's immediate surroundings, in this case, the heat bath. The second law defines all possible processes by the fact that if a process is possible, then the change in the entropy of the universe must be greater than or equal to zero. But now, I can write the change in entropy of the universe, as the change in entropy of the system, plus the change in entropy of its surroundings. And, since du equals TdS minus PdV, for a constant temperature, constant pressure process, we can rewrite the change in entropy as the change in internal energy divided by temperature, plus the change in volume times the pressure, and divide it by the temperature. So, we're left with this equation you see here.

By the way, this equation is called the fundamental equation for the entropy. And it's one we've seen before in our lecture on the second law. But now comes the crucial trick. Since we're working inside an isolated boundary, we know that both the internal energy, as well as the volume of the system plus its surroundings, must not change. This means that dU of the system, must equal negative dU of the surroundings for any process, and the same for the changes in volume.

Using these relationships, I can substitute into the equation from the second law. So, dS of the universe now can be written as dS of the system, minus dU of the system, divided by temperature, minus dV of the system, times pressure, divided by temperature. And according to the second law, all of this must be greater than or equal to zero for any possible process.

If I multiply through by negative T, then I finally arrive at the following equation: minus T times dS of the system, plus dU of the system, plus PdV of the system, must always be less than or equal to zero for a possible process. And this is exactly the kind of equation I needed, since it only has variables that depend on the system, as opposed to the surroundings, or worse, the whole universe. And also, it has variables that I can control experimentally.

As I discussed earlier, the Gibbs free energy G is defined as the internal energy plus pressure times volume minus temperature times entropy. With that definition, this equation becomes simply that dG for constant temperature and pressure, must be less than or equal to zero for all possible processes. This is because when you take the derivative of G to get dG at constant temperature and pressure, you wind up with exactly the expression we derived from our substitutions for entropy above. And this equation is, for practical calculations, perhaps the most important one in all of thermodynamics for predicting the behavior of materials.

In words, the second law demands that the Gibbs free energy must be lowered by the change happening to a system in order for that change to happen spontaneously at constant temperature and pressure. It also dictates that the Gibbs free energy at constant temperature and pressure is a minimum at equilibrium. But what is the Gibbs free energy, really?

Basically, if you look at the definition for it, what we've done is taken the total internal energy and subtracted off the thermal energy and the compressive energy, so, what's left? The answer is that what remains is the energy arising from other kinds of work. So far we've mostly focused on two sources of internal energy, compressive work from an applied pressure and thermal energy due to heat transfer. However, as I've mentioned there are many, many different types of work, such as the work of polarization, or chemical work, or magnetic work, and so on. In many practical problems in science and engineering, it is these other kinds of work in which we are most interested. So the Gibbs free energy is literally the amount of energy that is free to do these other types of work. That's why it's called a free energy.

Okay, hang on a minute. Are we all feeling our oneness with this stuff? Let's do a quick recap to be sure we've got this. Point number one: Thermodynamics deals with energy-like properties, such as the internal energy (U), which are closely connected to the first and second laws. Point number two: These two laws can be combined to give an equation that can then be used to show the conditions under which U is an equilibrium criterion. Point number three: the constraints that are required to make it a convenient criterion are not ones that are typically encountered in real laboratory conditions. For example, we rarely have an isolated system, and it's hard to measure directly the internal energy of a system. So, point number 4: By creating a new variable, the Gibbs free energy, and examining its differential form, and still using the combined statement of the first and second laws, we showed that G is a useful equilibrium criterion for systems in typical laboratory conditions, like constant temperature and pressure.

Now, before I move on to the last part of this lecture, I want to introduce another thermodynamic variable. I know, I know, it seems like I'm always introducing variables. I just can't help myself. But first, it's good to know about these different important variables because they're really important, and second, as we've discussed, in thermodynamics, we often rewrite one variable in terms of other ones in order to more easily solve the problem at hand.

The new variable I want to describe is called the enthalpy, and it's equal to the internal energy U, plus the pressure P, times volume V. This is a new energy state variable, since it only depends on other state variables. And we like to give it the letter H. As with the free energy, Enthalpy is not a new independent variable, but rather, it is a function of other state variables we have already discussed, in this case, the internal energy, pressure, and volume. The enthalpy is useful when discussing constant pressure processes, since in that case, the change in enthalpy, or dH, is related to the heat of the system if only mechanical work is involved.

Let's take a look at why that is the case. If I write down that equation from the first law that we know and love, then I get the fact that any change in internal energy of a system is equal to the heat added to the system plus the work done on the system, or dU equals δQ plus δW. Now, remember that U being a state variable means that its value for a system under given conditions cannot depend on the path taken to get to those conditions, only on the state of the system at that point. Also remember that heat and work are both not state variables, although, as we learned in Lecture 6, the addition of the two is a state function.

But what if I wanted the heat content for a process to be a state variable? We know that Q cannot be a state variable. However, suppose that the only work that could be done in a process were the work of expansion, or pressure times a change in volume? If that's the case, then the first law expression becomes dU equals δQ minus P times dV. Remember, our sign convention is that work done on the system is positive, and so we have the negative sign in front of the PdV here. Now, let's bring the PdV over to the other side, so we have dU plus PdV equals δQ, for a process where the only kind of work is that of expansion, or compression, for that matter, just depending on which way the volume change goes.

And this is where enthalpy enters in. Since H equals U plus PV, that means that dH equals dU plus *d of PV*. And when we carry the differential through, we have dH equals dU plus dP times V plus dV times P. And now we can see that if we go back to our first law expression where the only type of work is that of expansion or compression, then we have dU plus PdV equals δQ, but I can substitute dH in there, and I get dH minus dP times V equals sqiggly-*d Q*.

Okay, and here's the punch line. If the process is a constant pressure process, then that dP term is zero, and dH simply equals sqiggly-d Q. So, for a constant pressure process, the change in enthalpy is exactly equal to the heat transfer involved in the process. And, it's a state variable. That means that, unlike the general variable Q, which is path dependent, the enthalpy can mean the same thing but is path Independent, as long as pressure is held fixed.

Unlike the internal energy, it's hard to give a simple, intuitive physical description of enthalpy. As I just described, a change in enthalpy is the heat gain or loss for a constant pressure process. And it turns the variable for heat energy Q into a state variable, since enthalpy is a state function. This is why enthalpy is also referred to as the "heat content."

And how common is it for pressure to be held fixed? Well, I certainly hope by now, from this lecture, that you're feeling like it's a pretty common thing, since that was also an important part of our discussion of the Gibbs free energy. As with the Gibbs free energy, the enthalpy is useful because of the fact that volume is not the most convenient variable to hold constant in an experiment; it's much easier to control the pressure, especially if the system is a solid or a liquid.

That's of course not true all the time, just think of combustion engines where the pressure is changing, but let's take the case of a beaker filled with some chemical, open at the top. If I add another chemical to this beaker to initiate a reaction, then as long as the top of the beaker remains open, the whole reaction is going to take place at constant pressure—one atmosphere of pressure if we're at sea level and the top is open to the air. That's a pretty common situation both in laboratory experiments, as well as in nature in general. So indeed, the thermodynamics of constant pressure environments is extremely important, and enthalpy is a very useful variable. You can think of it as being a thermodynamic variable that contains all the information that is contained in the internal energy but can be controlled by controlling the pressure.

And notice the connection to the Gibbs free energy. If we look at the expression for G, we see that we can rewrite it as H minus TS. The Gibbs free energy is a state function useful for determining equilibrium in the most common type of experimental system, one where pressure and temperature are maintained at a constant value. And as we just learned, a system always tries to make its G as low as possible. You can see from this equation that what this means is that it will always be a compromise between trying to have a low enthalpy and a high entropy, with temperature playing an important role in determining which of these is more dominant.

So, to wrap up this lecture, I'd like to return to the point of what is free in the free energy in the first place. Let's take as an analogy a ball rolling around on a track. If the track has a part up high that is curved a little so the ball can get stuck up there, then we'd call that a metastable equilibrium for the ball. But it means that the ball is capable of doing work, as it rolls down to the lower point on the track, well, at least once it's pushed over the barrier to get it unstuck from that point higher up. The maximum work that the ball can do is exactly equal to the minimum amount of work it would take to push the ball back up to its higher, metastable point.

The Gibbs free energy is similar conceptually; it's equal to the maximum amount of useful work that a system can do as it changes from one state to another more energetically favorable state. Remember from our lecture on the second law that the missing ingredient to understanding why a given process goes in one way and not another is entropy. Entropy and the second law give us the arrow of time, since entropy always increases in spontaneous processes in isolated systems. But a parameter that's useful only in isolated systems is not of much practical use. So, today, we defined another state variable, the Gibbs free energy, that always decreases for a spontaneous process where the system is held at constant temperature and pressure. Mathematically, entropy and the Gibbs free energy are two sides of the same coin; one implies the other. But, for solving problems of practical interest, G is the parameter we have been looking for.

Mixing and Osmotic Pressure

Lecture 11

The last lecture introduced the Gibbs free energy, which allows us to determine equilibrium for systems under realistic experimental conditions. In this lecture, you will examine what happens when two different materials come together—and how the chemical gradient plays a fundamental role as a driving force to get to equilibrium. Along the way, a certain type of pressure known as osmotic pressure can build up if a barrier is present, so this lecture will also introduce you to the concept of osmosis.

An Open, Multicomponent System

- Let's examine a system composed of freshwater on one side and saltwater on the other, with a membrane in between these two types of water. If the membrane is impermeable, then no water from either side can flow back and forth. But what happens when the membrane is semipermeable, meaning that the membrane can let water pass but not salt? Can we predict the equilibrium state for this system?

- In that case, the simple form of the Gibbs free energy used in the previous lecture is not enough to determine equilibrium. But this simply introduces an additional degree of freedom that the system can use to further lower its free energy.

- When we remove the constraint of "no mass transport" across the boundaries, the system finds the additional degree of freedom of water transport and uses this freedom to lower its free energy. The modified equation for the change in free energy includes this additional degree of freedom—which is just another work term—and it is written as follows: dG equals VdP minus SdT (that's the same as before), plus the chemical potential mu (μ) for a given component in the system times the change in the number of moles of that component.

- The last term is a sum over all the different components in the system, and by including this term, we have now accounted for contributions to the free energy value by way of the transport of each component. From this equation, you can see that if the temperature and pressure are kept constant for a given process, then the free energy is governed solely by the last term.

- For an open, multicomponent system, in addition to the state variables such as pressure, temperature, and volume, each additional component adds a degree of freedom to the system. This degree of freedom is represented in terms of the number of moles of each component.

- So, in the case of this saltwater and freshwater system, dG simply equals the sum of the chemical potential of each component times its change in number of moles, with the sum carried out over all components in the system.

There is a great release of energy at the point at which a freshwater river meets the ocean.

- In this example, the same exact molecule, water, is on both sides of the membrane. However, on one side, the water has salt in it, and on the other side, it doesn't. The number of moles of water on each side can vary and, in fact, will vary until the chemical potentials of the waters are equal, at which point the system has attained chemical equilibrium.

- The difference in the chemical potential from the freshwater side of the membrane to the saltwater side is the thermodynamic force that drives the flow of water. This difference is called the chemical potential gradient, and it can be thought of as a general thermodynamic driving force for mass transport of a particular species. Water flows down this chemical gradient until the chemical potential on both sides are equal and this gradient no longer exists.

- How do we figure out these chemical potentials? Sometimes in thermodynamics we can simply look stuff like this up in tables. But when we have the exact same species—in this case, water—there are other ways to figure out the change in chemical potential of that species upon mixing it with another species.

The Entropy of Mixing

- The following model for the entropy of mixing is a model based on statistical mechanics, which is a set of tools for calculating macroscopic thermodynamic quantities from molecular models.

- The molar entropy of mixing is equal to $-R \times (X_A \log X_A + X_B \log X_B)$. This important result, where we have a logarithm dependence on the mole fraction, is quite accurate for a wide range of materials and systems.

- In terms of the flow of water across a semipermeable membrane between freshwater and saltwater, we are at the point where, based on the need for the system to lower its free energy and find equilibrium, it is going to try to make the chemical potentials on both sides equal. That is the chemical driving force for why water flows at all in such a setup.

- Returning to our free energy expression, we can write that the change in free energy upon mixing two components is equal to the change in enthalpy minus T times the change in entropy. Recall that enthalpy is a state variable defined as the internal energy plus pressure times volume. For a constant-pressure process, enthalpy is equal to the heat transfer.

- In this simple example, let's ignore the change in the enthalpy term; we'll make that zero for now. You can think of this as meaning that our two components mix randomly—that there are no specific interactions that favor the different components to be closer or farther apart.

- For the case of water, while this enthalpy-of-mixing term is not zero, it is smaller than the entropy-of mixing-term, which especially at higher temperatures will completely dominate the behavior. So, we're left with the change in free energy upon mixing being equal to $-T$ times change in entropy upon mixing.

- How do we get to the chemical potential from here? In the expression for a system held at constant temperature and pressure, the change in free energy (dG) is equal to the sum over all the components in the system of the chemical potential of each component times any change in the number of moles of that component.

- From this equation, we can see in fact that the derivative of the free energy with respect to a single component, while holding all the others fixed (that makes it a partial derivative), is exactly equal to the chemical potential of that component. In other words, the chemical potential of a given component in a system is equal to the partial molar free energy for that component.

- The partial molar free energy is the change in total free energy that occurs per mole of component i added, with all other components constant (at constant T and P). That means that the chemical potential of i measures the change in free energy caused by adding more of species i to the system.

- So, the chemical potential for a given component of the freshwater-saltwater solution will be equal to the partial molar free energy of solution for that component, holding temperature, pressure, and number of moles of all other components fixed.

- We now have a way to predict, starting from values of the chemical potential of a pure component that can be looked up, how much it changes when you mix it with something else. That change will tend to decrease the chemical potential of a component when it mixes with another one, although note that the magnitude of this effect is highly dependent on temperature.

- The logarithm term can be thought of as the excess free energy of that component in the solution. Keep in mind, thought, that this is only the excess free energy upon mixing due to entropy because we have ignored enthalpy, at least for now.

Osmosis and Reverse Osmosis

- When considering how the chemical potential gradient looks when we put freshwater on one side of a membrane and saltwater on another, remember that the membrane only allows water to pass across, not salt.

- What we've set up here is the process of osmosis, which by definition is the net movement of solvent molecules through a partially permeable membrane into a region of higher solute concentration, in order to equalize the chemical potentials on the two sides.

- Osmotic pressure is the pressure required to prevent this passive water passage—the pressure that builds up when we have two materials with different chemical potential on either side of a membrane. How big are these pressures?

- A technology that some researchers are working on uses osmotic pressure as an energy resource. Extracting clean, fresh water from salty water requires energy. The reverse process (called reverse osmosis)—mixing freshwater and salty water—releases energy. Instead of desalination, we call this simply salination, and there's quite a bit of energy involved.

- At every point in the world where a freshwater river meets the ocean—and that's a whole lot of places—there is a great release of energy, not from the flow of the water but simply by the mere fact that freshwater is mixing with saltwater.

- To give you a sense of just how much, by some estimates, the energy released by the world's freshwater rivers as they flow into salty oceans is comparable to each river in the world ending at its mouth in a waterfall 225 meters (or 739 feet) high.

Suggested Reading

DeHoff, *Thermodynamics in Materials Science*, Chaps. 5 and 7.1.

Porter and Easterling, *Phase Transformations in Metals and Alloys*, Chaps. 1.1–1.2.

Questions to Consider

1. Why do we need another thermodynamic potential, in addition to the internal energy?

2. What governs equilibrium when a liquid such as water is mixed with a solute such as salt?

3. Using our equation for the chemical potential of mixing, can you derive the minimum thermodynamic work (energy) required to desalinate seawater?

4. How does this energy (for a process called reverse osmosis) compare to that of distillation (boiling a given volume of water)?

Mixing and Osmotic Pressure
Lecture 11—Transcript

In the last lecture, we got deep about the notion of equilibrium by introducing the Gibbs free energy, which allows us to determine equilibrium for systems under realistic experimental conditions. In this lecture we'll look in greater detail at just what happens when two different materials come together and how the chemical gradient plays a fundamental role as a driving force to get to equilibrium.

Along the way, a certain type of pressure known as osmotic pressure can build up if a barrier is present; this pressure is extremely relevant in the life sciences, since biological systems are regulated at the cellular level by osmotic pressure levels. It's the reason why not drinking enough water and drinking too much water can be bad for you. Osmosis is also a great concept to use for getting up nice and close to the concept of the chemical potential, which we learned about a couple of lectures ago.

Remember from the last lecture that the Gibbs free energy is a state function, useful for determining equilibrium in the most common type of experimental system, one where the pressure and temperature are maintained at a constant value. But, before we go into the concepts, let's start by taking a look at osmosis in action. Now, I had wanted to do this as a live demo, but unfortunately, the process of osmosis is based on diffusion, which is kind of slow. And since I didn't want to make you wait the whole lecture to see it in action, we took some video and sped it up.

So here I have an egg that I've soaked in vinegar overnight. What that does is it eats away at just the outer shell. To be more chemically specific, the acetic acid of the vinegar dissolves away the calcium carbonate of the shell. What's left is still the egg, but now it has only a very thin membrane holding it together. And the point of doing this is that the outer membrane of the egg that is left is now permeable to water, which means that water can pass through it.

So in the next step, I place the eggs in corn syrup and immerse them fully. I could just as easily have used highly concentrated salt water, but corn syrup works just as well, and in fact, a little bit faster. I wait. Again, osmosis is kind of slow. But after soaking overnight, check out these eggs. They look all shriveled up and deflated. The reason is that the fresh water that was in the egg flowed across the outer membrane into the salt water.

And finally, we can re-inflate the egg if we want. This time, I put the eggs in a bowl with fresh water, and just for fun, I'll put some green dye in the water. Again, it takes some time, but after a while, the egg becomes completely filled with water. Actually, much more so than before. The water was driven both out of and into the egg, because of a difference in chemical potential, as we'll learn in this lecture.

So, to conclude, the question might be, what do I do with a swollen green egg? Well, I'd say one of two things. You could pop it, like this, and a make a big mess. Or, if you want, you could fry it up, and together with some ham, serve up some truly green eggs and ham, thanks to your knowledge of thermodynamics!

So, that was pretty cool, but how did it work? We'll answer this question today by building up the concept of chemical equilibrium and adding some new concepts along the way. So let's get going so we can put some rigorous thermodynamics onto our inflatable eggs. In order to dig in, let's examine a system composed of fresh water on one side and salt water on the other. We're going to place a membrane in between these two types of water. If the membrane is impermeable, then of course, nothing happens, no water from either side can flow back and forth. But what happens when the membrane is semi-permeable? For example, as in the experiment I just showed you, what if the membrane can let water pass but not salt? Can I predict the equilibrium state for this system?

In that case, the simple form of the Gibbs free energy I defined in the last lecture is not enough to determine equilibrium, because, in the last lecture, I did not allow mass transport across the boundaries in the examples I used. But this is fine, and it simply introduces an additional degree of freedom that the system can use to further lower its free energy.

When we remove the constraint of no mass transport across the boundaries, the system finds the additional degree of freedom of water transport, and it uses this freedom to lower its free energy. The modified equation for the change in free energy includes this additional degree of freedom, which in the end is just another work term, and it's written as follows: *dG* equals *V* times *dP* minus *S* times *dT*—that's the same as before—and now plus the chemical potential μ for a given component in the system times the change in the number of moles of that component. The last term is a sum over all the different components in the system, and by including this term, we have now accounted for contributions to the free energy value by way of the transport of each component. From this equation you can see that if the temperature and pressure are kept constant for a given process, then the free energy is governed solely by the last term. The last term was, again, chemical potential μ, for a given component, times the change in number of moles of that component.

Remember from Lecture 9 for an open multicomponent system, in addition to the state variables, like pressure, temperature, and volume, each additional component adds a degree of freedom to the system. This degree of freedom is represented in terms of the number of moles of each component. So in the case of this salt water and fresh water, *dG* simply equals the sum of the chemical potential of each component times its change in number of moles, with the sum carried out over all components in the system.

In the example here, I mentioned that I'm using a membrane that allows the transport of water across, but not salt. This means that, effectively, the only mobile species is water. Let's refer to the side with pure water as side 1 and the side with saltwater as side 2. I obtain an expression for the change in free energy, which is equal to the chemical potential of the water on side 1, that's the fresh water, times any change in its number of moles. We'll call that μ1 times *dn*1, plus the chemical potential of the water on side 2, that's the salt water, times any change in its number of moles, call that μ2 times *dn*2.

But, as long as there's no leaking, then the change in number of water molecules on one side must equal minus the change on the other side, so we can write that dn1 equals negative *dn*2, and if we then make a substitution we obtain that *dG* equals dn1 times, in parentheses (μ1 minus μ2). And remember from last lecture that one of the single most important aspects of determining the free energy change for a given process is that I can use this to find equilibrium for that process. When the change in free energy is zero, then the system is in equilibrium.

This equation shows, therefore, that at constant temperature and pressure, the equilibrium condition is dictated by the chemical potential of the mobile species, here, water. In this case, equilibrium is attained when the chemical potential of water on both the sides of this membrane is equal. This is called chemical equilibrium.

Let's summarize for a moment here; we have the same, exact molecule, water, on both sides of this membrane. But, on one side, that water has salt in it, and on the other side, it doesn't. What we just learned is that the number of moles of water on each side can vary, and in fact, will vary, until the chemical potentials of the waters are equal.

But what is the thermodynamic force that drives the water to flow, in this case, it will flow from side 1 (fresh water) to side 2 (salty water)? The difference in the chemical potential ($\mu 1 - \mu 2$) is what drives the water to flow. This difference is called the chemical potential gradient, and it can be thought of as a general thermodynamic driving force for mass transport of a particular species.

Water flows down this chemical gradient until the chemical potential on both sides are equal and this gradient no longer exists. You can think of it as analogous to a ball rolling down a hill, but in this case, instead of the force of gravity acting on mass to cause a change in height, we have the force of chemistry acting on a species to cause a change in its number of moles. The chemical potential difference tells us how high the hill is.

And now, we come to a very important question. How do we figure out these chemical potentials? I've already mentioned in previous lectures that sometimes in thermodynamics we can simply look stuff up. But when we have the exact same species, in this case, water, there are other ways to figure out the change in chemical potential of that species upon mixing it with another species. And, to begin with, we'll start with good old entropy. Actually, you may remember that when I first talked about entropy, I gave salt dissolving in water as an example of a process that is dominated by the entropy change. That's because what happens is that when the water is surrounding salt ions—sodium and chlorine atoms if it's just regular table salt—then those salt ions have many more ways in which they can arrange themselves compared to when they were locked up in a crystal of the salt.

Since entropy is a measure of the number of degrees of freedom, or more formally, the number of different possible microstates for a given macrostate, you can see that the dissolved salt ions have a much higher entropy, so much higher, that it dominates over other effects, like the fact that the interaction energy between salt and water is different than between salt and itself. In fact, the interactions are weaker when the salt is dissolved, which without entropic effects, would have us conclude that it wouldn't dissolve, but entropy takes over and wins the day.

So again, you may remember this salt dissolution example from before, but what I didn't do before is quantify the changes. We're now in a position to do so, by introducing our first model for the entropy of mixing. This is going to be a model based on statistical mechanics, which I introduced way back in the beginning lectures as a set of tools for calculating macroscopic thermodynamic quantities from molecular models. And in the one I'm about to present, we can use simple, so-called coarse-grained lattice models to capture the important aspects of the material's behavior.

So, what is a lattice model? Well, we start with, you guessed it, a lattice. This is nothing more than a separation of space into different compartments. For this simple derivation, I'll use a two-dimensional lattice, where each space is a square. We could come up with any shape or size lattice we want to, but, for this example, let's use a 6 by 3 set of squares. Now, consider that we have two different components, call them component A and component B. And we're going to fill the lattice squares in with one or another component, like you see here. Now, remember, our goal is to use this simple picture to come up with a model that can predict the change in entropy of the system upon mixing. So, what is the entropy?

Well, remember from our lecture on entropy, that it can actually be quantified, and that it's equal to Boltzmann's constant (or k_B) times the natural log of the number of available microstates. S equals k_B times log of W, where W is the variable we use for the number of states. So let's look back at our lattice. Suppose all of the squares were filled with only one component. How many different ways can this be arranged? Well, you can see that this is, in fact, the only one. I can only have one way for this to occur, namely, all molecules of that component occupy a square, and any changes do not change the nature of this system. So for this macrostate, there exists only one possible microstate.

But now, suppose that I have N_A A molecules and N_B B molecules. The total number of molecules is $N = N_A + N_B$, and it will be equal to 18 for this example, since as I already mentioned, I have a 6 by 3 grid that will be fully filled up. The number of states for this system is defined by the number of unique ways these components can be arranged on the lattice. It turns out that this is a fairly simple problem, since it just involves counting up possibilities. Suppose I have an empty grid and I want to know how many ways I can place an A-type component? Well, the answer is simply N, since there are N spaces available. And if I then place that one down and ask how many ways can I place the next one, the answer is $N - 1$, since that's how many free spaces are left. And so on.

And if you have taken some pre-calculus, then you may be seeing the formation of a factorial here, which is exactly where we come to. Skipping over a little bit of math, the total number of ways in which I can arrange N_A and N_B components on the N lattice sites is exactly equal to N factorial divided by N_A factorial times N_B factorial. I won't go through the detailed math on how we arrived at this relationship, but you can try out a few possibilities on your own and see that it works.

Now, two quick points. First, remember that a factorial just means that we multiply that number times one smaller times one smaller, and so on, all the way down to the number 1. Second, this expression assumes that the components of a given type are indistinguishable from one another. So, if I swap two A-type molecules, we cannot tell the difference in the microstate.

Okay, so, now that we have an expression for the number of possible states W, we can determine the entropy of mixing. Remember, the entropy is equal to Boltmann's constant k_B times the natural log of the number of microstates.

So if the two components are mixed on a lattice, as I just described, this expression becomes *k-B* times the natural log of N_A factorial times N_B factorial divided by *N* factorial.

By the way, just as an aside for those of you who may not have taken a logarithm in a while, the logarithm of a number is the exponent to which another number, called the base, must be raised to produce that number. For example, the logarithm in base 10 of 1000 is 3, because 1000 is 10 to the power 3. The natural logarithm is a logarithm taken in a different base, that is, the base of the number *e*. And in case you haven't seen that before, the number *e* is a constant that is quite useful in mathematics; it's roughly about 2.7.

Okay, so, what we do next is a little bit more math on this logarithm expression, which I won't go through here, but what we get to is the following expression for the entropy of mixing. It's that the molar entropy of mixing is equal to minus *R* times the sum X_A log X_A plus X_B log X_B. Notice that I've converted those *N*'s into *X*'s, which means I've gone from total number of particles into mole fractions. As I mentioned in our lecture on molar quantities, it's easier to work with mole fractions. I know, this seems like it was derived from a pretty simple model. So how could it be very accurate? Well, trust me; this is an extremely important result that is quite accurate for a wide range of materials and systems.

As a quick side point, how many possible states do you think there are for, say, a mole of water around its freezing point? It's actually astounding. There are, give or take a few, roughly 10 to the power 2 followed by 24 0's microstates for a mole of water at that temperature. And when the water is heated by just a single degree? Well that number gets increased by 10 to the 22 times. For comparison, the number of atoms in the whole universe is only about 10 to the 70th, and a lowly old googol is just 10 to the 100th power. So when we talk about the number of microstates accessible to a system in thermodynamics, we are dealing with very large numbers.

Okay, let's take a look at this function; here's a plot of the molar entropy of mixing. By the way, don't let the word molar throw you off, it's just the entropy of mixing per mole. In this plot, we vary the mole fraction of component B along the x-axis and plot the molar entropy of mixing on the y-axis. Also, keep in mind that X_A equals 1 minus X_B, and vice versa. So we only need to examine this as a function of one or the other components. But you can see from the plot that the entropy of mixing increases as two components are mixed, and it reaches a maximum for a 50/50 mixture. Now, I know that was a bit of math, but it's a nice example of how we build simple models for our thermodynamic variables. And, the result we came up with, where we have a logarithm dependence on the mole fraction, is something we use a lot in thermodynamics.

But now, let's bring this back to the topic that I started with, namely, the flow of water across a semipermeable membrane between fresh water and salty water. We were at the point where, based on the need for the system to lower its free energy and find equilibrium, that meant that it was going to try to make the chemical potentials on both sides equal. That is the chemical driving force for why water flows at all in such a setup.

Returning to our wonderful free energy expression, we can write that the change in free energy upon mixing two components is equal to the change in enthalpy minus T times the change in entropy. And remember that enthalpy is a state variable defined as the internal energy plus pressure times volume. For a constant pressure process, enthalpy is equal to the heat transfer. In this simple example, let's ignore the change in enthalpy term; we'll call that zero for now. You can think of this as meaning that our two components mix randomly, that there are no specific interactions that favor the different components to be closer or farther apart. As I mentioned for the case of water, while this enthalpy-of-mixing term is not zero, it's smaller than the entropy of mixing term, which especially at higher temperatures will completely dominate the behavior. So, we're left with the change in free energy upon mixing being equal to minus T times the change in entropy upon mixing.

Okay, now, how do we get to the chemical potential from here? Well, in the expression I derived earlier in this lecture for a system held at constant temperature and pressure, we had that the change in free energy, dG, is equal to the sum over all the components in the system of the chemical potential of each component times any change in the number of moles of that component. And from this equation, we can see, in fact, that the derivative of the free energy with respect to a single component, while holding all the others fixed (that makes it a partial derivative), is exactly equal to the chemical potential of that component.

That's a really big deal, so I'd like to make it my repeat statement for the day. The chemical potential of a given component in a system is equal to the partial molar free energy for that component. That's pretty awesome, and it's very powerful. In our lecture on partial molar quantities, we talked a lot about the chemical potential, and we related it to the partial molar internal energy. But that was for a situation in which the entropy and volume are held fixed. And as we discussed in our last lecture, this is very hard to do in real systems. On the other hand, keeping temperature and pressure fixed is much easier to do, and in that case, we now know that the chemical potential is equal to the partial molar free energy.

In words, the partial molar free energy is the change in total free energy that occurs per mole of component added, with all other components constant at constant temperature and pressure. That means that the chemical potential of say, component I, measures the change in free energy caused by adding more of species I to the system. So the chemical potential for a given component of the mixed solution we've been talking about, will be equal to the partial molar free energy of solution for that component, holding temperature, pressure, and number of moles of all other components fixed.

Okay, so for component A, that will be equal to the chemical potential of the unmixed species of A, which we write as μ naught A, plus RT log X_A. Remember when we talked about chemical potentials how I said you could just look them up? Well that's exactly how we found values for μ naught. We did it before in an example where we wanted to know if a given reaction was going to happen or not. But in this case, we now have a way to predict, starting from those looked up values of the chemical potential of a pure component, how much it changes when you mix it with something else.

And as you can see from this simple plot, that change will tend to decrease the chemical potential of a component when it mixes with another one, although, note that the magnitude of this effect is highly dependent on the temperature, since you can see it's a linear relationship with that T out in front.

This logarithm term, the one we've spent some time here to derive, can be thought of as the excess free energy of that component in the solution. And I'll remind you again, just for good measure, that this is only the excess free energy upon mixing due to entropy, since I've ignored enthalpy, at least for now.

Okay, and finally, after all of that, we are now in a position to go back to our initial question. That was, how does the chemical potential gradient look when I put fresh water on one side of a membrane and salt water on another? Remember, the membrane only allows water to pass across, not salt. The difference between the chemical potential of pure water minus that of salt water is simply equal to μ of pure water minus μ of salt water. In the pure-water case, μ equals μ naught, since there's no additional mixing term, whereas, in the saltwater case, we've mixed in another component, namely salt, so μ equals μ naught plus *RT* times the natural log of the mole fraction of salt water.

Let's suppose here that we're dealing here with seawater, for which the salinity varies quite a bit, but on average it's around 3.5%. Or in other words, there are about 35 grams of salt for every kilogram of seawater. Given that there are around 58 grams of salt per mole of salt, and 18 grams of water per mole of water, this gives us a mole fraction of around 0.01 for the salt in seawater, and 0.99 for the water in seawater. If we do the math, we see that the chemical potential difference between the two at room temperature is equal to −8.314 times 300 times the natural log of 0.99, and the units here are joules per mole. Multiplied out, we arrive at a chemical potential difference of around 25 joules per mole, or in this case, around 1400 Joules per liter of water.

Let's think about that for a moment. How much of a driving force is that? What I've set up in this problem is, in fact, the process of osmosis, which is, by definition, the net movement of solvent molecules through a partially permeable membrane into a region of higher solute concentration, in order to equalize the chemical potentials on the two sides. Osmotic pressure is the pressure required to prevent this passive water passage.

So water flows from fresh to salty water until the two sides have the same chemical potential. But I still haven't gotten a feel for how much energy is at play here. Well, one way to look at this is to suppose I want to reverse this process. That, in fact, is the leading technology today for creating fresh water from salt water, and it's known as reverse osmosis. And now we know how much energy that requires. At a minimum, we need to fight against this chemical potential gradient, so we'll need to put in 1400 Joules per liter. That means, if you run at 1.4 kilowatts of power—that's enough power for 23 60-watt light bulbs, or else a single hairdryer—then in one hour you could desalinate around 3600 liters of water. That's actually quite a bit when you think about it. Now, unfortunately today's best RO desalination plants do not operate near this theoretical limit, at least not yet. But power consumption has been dramatically improved over the past 40 years, and today, we're only about a factor of 3 to 5 away from this lowest possible value.

So, I mentioned that osmotic pressure is the pressure that builds up when we have two materials with different chemical potential on either side of a membrane. How big are these pressures? I'll answer that by finishing this lecture with a technology some researchers are working on that uses osmotic pressure as an energy resource. What I just described is the fact that extracting clean, fresh water from salty water requires energy. But the reverse process—mixing fresh water and salty water—releases energy.

Instead of desalination, we call this simply salination, and there's quite a bit of energy involved. You can just imagine, at every point in the world where a fresh-water river meets the ocean—and that's a whole lot of places—there is a great release of energy, not from the flow of the water, but simply by the mere fact that fresh water is mixing with saltwater. To give you a sense of just how much, by some estimates, the energy released by the world's freshwater rivers as they flow into salty oceans is comparable to each river in the world ending at its mouth in a waterfall 225 meters high!

I really like that image as a way to end today's lecture. It illustrates the power of osmotic pressure, and it gives a nice, visual way to think of all of that energy that comes from the thermodynamic concept we learned here today, namely, the entropy of mixing.

How Materials Hold Heat

Lecture 12

You have already learned all about heat and how it is a fundamental part of the first law—how it can be made into the state variable entropy by dividing by temperature or the state variable enthalpy by holding pressure constant. But in this lecture, you are going to learn the nature of how materials keep thermal energy inside of them. This is called heat capacity, because it is the capacity of a material to absorb heat. Heat capacity is an extremely important property, both because it tells us how a material holds onto thermal energy, and because it connects us, through very simply laboratory measurements, with other thermodynamic variables.

Heat Capacity: Monatomic Ideal Gas

- The ratio of the heat transfer divided by temperature change is the quantity that measures the ability of a material to store energy transferred as heat. In general, the heat capacity of the material is equal to $\delta Q/dT$, or the ratio of the differential heat to differential temperature.

- Physically, this means that a higher heat capacity material can absorb a higher quantity of heat without changing temperature very much, whereas a low heat capacity material will take less heat absorbed before changing the same amount of temperature. Because Q has units of joules and T has units of kelvin, the heat capacity units are joules per kelvin.

- Different materials all have different capacities for holding thermal energy. What is heat capacity, more fundamentally, and what is it about a material that gives it a high or low heat capacity? To answer this question, let's start by traveling down to the scale of the atom, where everything is in motion. At room temperature, atoms and molecules fly around at speeds nearing a kilometer per second.

- This motion is the kinetic energy that gives rise to the temperature of a material. And when there is more than one atom, such as in a molecule, then the kinds of motion can be more than just translational. For example, the molecule could rotate, and different atoms within it could also vibrate. All of these different types of movements taken together give rise to the temperature of the material.

- When a material absorbs heat energy, that energy goes right into these atomic-scale motions. But which of these different kinds of motion in particular does the heat energy go into? The more complicated and the greater the number of degrees of freedom, the more complicated the question of where heat flows becomes. In addition, there is a big difference between the different phases of matter: gases, liquids, and solids.

- In terms of talking about heat and temperature, the very simplest material is a monatomic ideal gas. The assumption with an ideal gas is that the atoms are noninteracting—so, they can bump into one another and collide, but they don't have any inherent attraction or repulsion between them.

- This is a pretty good approximation in many cases, but the more noninteracting the atoms are, the more accurate the ideal gas model will be. For noble gases like helium or neon, this is an excellent model.

- The first law tells us that a change in the internal energy is equal to the change in energy from heat plus change in energy due to work. This allows us to derive two versions of the heat capacity: one that holds for a constant-volume process and another that holds for a constant-pressure process.

- The constant-volume heat capacity is defined in terms of internal energy: It's the partial derivative with respect to temperature. And the constant-pressure heat capacity is equal to the partial derivative of the enthalpy with respect to temperature. From the first law, we obtain two different definitions for heat capacities in terms of thermodynamic state functions, which are really handy because they are path independent.

- For the case of the monoatomic ideal gas, we know that the internal energy is U equals 3/2 times the number of moles of gas n times the ideal gas constant R times the temperature of the gas. For constant volume, the heat capacity is the partial derivative of U with respect to temperature, which gives 3/2 times number of moles of gas n times the gas constant R.

- Heat capacity is an extensive quantity. If the system size increases, so does the heat capacity. As is often the case, it's useful to use the intensive version of an extensive variable. In this case, it could either be the molar heat capacity, where we divide by the number of moles, or what is known as the specific heat capacity, where we divide by the mass of the system.

- The constant-pressure heat capacity for an ideal gas can be obtained in a similar manner. The heat capacity at constant pressure equals heat capacity at constant volume plus n times R. Written as molar heat capacities, the constant-pressure molar heat capacity equals constant-volume molar heat capacity plus the ideal gas constant R. Note that the heat capacity at constant pressure is always greater than the heat capacity at constant volume.

- So, the molar heat capacity for an ideal gas at either constant volume or constant pressure is simply a constant. Therefore, all ideal gases have the same exact heat capacity—at constant volume, anyway—of 12.5 joules per mole kelvin, regardless of the gas and regardless of the temperature.

Heat Capacity: Diatomic Ideal Gas

- Let's move up one degree of difficulty in terms of the system to a diatomic ideal gas. These are pairs of atoms, sometimes also called dimers. Because it's still an ideal gas, they do not interact with one another, but they have a very important difference from the single-atom, monatomic case. As they are translating around at those tremendous speeds, the molecules can now also rotate as well as vibrate.

- These rotational and vibrational motions of the molecules can be activated by collisions, so they are strongly tied in with the translational degrees of freedom of the molecules. All of this energy that was getting bounced around in the form of translation motion along different directions is now—on average—shared equally by each independent degree of freedom. This is known as the equipartition theorem, which states that each degree of freedom contributes the Boltzmann constant k_B times the temperature T of energy per molecule.

- Not counting vibration for now, for diatomic molecules, there are five degrees of freedom from translation and rotation: three associated with the translational motion and two associated with the rotational motion.

Physicist Ludwig Boltzmann (1844–1906), pictured at center, used his knowledge of thermodynamics to develop the branch of physics known as statistical mechanics.

- From the equipartition theorem, each degree of freedom contributes, on the average, $k_B T$ of energy per molecule, which means that the internal energy for this case is equal to (5/2)nRT. This gives us a value for the constant-volume molar heat capacity of (5/2)R, or 20.8 joules per mole kelvin.

- However, ignoring vibrational energy is not correct. And vibrational motion adds two more degrees of freedom: one coming from the kinetic energy and the other coming from the potential energy associated with the vibrations.

- The equipartition theorem should really include all three types of motion, and if we add this vibrational part, we get an extra two (1/2)R contributions, giving a constant-volume molar heat capacity of (7/2)R.

- The model we derived based on the ideal gas law and the equipartition theorem is purely classical in nature. It predicts a value of the molar heat capacity for a diatomic gas that only agrees with experimental measurements made at very high temperatures.

Heat Capacity: Solids

- In a solid, all of the atoms are arranged on a regular lattice. In the simplest picture, you could imagine all of these atoms held together by a kind of spring that bonds them, just as in the case of a diatomic molecule. And as you're thinking about this lattice of atoms held together by springs, think about which of the three degrees of freedom solids can have: translational, rotational, or vibrational.

- Unlike atoms and molecules, solids can only have one single type of freedom, and that would be vibrations. The atoms in the solid don't translate, and they don't spin around because they're pretty much stuck in their crystalline position. So, when heat energy is transferred to a solid, it all goes into those vibrations of the atoms.

- In a solid, there are six degrees of freedom. Each one brings an amount of $(1/2)R$ to the heat capacity, so according to this analysis, the molar heat capacity will be $3R$ for a solid. And this is in fact exactly what happens for most solids, as long as the temperature is high enough. This result of $3R$ for the constant-volume molar heat capacity of solids is known as the Dulong–Petit law.

- However, just as in the case of the diatomic molecule, there is a very strong temperature dependence for solids. For some metals, the variation from $3R$ occurs below room temperature, but for many materials, it occurs at temperatures much higher, meaning that for quite a range of temperatures, the behavior is much more complicated.

Suggested Reading

Callister and Rethwisch, *Materials Science and Engineering*, Chap. 19.

DeHoff, *Thermodynamics in Materials Science*, Chaps. 2.2 and 4.

Serway and Jewett, *Physics for Scientists and Engineers*, Chap. 20.

Questions to Consider

1. Why do you burn your tongue on the cheese of a piece of hot pizza but not the crust?

2. Both heat capacity and entropy seem to have to do with the number of degrees of freedom in a material—so what's the difference between them?

3. Which of the following has more degrees of freedom: a monatomic gas, a diatomic gas, or a polyatomic gas?

How Materials Hold Heat

Lecture 12—Transcript

Hi, today we'll be talking more about heat. I mean, our subject is, after all, "the motion of heat," so it's a pretty important topic. Now, we've already learned all about heat and how it is a fundamental part of the first law, how it can be made into the state variable entropy by dividing by temperature, or the state variable enthalpy, by holding pressure constant. But today, we're going to discuss something else about heat, namely, the nature of how materials keep thermal energy inside of them. This is called heat capacity, since it is, literally, the capacity of a material to absorb heat.

Now, remember, we would never say store heat, not students of this course anyway. We can't say store, since heat is not just energy; it's energy transfer. But just how much energy can be transferred to or from a given material in the form of heat? Let's pop some balloons for our demo to help us see this heat capacity in action.

So, the following is a pretty simple demo, but, it's a great way to illustrate the concept of heat capacity. Here I have two different balloons. And for each one, I'm going to try to pop them with a flame. So, first let me make my flame, and I hope you notice the candle holder that I'm using; I'm very excited about that one. And then, I'm going to take, once that flame gets going, I'm going to take this balloon here; this balloon is filled with air, so, what do you think is going to happen when I put this over the candle flame? Well, let's take a look.

As you can see, it popped. And actually, the popping put out the candle. The reason it popped is because the fire damaged the polymer and the plastic sort of polymers that make up the balloon material. Okay? And heat from the fire went straight into damaging that material, and a hole was created, and the air burst out.

But, what if I take this second case? Here, I have a balloon that's filled with water, instead of air. Now, I'm going to hold this one over the same flame. Okay, we'll get it nice and close. By the way, did you know that a flame is around 1400 degrees Celsius? Pretty hot, even a small one like this. Now, you can see that smoke is coming, but, even after 5, 10 seconds, and even if kept holding it here, this balloon will not pop.

You can see that what happened was it got dark, it got charred. So it looks like maybe I am damaging the balloon, but in fact, that char, that will just wipe off. That's just from the smoke of the candle. So that same material that got damaged before didn't even get scratched this time. Why is that? Well, it has to do with heat capacity. The heat capacity of the water is so much higher than the heat capacity of air. That means that the water wants that thermal energy so badly, so badly, in fact, that it pulls it away from the candle and away from the liner of the balloon, the material of the balloon, instantly. And it grabs the heat, and it stores it inside of its internal degrees of freedom.

The heat capacity of the water is so high that I could have held this over the flame, literally, until the water started to boil, and no damage would have happened to the balloon. And in fact, I am so confident in the high-heat capacity of water, that I want to show you one more thing. It turns out that I happen to have the 50/50 mixture of ethanol and water from the last demo.

Now, I'm so confident in the laws of thermodynamics and the role of heat capacity that I'm going to take a $100 bill, and what I'm going to do is, I'm going to dip it into this 50/50 mixture of water and ethanol. Okay? I'm going to get it nice and soaking wet. And now, I'm going to take the bill, and I'm going to light it on fire. Let's see what happens.

You can see that you get a nice, blue flame, and you can see that it's on fire, but no damage was done. And the reason for that is that, again, the heat capacity of the water is so high, that it pulls all of the fire's energy away from the material that makes up the bill before it can cause any damage to that material. So, what is it that determines heat capacity? Well, let's get back to the lecture to find out. Let's dive into some thermo to understand this.

If we ask the question, what quantity would measure the ability of a material to store energy transferred as heat? Then we might think that it has to have something to do with heat transfer, or the variable we like to call δQ, as well as temperature, since a change in the temperature of the material could be related to its loss or gain of thermal energy. In fact, it is the ratio of the heat transfer divided by temperature change that is the quantity we seek. In general heat capacity of the material, is equal to δQ divided by dT, or, the ratio of the differential heat to differential temperature.

Now, physically this means a higher heat-capacity material can absorb a higher quantity of heat without changing temperature very much, whereas a low heat-capacity material will take less heat absorbed before changing the same amount of temperature. And as always, units are very important to know about. Here, you can see that since Q has units of joules and T has units of Kelvin, the heat capacity units will be joules per kelvin.

Now, we've all felt heat capacities in real life in many different ways. I know most of us don't go around trying to light water-filled balloons on fire, but beyond that, heat capacity is everywhere. If you have two different outdoor chairs, one made of wood and one made of metal, and you leave them out in the sun on a hot day, have you noticed that picking up the wood chair is comfortable, while picking up the metal one feels much hotter? Or how about how long it takes to heat a cast iron pan over the same flame that much more rapidly gets a copper pan hot? Or how much longer it takes to heat your soup in the microwave than say, a croissant? Or how much hotter sand on the beach gets out in the sun compared to say, a small pool of water. And last, my favorite, heat capacity is why we can burn ourselves on the cheese of a piece of pizza but not the crust. They both came out of the oven at the same temperature, but the cheese stored a whole lot more thermal energy than the crust.

The biggest and most important example of heat capacity is that of just plain, old water. It's got a very high heat capacity, and trust me, it's a really good thing that it does, because the oceans of our great planet are what regulate its temperature more than anything else. Without the high heat capacity of water, the oceans would not be able to serve as the ultimate moderator, preventing the planet's temperatures from going through wild oscillations that would threaten life as we know it. So we have all these different materials, all with different capacities for holding thermal energy. Let's dig a bit into our thermo to understand what's going on under the hood in these materials. What is heat capacity, more fundamentally, and what is it about a material that gives it a high or low heat capacity?

To answer this question, let's start by traveling all the way down to the scale of the atom. The world is pretty awesome down here, where we can see that everything is in motion. At room temperature, atoms and molecules are flying around at speeds nearing a kilometer per second, and this motion is the kinetic energy that gives rise to the temperature of a material. And we notice that when there is more than one atom, such as in a molecule, then the kinds of motion can be more than just translational. For example, the molecule could rotate, and different atoms within it could also vibrate. All of these different types of movements taken together give rise to the temperature of the material.

So, why am I all the way down at this scale, looking at how atoms and molecules move, twist and vibrate? Well, it's because I'm in search of an answer to the following question. When a material absorbs heat energy, where does that energy go? Well, from our discussions and definition of temperature, you probably have a pretty intuitive feeling already for this. It goes right into these atomic-scale motions. But, take a look at all of these different kinds of motion. Which ones, in particular does the heat energy go into? Does it just get spread evenly over all of them? Or, is there some preferential way in which energy populates atomic motion? And how, in the end, does that relate to the thermodynamic variables we know and love, like temperature, internal energy, or entropy?

You might be able to guess from looking at materials at this atomic scale that the more complicated and the greater the number of degrees of freedom, the more complicated the question of where heat flows becomes. You can also probably guess that there would be a big difference between the different phases of matter—gases, liquids, and solids. What we're going to do is start with the simplest case and then build up our understanding from there.

So, in terms of talking about heat and temperature, the very simplest material is a monatomic, ideal gas. Remember from our discussion of the ideal gas in Lecture 5, that the assumption here is that the atoms are non-interacting. So they can bump into one another and collide, but they don't have any inherent attraction or repulsion between them. This is a pretty good approximation in many cases, but the more non-interacting the atoms are, the more accurate the ideal gas model will be. As we discussed previously, for noble gases, like helium or neon, this is an excellent model. So let's start with some neon gas atoms.

Imagine that we have a container with atoms flying around inside with a given speed—that would be very fast—and bumping into the container walls, as well as one another. Think of an ideal gas as being like a whole lot of ping-pong balls crashing into one another at millions of miles per hour but never getting damaged or dented. Notice that in this simple material we have only one way in which the atoms can move, namely, by translating, since we just have single atoms. The atom doesn't rotate or vibrate around itself. So, that means that if I add thermal energy to this material, the only place where it can go, is into the translational motion of its atoms. Thermal energy in a monatomic ideal gas comprises only translational motions.

Now, in order to get from this picture to an expression for the heat capacity, we turn to the first law of thermodynamics. Remember, that the first law tells us that a change in the internal energy is equal to the change in energy from heat plus change in energy due to work. With just a little bit of math, this allows us to derive two versions of the heat capacity, one that holds for a constant-volume process, and one that holds for a constant-pressure process. The constant-volume heat capacity is defined in terms of internal energy; it's the partial derivative with respect to temperature. And the constant-pressure heat capacity is equal to the partial derivative of the enthalpy with respect to temperature.

So, this is a good thing. From the first law, we obtain two different definitions for heat capacities in terms of thermodynamic state functions, which remember, are really handy, since they are path independent. For the case of the monoatomic ideal gas, we know that the internal energy is *U* equals $^3/_2$ times the number of moles of the gas n, times the ideal gas constant *R*, times the temperature of the gas. For constant volume, the heat capacity is the partial derivative of *U* with respect to temperature, which gives $^3/_2$ times number of moles of the gas n times the gas constant *R*. Now, an important observation I'd like to make is that you can see that heat capacity is an extensive quantity. If the system size increases, so does the heat capacity, and that makes sense if you think about it; the more material there is in your system, the more heat you'll be able to put into it.

As is often the case, it's useful to use the intensive version of an extensive variable. In this case, it could either be the molar heat capacity, where we'd divide by the number of moles, or what is known as the specific heat capacity or specific heat for short, where we divide by the mass of the system. Okay, now, the constant-pressure heat capacity for an ideal gas can be obtained in a similar manner. I'll let you all at home have fun with the math steps involved; it's just a few partial derivatives, but here I'll skip right to the answer, which is that the heat capacity at constant pressure equals heat capacity at constant volume plus n times R.

Written as molar heat capacities, we have that the constant pressure molar heat capacity equals constant volume molar heat capacity plus the ideal gas constant R. Note that the heat capacity at constant pressure is always greater than the heat capacity at constant volume. So, the molar heat capacity for an ideal gas at either constant volume or constant pressure is simply a constant! So all ideal gases have the same exact heat capacity, well, at constant volume anyway, and that is 12.5 Joules per mole kelvin, regardless of the gas and regardless of the temperature. It seems kind of simple, Could this possibly be right?

In fact, it's really quite a good description. But, let's now move up one degree of difficulty in terms of the system; let's go for a diatomic ideal gas. So these are pairs of atoms, sometimes also called dimers, which since it's still an ideal gas do not interact with one another, but they have a very important difference from the single atom, monatomic case. As they are translating around at those tremendous speeds, the molecules can now also rotate as well as vibrate. These rotational and vibrational motions of the molecules can be activated by collisions, and so, they are strongly tied in with the translational degrees of freedom of the molecules.

But the key point here is that all of this energy that was getting bounced around in the form of translational motion along different directions is now, on average, shared equally by each independent degree of freedom. This is known as the equipartion theorem, which states that each degree of freedom contributes ½ times the Boltzmann constant, k_B, times the temperature T of energy per molecule.

For a diatomic gas molecule, by the way, that would be most of what we breathe, since air consists primarily of N_2 and O_2 diatomic molecules. For these molecules, the shape looks kind of like a dumbbell. If you take a look, you can see that in addition to being able to translate around along the x, y, or z axes, the molecule can also rotate about any of these axes. But if the dumbbell is pointing along, say, the z-axis, then we can neglect the rotation about the z-axis, because the molecule's moment of inertia and rotational energy about this axis are negligible compared to the x and y axes.

So, not counting vibration for now, for diatomic molecules, there are five degrees of freedom from translation and rotation—three associated with the translational motion and two associated with the rotational motion. From the equipartition theorem, as I mentioned, each degree of freedom contributes, on the average, $\frac{1}{2} k_B T$ of energy per molecule, which means that the internal energy for this case is equal to $^5/_2\ n\ R\ T$. This gives us a value for the constant volume molar heat capacity of $^5/_2$ times R, or 20.8 Joules per mole Kelvin.

By the way, you may have noticed that I'm going back and forth between using the Boltzmann constant *k-B* and the ideal gas constant *R*. That's because they're very much related. The number of moles in a system times the ideal gas constant is equal to the number of particles in a system times the Boltzmann constant. This means that the Boltzmann constant is exactly equal to the gas constant divided by Avogadro's number, which is another constant, roughly equal to 6.022 times ten to the 23rd.

Take a look at the experimentally measured values for the molar heat capacity of some diatomic molecules at room temperature. Notice that they are all very close to this value. So it seems that our model is working well! But, and you could probably sense that there was a but, there is a little problem here. Remember how I just ignored the vibrational energy? Well, that's not correct. In fact, it's a bit troubling how well the model agrees with the experiment, given that these entire other degrees of freedom have been ignored.

And vibrational motion adds not just one more degree of freedom, but rather, two more degrees of freedom, one coming from the kinetic energy and the other from the potential energy associated with the vibrations. You can think of this as two balls connected by a spring, for a simple picture; if you put energy into it, it could change the distance the balls are apart, which would be a change in the potential energy, or, it could change the speed at which the balls are moving, which changes the kinetic energy. Therefore, there are two additional degrees of freedom from this single vibration.

The equipartition theorem should really include all three types of motion, and if we add this vibrational part, we get an extra two ½ R contributions, giving a constant volume molar heat capacity of $^7/_2\ R$. A whole extra R, but that's not what we see in experiments, or is it? I'm very excited about what I'm about to show you, because historically, it was truly one of the most confusing failures of classical physics, and it helped lead to the development of quantum mechanics. Now, don't worry, this isn't a course on quantum, and I don't plan to go all quantum on you here, but in order to understand what's happening with these heat capacities, we have to look at the behavior of heat capacity as a function of temperature.

Take a look at this plot showing the heat capacity for the hydrogen dimer, the H_2 molecule, from 20 to 10,000 Kelvin. Notice that it's not at all a constant, but rather, appears to have three different plateaus. And this is quite remarkable; each plateau occurs at exactly the values we have already predicted for different scenarios. For the lowest temperature range, we get the value $^3/_2\ R$, same as for a monatomic ideal gas. As the temperature rises to room temperature, the molar heat capacity goes up to $^5/_2\ R$, the value we obtained for a diatomic ideal gas, including rotational energy, but without vibrational energy. And finally, at high enough temperatures, the measured heat capacity is consistent with our model for a diatomic ideal gas, including all types of motion.

So, the key to understanding what's going on here is that the model we derived based on the ideal gas law and the equipartition theorem is purely classical in nature. It predicts a value of the molar heat capacity for a diatomic gas that, according to the figure I just showed you, only agrees with experimental measurements made at very high temperatures. In order to understand this, we have to go beyond classical physics and bring quantum physics into the model. However, I did promise that I won't go all quantum here, and I plan to keep that promise!

But I do want to give you a good intuitive feeling for what's happening. Quantum mechanics tells us that the energies of atoms and molecules are, well, they are quantized. This means that they can only take on discrete, specific values, as opposed to any old values they want. And this means that the energies of those rotational and vibrational degrees of freedom, they are quantized and can only have certain values. And the vibrational energies that are allowed tend to be highest, with rotational next, and translational lowest.

So, at very low temperatures, molecules are colliding about, but the energy they can gain from these collisions is generally not enough to raise it to the amount needed to occupy either rotations or vibrations. That's why, even though rotation and vibration are allowed according to classical physics, they don't occur at low temperatures. And that's why we get the same heat capacity as we did for a single monatomic ideal gas. The pair of atoms behaves just like a single atom, because it cannot rotate or vibrate at low temperatures; it can only translate. The number of degrees of freedom is back to only three.

But as the temperature is raised, the average energy of the molecules increases, and now, enough energy is present to reach the amount needed for the first allowed rotational modes, in which case the rotational degrees of freedom become accessible to the system, and as a result, rotation begins to contribute to the internal energy, and the heat capacity increases.

To be able to access those vibrational states we need to go to much higher energies, again, because quantum mechanics tells us that only certain energies are allowed, as opposed to any energy, and in the case of vibrations, the first allowed energy happens to be quite high. As you can see in the data for H_2, these vibrational modes start getting occupied between 1000 and 10,000 degrees Kelvin. By 10,000 Kelvin, vibrations are fully contributing to the internal energy and the heat capacity has the value we predicted for this case, namely $^7/_2\,R$. So in a way, where we've come to is pretty cool, since the equipartition theorem still works very well; it's after all what gives us those plateau values.. It's just that which degrees of freedom are allowed to be equipartitioned over is more complicated, and that dictated by quantum mechanics.

Now, at this point, I'd like to take a step back for a moment to be sure that we're feeling our oneness with all of this. Remember that our original goal of this lecture was to understand how thermal energy is stored in materials. And we went from that goal to the definition of a new thermodynamic variable, the heat capacity. The variable gives a quantitative measure of the heat that can "fit," in a sense, into a material, per change in the temperature of the material. And then, we talked about the fact that the word "fit" here means degrees of freedom at the atomic scale. How many different places are there for the heat energy to go? And that brought us to the calculation of how many degrees of freedom there are in a given system, starting from the simplest case of an ideal gas. The equipartition theorem states that the thermal energy will be spread evenly over all of the degrees of freedom. But, as we just saw, quantum mechanics can get in the way and freeze out some of them, depending on the temperature. So, as you can see, the heat capacity of a material is indeed quite a complicated variable!

Okay, so, we've worked our way up the chain, all the way from a single atom to a two-atom molecule. You might think we have a whole lot more to go, but actually, there's only one more case I want to discuss, namely that of solids. In a solid, all of the atoms are arranged on a regular lattice, like the one shown here. In the simplest picture, you could imagine all these atoms held together by a kind of spring that bonds them together, just as in the case of a diatomic molecule. And as you're looking at this lattice of atoms held together by springs, think about which of the three degrees of freedom solids can have. Is it translational, rotational, or vibrational? Unlike atoms and molecules, you can see that, in fact, solids can only have one single type of freedom, and that would be vibrations. The atoms in the solid don't translate, and they don't spin around, since they're pretty much stuck in their crystalline position. So when heat energy is transferred to a solid, it all goes into those vibrations of the atoms.

And we can decompose each atom in a solid into having some amount of vibration along each axis—x, y, and z. For each axis, we have the two degrees of freedom that a vibration gives. Taken together, this means that in a solid there are six degrees of freedom. Each one brings an amount of ½ R to the heat capacity, so, according to this analysis, the molar heat capacity will be $3R$ for a solid. And this is, in fact, exactly what happens for most solids, as long as the temperature is high enough. This result of $3R$ for the constant-volume, molar-heat capacity of solids is known as the Dulong–Petit law.

However, just as in the case of the diatomic molecule, there is a very strong temperature dependence for solids as well. For some metals, the variation from $3R$ occurs below room temperature, but for many materials, it occurs at temperatures much higher, meaning that for quite a range of temperatures, the behavior is much more complicated. Take a look at the temperature-dependence of the molar heat capacities for these solids.

And to close today's discussion, I'd like to impress upon you why this is all so important, why I wanted to be sure to have an entire lecture devoted just to heat capacity. As I've mentioned in other lectures, absolute values of thermodynamic functions, like internal energy and entropy, are not as useful as measured changes in these quantities. But how can the changes be measured easily?

It is the knowledge of heat capacities that allows us to calculate changes in internal energy and entropy occurring in real processes from simple experimental measurements. We already know that internal energy is directly related to heat capacity, and with a little help from the second law, we can rewrite the heat capacity as TdS divided by dT, giving us a simple way to measure changes in entropy for real processes.

Be sure not to confuse heat capacity with entropy. Remember our physicalnotion of entropy as a measure of the degrees of freedom available to the molecules in the material, this implies that materials with more molecular degrees of freedom, that is, the ability for molecules to translate, vibrate, or rotate, have higher heat capacities. But, entropy can only go up with temperature, not so for heat capacity.

Here's a plot of the heat capacity of the chlorine dimer as a function of temperature, going through each of its difference phases from solid to liquid to gas. Note that the heat capacity can either go up or down as temperature increases, due to all of the things we have been discussing today, as well as the complexity inherent to the material changing from one phase to another.

While the entropy of a gas would always be larger than the solid form of the same material, the heat capacity is actually lower!

So, in the end, heat capacity is an extremely important property, both because it tells us how a material holds onto thermal energy, but also, because it connects us, through very simple laboratory measurements, with other thermodynamic variables. Take any material, carefully add a controllable amount of heat while measuring the temperature, and you have the heat capacity.

How Materials Respond to Heat

Lecture 13

The heat capacity gives us a crucial way to correlate energy flow with temperature, but what about thermal energy flow and volume? When heat flows into or out of a material, how does it expand or contract in response? How does that depend on the type or phase of the material, or its temperature? That will be the topic of this lecture. The fact that materials expand when heated is an essential component of life and a key ingredient in thermodynamics.

The Fundamental Basis for Thermal Expansion

- Temperature is related to the average translational kinetic energy of the atoms or molecules that make up the material, or 1/2 mass times average velocity squared: $1/2mv^2$. At this microscopic scale, temperature is intimately connected to the motions of the particles.

- There are three different types of degrees of freedom: translational, rotational, and vibrational. For monatomic gases, it's all about translational motion because that's the only kind of motion possible, and temperature is therefore entirely determined by the kinetic energy from translation of the atoms. In the case of solids, there is only one possible degree of freedom for the atoms or molecules: vibration. Rotation and translation are not possible in a solid.

- Think about a solid as consisting of atoms in a lattice held together by springs. These springs are a simple but helpful way to represent the bonding that occurs between the atoms, which we can describe as interacting through some interatomic potential.

- This potential is such that when the atoms are far apart, the forces tend to be attractive, and when they are very close to each other, they tend to be repulsive. The attractive and repulsive forces compensate each other at an equilibrium distance that represents the lowest potential energy point and stable equilibrium of the system.

- Let's consider the simplest case of just two atoms held together by a single spring. In this case, they would oscillate back and forth around some central point, the point at which no force is pushing or pulling on them and the spring has no tension on it.

- When we have a spring system, the motion of the atoms back and forth is called harmonic. If we are approximating the vibrations of atoms in a solid as connected to springs, then we can think of these vibrations as being roughly harmonic in nature—no tension, no force pushing or pulling.

- Let's incorporate temperature into the picture. At zero temperature, the atoms won't be moving at all, so they will just sit in that equilibrium position where the spring likes to be when it has no tension.

- As we slowly heat up this harmonic oscillator, the atoms move together and apart, together and apart, oscillating about that same equilibrium position as they trade back and forth kinetic with potential energy.

- As we continue to increase the temperature, the total energy increases, and the atoms will vibrate very quickly now, stretching out farther and farther and traveling with more speed, still about that same equilibrium point.

- The higher the temperature, the faster the vibrations, but the center of the vibration is the same. That means that, on average, these atoms are not going to be any farther apart from one another—so, on average, the material cannot have expanded with temperature.

- This means that the simple model of atoms vibrating as if attached by springs is not good enough to explain thermal expansion. It turns out that it is a pretty good description of real materials, but only when the atoms are very, very close to their equilibrium position.

Thermal-expansion joints are built into bridges to account for the amount of thermal expansion the bridge endures.

- Instead, for a real material, at the minimum of the potential energy, it is pretty harmonic looking, but as we move in to closer distances, the potential shoots up very quickly. And as we move out to farther distances, it goes up much, much more slowly.

- The key point is that the potential energy between atoms in a real material is asymmetric about the equilibrium point. And that is what causes thermal expansion.

- If we now go back to zero temperature, we'd be at the same equilibrium point and have no motion. As we increase the temperature, just as before, the atoms move both more quickly as well as farther apart with each oscillation.

- But now, as temperature increases, the oscillations lead to much more change in how far apart the atoms can get compared to how close they can get (which changes very little). The equilibrium distance shifts as the atoms spend more time at distances greater than the original spacing. The repulsion at short distances is greater than the corresponding attraction at farther distances.

- So, the average position of the oscillating atoms increases as the oscillations become larger. This is the atomic-scale picture of thermal expansion, and it occurs only because of the asymmetry in the way atoms interact with one another.

Quantifying Thermal Expansion

- Thermal expansion is very important in our daily lives. It may not seem like a material is expanding very much compared to its original size, but especially for nonelastic materials like steel, concrete, and bricks, even just a small expansion can cause massive changes. That's why when buildings, concrete highways, railroad tracks, most types of walls, and bridges are designed, so-called thermal-expansion joints must be used.

- Thermal expansion is such an important property that it has its own variable, called the thermal expansion coefficient. For many situations, to a good approximation, the thermal expansion along any dimension of an object is linearly proportional to the temperature change.

- If an object has an initial length L along some direction at some temperature T, that length increases by an amount ΔL when the temperature changes to T plus ΔT. The coefficient of linear expansion is called alpha (α), and it is defined as follows: $\alpha = \frac{1}{L}\frac{dL}{dT}$. It is the fractional change in length per change in temperature.

- Often, we'll see this written in terms of the change in length, so we have $\Delta L = \alpha L \Delta T$. In many experiments, we observe that alpha is only weakly dependent on temperature, at least for small changes in temperature, so often alpha is just taken as a constant for a given phase of a given material. In many cases, we are only interested in the net volume change of the material, as opposed to its change in each direction.

- Similar to the linear expansion coefficient, we write a volume expansion coefficient as follows: $\alpha = \frac{1}{V}\frac{dV}{dT}$. It is the fractional change in volume divided by the change in temperature. The change in volume can then be written as $\Delta V = \beta V \Delta T$.

- For a solid, the average coefficient of volume expansion is roughly three times the average linear expansion coefficient, so $\beta = 3\alpha$. This assumes that the average coefficient of linear expansion of the solid is the same in all directions, or that the material is isotropic.

- Regardless of this relation between linear and volumetric thermal expansion, if we're talking about only very small changes, then we can write our expression as a differential. So, the volumetric thermal expansion coefficient is defined as 1 over volume times the partial derivative of volume with respect to temperature, and it's a partial derivative because we take it while holding other thermodynamic variables like pressure or number of moles fixed: $\beta = \frac{1}{V}\left(\frac{\partial V}{\partial T}\right)$.

- From this definition of thermal expansion as a derivative, you can see that it is simply equal to the slope at any given point of a plot of the volume versus temperature of a material. It's also now easy to take into account any temperature dependence of β, because the way we would compute the volume change is by integrating this expression over the temperature change.

- In a plot of molar volume versus temperature, not only is the volume higher for a liquid than for a solid, but the slope is higher as well, indicating that the expansion coefficient is larger for the liquid than for the solid. In fact, the difference is often as much as a factor of 10 or even 100 times larger for the liquid.

- We can understand this difference by thinking back to our atoms and degrees of freedom. In the liquid, unlike the solid, we now have accessible all three types of degrees of freedom: vibration, rotation, and translation. As the temperature of the liquid increases, all of these degrees of freedom increase in energy, and in particular, the presence of the translational motion leads to larger average volume expansion.

Compressibility

- Another important thermodynamic property that is quite similar in spirit to thermal expansion is compressibility. Remember that we wanted to hold pressure fixed in order to understand the effects of increasing temperature on volume. But what is the effect of increasing pressure?

- Intuitively, we feel that if we squeeze on something, it will compress. And, indeed, mathematically we can define this effect in the same manner as the expansion coefficient. The isothermal compressibility is defined as −1 over the volume times the partial derivative of the volume with respect to pressure, holding everything else, like temperature and number of moles, fixed: $= -\frac{\;}{V}\left(\frac{\partial}{\partial p}\right).$

- The compressibility of a material is its fractional decrease in volume that occurs with pressure at constant temperature. That means that if we plot the volume versus pressure for a material, the compressibility is equal to the slope of that curve (just as thermal expansion was the slope of the volume versus temperature).

- The compressibility of materials often tends to be quite small. It takes a lot more energy to push these atoms close together than it is to pull them apart.

Suggested Reading

Callister and Rethwisch, *Materials Science and Engineering*, Chap. 19.

Halliday, Resnick, and Walker, *Fundamentals of Physics*, Chap. 18.6.

Questions to Consider

1. How can you store energy using the concept of thermal expansion?

2. Why do different solids have different thermal expansion coefficients?

3. How would thermal expansion be different between a gas and a solid and a liquid?

4. How is the macroscopic quantity temperature connected to atomic quantities like the average separation between atoms in a structure?

5. Why does zinc have a higher thermal expansion coefficient than tungsten?

How Materials Respond to Heat
Lecture 13—Transcript

In the last lecture, we talked about how different materials, or even just different phases of the same material, can absorb heat quite differently from one another. The heat capacity, as we discussed, is a measure of how much heat a material absorbs or loses per degree of temperature increase or decrease. It gives us a crucial way to correlate energy flow with temperature, and we understood it from the point of view of degrees of freedom of the atoms and molecules that make up the material. So that's all nice and good. Thermal energy flow and temperature are nice and connected now. But, what about thermal energy flow and volume? When heat flows in or out of a material, how does it expand or contract in response? How does that depend on the type or phase of the material, or its temperature? That will be the topic of this lecture.

We've already touched upon thermal expansion in earlier lectures, for example, when we discussed thermometers, we saw how the expansion of a gas, liquid, or solid, all corresponding to different types of thermometers, can lead to an accurate temperature measurement. But in this lecture, we're going to dig deeper into how and why the volume of the material changes with changes in temperature.

I'll start with a question. One of the buildings on MIT's campus happens to be quite tall and similarly rectangular on all sides, going straight up to a height of around 50 meters. This building also happens to face exactly in line with the sun's trajectory, so that in the morning fully one side of the building is exposed to the sun, while the other side is dark, while in the afternoon, it's the opposite. Now, sadly, in Cambridge Massachusetts we don't get all that much sun overall, but in the summer time we can get as much as five hours per day. And the peak of that nice, warm sunshine comes right around 2:00 pm during the summer. So I have the following question: At 2:00 pm on a hot summer day at MIT, what is the shape of this building?

As you may have surmised, the building has changed, because at midday, the sun gave quite a bit of thermal energy to only one side of the building, not the other. One side is quite hot, the other still cold. And because of thermal expansion, the building is literally lopsided, not by all that much, but still, one side is taller and wider than the other. We'll calculate just how much a bit later in the lecture.

Okay, enough warming up—no pun intended, I swear. Let's dive into some thermo to help us understand the concept of thermal expansion. In order to begin to understand how materials expand with heat, just as with the heat capacity, we head all the way down to the scale of the atom. As you may recall, temperature is related to the average translational kinetic energy of the atoms, or molecules, that make up the material, or ½ mass times average velocity squared. At this microscopic scale, temperature is intimately connected to the motions of the particles.

Now, in our last lecture on heat capacity, I discussed the three types of degrees of freedom—translational, rotational, and vibrational. For monatomic gases, it's all about translational motion, since that's the only kind of motion possible, and temperature is therefore entirely determined by the kinetic energy from translation of the atoms. Now, during this lecture, I'll be focusing on liquids and solids, and I'd like to begin with solids, since in that case, there is only one possible degree of freedom for the atoms or molecules, namely, that of vibration. Rotation and translation are not possible in a solid.

Let's think about a solid as consisting of atoms in a lattice held together by springs. These springs are a simple but helpful way to represent the bonding that occurs between the atoms, which we can describe as interacting through some interatomic potential. This potential is such that when the atoms are far apart, the forces tend to be attractive; and when they are very close together, the forces tend to be repulsive. The attractive and repulsive forces compensate each other at an equilibrium distance that represents the lowest potential energy point and stable equilibrium of the system. So you can kind of see why springs are a good analogy for what holds atoms together in a solid.

This picture encourages us to think of the atoms vibrating back and forth as soon as they have any sort of thermal energy. By the way, how fast are the atoms vibrating, on average, in a solid, and how far do they travel during the vibration? Well, the atomic scale is set at the scale of the Angstrom, or ten to the minus ten meters, and a typical bond distance between atoms is around a few angstroms, depending on the material. When the atoms vibrate at a given temperature, they typically don't move as far as a full bond distance, but rather, on the order of one tenth that distance. So the distance an atom actually travels during its vibration is roughly on the order of ten to the minus eleven meters, or a tenth of an angstrom.

And the speed of the atoms during these vibrations? Remember that for a gas at modest temperatures, the atoms or molecules are kicking around at very high speeds, thousands of miles per hour. In a solid, you might first think the atoms are moving much slower, since they're not as free. But, if the springs are stiff, that is, if the bonding is strong, which is the case for most solids, then motion is actually extremely fast. In a typical solid, even at just room temperature, the atoms are vibrating at a frequency of ten to the 13th hertz. That means it stretches out and in ten trillion times each and every second. That's fast!

Okay, let's consider the simplest case of just two atoms held together by a spring. In this case, they would oscillate back and forth around some central point, the point at which no force is pushing or pulling on them and the spring has no tension on it. You may have learned this in an introductory physics course; when we have a spring system like the one shown here, the motion of the atoms back and forth is called harmonic. That means that if I draw a plot of the kinetic energy vs. the distance between the atoms, I'd get a picture that looks like this. It's a relationship that goes as y = x squared, and it's the basic kinetic and potential energy functions for a spring. So, that means, if I'm approximating the vibrations of atoms in a solid as connected to springs, then I can think of these vibrations as being roughly harmonic in nature—no tension, no force pushing or pulling.

Let's take this picture and incorporate temperature into it. If I'm at zero temperature, then the atoms wouldn't be moving at all, so they just sit in that equilibrium position where the spring likes to be when it has no tension. As I slowly heat up this harmonic oscillator, the atoms move together and apart, together and apart, oscillating about that same equilibrium position as they trade back and forth kinetic and potential energy. As I continue to increase the temperature, the total energy increases, and the atoms will vibrate now very quickly, stretching out farther and farther and traveling with more speed, still about that same equilibrium point.

But wait a minute. I thought this model was going to tell me something about thermal expansion.

If I look at what's happening, the higher the temperature, the faster the vibrations, but, the center of the vibration is the same. That means that, on average, these atoms are not going to be any further apart from one another. So on average, the material cannot have expanded with temperature. What does this mean? Well, it means that the simple model of atoms vibrating as if attached by springs is not good enough to explain thermal expansion. If we plot the potential energy of those two atoms undergoing harmonic motion on a spring, we would see that y equals x squared dependence. And actually, it turns out that this is a pretty good description of real materials, but only when the atoms are very, very close to their equilibrium position.

Instead, the potential energy of atoms in a real material looks something like this. See, at the minimum of the potential it's pretty harmonic looking, but, as soon as we move in to closer distances, the potential shoots up very quickly. And as we move out to farther distances, it goes up much, much more slowly. So the key point here is that the potential energy between atoms in a real material is asymmetric about the equilibrium point. And that is what causes thermal expansion.

If I now go back to zero temperature, we see that, once again, we'd be at this same equilibrium point and have no motion. But now, as I increase the temperature, just as before, the atoms move both more quickly, as well as farther apart, with each oscillation. But now notice what's happening. As temperature increases, the oscillations lead to much more change in how far apart the atoms can get compared to how close they can get, which changes very little. The equilibrium distance shifts as the atoms spend more time at distances greater than the original spacing. The repulsion at short distances is greater than the corresponding attraction at far distances. So, the average position of the oscillating atoms increases as the oscillations become larger. This is the atomic-scale picture of thermal expansion, and as you can see, it occurs only because of the asymmetry in the way atoms interact with one another.

Okay, now that we've spent some time hanging out with the atoms in a solid to understand the fundamental basis for thermal expansion, let's move on to some quantifying. I want you to get an appreciation for how important thermal expansion is in our daily lives. It may not seem like a material is expanding very much compared to its original size, but especially for non-elastic materials, like steel, concrete, and bricks—that would be, much of the stuff we use to build things—even just a small expansion can cause massive changes. That's why, when buildings are designed, so-called thermal-expansion joints must be used, and the same for concrete highways, railroad tracks, most types of walls, and bridges, just to name a few important examples.

So, how do we quantify and calculate the thermal expansion of materials? Well, we've understood why it happens, but now, how do we incorporate it into our thermodynamic world? First off, it's such an important property that we give it its own variable, called the thermal expansion coefficient. For many situations, to a good approximation, the thermal expansion along any dimension of an object is linearly proportional to the temperature change.

If an object has an initial length L along some direction at some temperature T, then that length increases by an amount delta-L when the temperature changes to T plus delta T. The coefficient of linear expansion we call alpha, and it's defined as delta-L divided by L, over delta T. It's the fractional change in length per change in temperature. Often, we'll see this written in terms of the change in length, so we have delta-L equals alpha times L times delta-T. In many experiments, we observe that alpha is only weakly dependent on temperature, at least for small changes in temperature, so often, alpha is just taken as a constant for a given phase of a given material. I'll come back to this point in a bit.

In many cases, we are only interested in the net volume change of the material, as opposed to its change in each direction. So, similar to the linear expansion coefficient, we write a volume expansion coefficient. And we write that with the variable beta; beta equals the fractional change in volume divided by the change in temperature. The change in volume can then be written as beta times the initial volume, V, times the change in temperature.

For a solid, the average coefficient of volume expansion is roughly three times the average linear expansion coefficient, so beta equals 3 times alpha. This assumes that the average coefficient of linear expansion of the solid is the same in all directions, or as I just discussed, that the material is isotropic.

Okay, but regardless of this relation between linear and volumetric thermal expansion, if we're talking about only very small changes, then we can write our expression as a differential. So, the volumetric thermal expansion coefficient is defined as 1 over volume, times the partial derivative of volume with respect to temperature. And it's a partial derivative, since we take it while holding other thermodynamic variables, like pressure or number of moles, fixed.

Now, holding these other variables fixed makes sense if you think about it. Take pressure for example; it would become quite complicated if the definition I just showed you included effects like increased pressure upon heating, which would, in turn, act to prevent the material from expanding as much. But from this definition of thermal expansion as a derivative, you can see that it's simply equal to the slope at any given point of a plot of the volume vs. temperature of a material. And it's now easy to take into account any temperature dependence of beta, since the way we would compute the volume change is by integrating this expression over the temperature change.

Let's take a look at the molar volume vs. temperature plot for a given material. Remember that when I say molar volume, it's just the volume per mole of the material. In this case, we're looking at a polymer known as polyethylene. It is the most basic plastic in the world today, with global production closing in on 100 million tons of the stuff per year. As we look at the volume vs. temperature plot for polyethylene, first of all, notice that there is a huge jump in volume; that's right at the temperature where the material melts, or, if you think of it as going from right to left, it's where it freezes. This is the temperature point at which the material undergoes a phase transition, a topic we'll be turning to in detail in our next several lectures. But for now, just note that a large volume change such as this often occurs at a phase transition.

Remember how we said that the thermal expansion coefficient is equal to the slope of this curve? Well, one thing you can see is that at a phase transition, the thermal expansion coefficient is simply not well defined, since the slope tends toward infinity. So, we like to use expansion coefficients to understand changes in a given phase of a material, rather than during phase changes. And within one particular phase, either the solid or liquid, you can see that, indeed, the volume is changing with respect to temperature. But, the expansion coefficient is pretty close to constant, since the curves are quite close to linear in each regime. When the curve is linear, then the slope of that curve, which is the thermal expansion coefficient, is of course, a constant.

Now, the last point I'll make about this plot of molar volume vs. temperature is that, as you can see, not only is the volume higher for the liquid than for the solid, but the slope is higher as well, indicating that the expansion coefficient is larger for the liquid than for the solid. In fact, the difference is often as much as a factor of 10 or even 100 times larger for the liquid.

We can understand this difference by thinking back to our atoms and degrees of freedom. In the liquid, unlike the solid, we now have accessible all three types of degrees of freedom—vibration, rotation, and translation. As the temperature of the liquid increases, all of these degrees of freedom increase in energy, and in particular, the presence of the translational motion leads to larger average volume expansion.

Okay, before moving on to a few example problems, I want to mention another important thermodynamic property that is quite similar in spirit to thermal expansion. Namely, compressibility. Remember that we said we wanted to hold pressure fixed in order to understand the effects of increasing temperature on volume. But what is the effect of increasing pressure? Intuitively, we feel that if we squeeze on something, it will compress. And indeed, mathematically we can define this effect in the same manner as the expansion coefficient. The isothermal compressibility is defined as minus one over the volume, times the partial derivative of the volume with respect to pressure, holding everything else like temperature and number of moles fixed.

The compressibility of a material is its fractional decrease in volume that occurs with pressure at constant temperature. That means that if we plot the volume vs. pressure for a material, the compressibility is equal to the slope of that curve, just as thermal expansion was the slope of volume vs. temperature.

Now, we won't dwell on compressibility, since I think you can understand it pretty easily now that we've worked through thermal expansion. But one point I'll mention is that the compressibility of materials often tends to be quite small, something we can understand by looking back at our potential energy plot for the oscillating two atoms. It's going to take a lot more energy to push these atoms closer together than it is to pull them apart.

Now, what we've talked about so far, regarding expansion with temperature, corresponds to the typical behavior of materials. But I couldn't possibly go through a whole lecture on thermal expansion without at least mentioning one highly unusual case. That would be the stuff that we humans, as well as the earth's surface, are mostly made of, namely water.

As I mentioned, liquids generally increase in volume with increasing temperature, and they have coefficients of volume expansion that at least ten times greater than those of solids. But liquid water near its freezing point is an exception. Take a look at a plot of the density vs. temperature for water, and notice that there is a peak in the plot around 4°C. That means that as the temperature of water goes from zero to four degrees, it actually contracts, leading to the density increase you see here. After 4°, water then expands with temperature, and the density starts to decrease.

This behavior of the density of water near its freezing point is extremely unusual, and it has to do with the way in which water molecules bond together. The particular type of bonding that occurs from one molecule to another for water, is known as hydrogen bonds, and combined with the shape of the water molecule, that makes it so that the molecules can pack in more closely as a liquid than as a solid at these near-freezing temperatures. This is in contrast to most other materials that exist.

But the weird thermal expansion behavior of water is nothing less than essential to life as we know it. Think about the way in which water in a pond freezes; we know that it freezes from the top first to the bottom. This is why for so many marine-life environments, life can survive during the winter. The surface freezes over, but in many cases, underneath an ice layer, remains liquid water, so those fishies don't turn into blocks of ice.

Why is this the case? Well, we can use the unusual thermal-expansion behavior of water to understand it. When the temperature outside drops from, say, 7°C to 6°C, the surface water also cools, but due to the density profile of water, the density increases during this cooling, which in turn, makes the surface water more dense than the water below it. As a result, the surface water sinks, and warmer water from below comes up to the surface to be cooled next. When the temperature outside goes to 4°C, the density of the surface water is at a maximum, but then, as the outside temperature goes from 4°C down to 0°C, the density starts to decrease, making the surface water expand as it cools. Eventually, the mixing of top layer with deeper layers stops, and the surface water freezes into ice. However, the ice remains on the surface because ice is less dense than water.

Once a nice, strong surface ice layer as formed, it serves as a fantastic thermal insulator from cold air, which dramatically slows down any subsequent cooling of the water. Meanwhile, the water near the bottom remains at a toasty 4°C, since that's the maximum density, a thermodynamic fact that makes the fish quite happy.

So, in the last minutes of this lecture, I'd like to show you a little thermal expansion, and then, we'll come back to that problem of the building on MIT's campus, as well as see some last examples of thermal expansion in our daily lives.

Let's look first hand at how heat can change the physical volume of a material. What happens when we add heat to a material? Let's take a look. Here, I have a ring, and I have a ball. And at room temperature, which is where we are right now, the ball can easily slide through the ring, back and forth. Okay?

Now, let's heat up the ball and see what happens. So I'm going to hold the ball above a flame, like this, and so we're raising the temperature of the material that makes up the ball, probably not by a whole lot, maybe by 50°C overall. So now I'm going to take the same ring, and I'm going to try to put the ball through it. And you can see, very clearly, that it simply doesn't fit. I cannot put it through this same hoop. Okay? The volume of the ball has expanded.

Now, that was a single material, so that's one material that undergoes thermal expansion. But what if we had two materials stuck together? Well, that's exactly what I have here. I have two different metals bonded together to form this strip here. Okay? This is called a bi-metal strip—simply two different metals stuck together. But the key here is that they have two different thermal expansion coefficients. So when heat enters into this material, they try to expand at different rates, and the combined strip that's bonded together will bend. Let's take a look.

Okay, so now you can see that when I add heat into this material, it doesn't just expand uniformly, but instead, the whole material wants to bend. That's because of the different thermal expansion coefficients of the two different materials. And as soon as the heat is taken away, you can see that it wants to start to bend back.

Now, we can use this kind of bending that happens in bi-metallic strips to actually do useful things, like, for example, measure the temperature in a thermostat, or make a system oscillate back and forth. Take a look at this switch that I built using simply the same technology, a bi-metal strip that reacts to heat and bends when it's heated. So I'm going to plug this in, and the light will go on, and the light is emitting heat. And that heat is going to make this bi-metallic strip bend just so slightly, which then turns off the circuit and turns off the light.

Then, when the light is turned off, the heat isn't present; the bi-metal strip cools back down, and it will reconnect with the circuit, and the light will go back on. So you can see now, the bi-metallic strip is just barely making contact, closing the circuit, opening the circuit. And what we have is a switch that reacts to heat. That's what's in a lot of thermostats in a lot of homes around the world.

So there you have it. What we're seeing is, at the atomic scale, how, when you have an asymmetry in the way atoms see each other, the way atoms are bonded together, that causes an expansion of the material when heat is added. And we've seen that with several different examples—a single material and two different materials bonded together.

Let's put some numbers to that problem about the rectangular building on MIT's campus that I mentioned in the beginning of the lecture. Well, that's actually pretty easy, since it's about the steel. For a temperature change of, say, 25°C, the thermal expansion of the side of the building facing the sun will be about one and a half centimeters. That may not sound like much, but it would be plenty to crack open the building if it's not constructed with thermal expansion in mind.

And there are so many other examples that we experience all the time, but perhaps, are not aware of. An airplane can expand by 10 to 20 centimeters during flight because of the heating it undergoes due to friction with the air. Running hot water over a tight metal lid on a glass jar makes it easier to open, because the thermal expansion coefficient of the metal is larger than that of the glass.

And here's a fun one, the old-fashioned pendulum clocks, which back in the day were all the rage—that would be "the day," as in hundreds of years ago. Does a pendulum clock in Phoenix run slower or faster than the exact same pendulum clock in Boston? I'll let you work that one out on your own. Or how about this one that most of us have heard, Never fill your gasoline tank up all the way, especially on a hot day. But why not? It's because of thermal expansion of course! When the cold gasoline is pumped into the car's tank, it first cools the steel of the tank's container. But then, both the container and the gas inside quickly heat up to the outside air temperature. Both expand, but, the thermal expansion coefficient of gasoline is much larger than that of the steel tank, so it expands much more and overflows if the tank is full.

Thermal expansion is also why the reliability of the gasoline gauge itself can vary from summer to winter. When you see that fateful gasoline light and hear the beep in your car, it means you only have a little bit left. But the gauge reads the volume of the tank, and in the summer, the same mass of gasoline will take up a lot more volume, due to thermal expansion, compared to the winter. So when you see the light come on, you've probably got a lot less time until you're stranded in the summer than in the winter.

So those are just some of the many examples of thermal expansion in our lives. And it's even more literally in our lives when we consider medical technology. Most hip or knee implants need to be replaced after some time, partly because the metal doesn't bond well enough with the bone. A lot of research is going on to try to find better materials that could coat the metal and provide a stronger bond to the bone. However, a major challenge in this research is to find coatings that also have a thermal expansion coefficient similar to the metal. If not, the coating will crack and the whole implant would crack off of the bone and fail.

And of course, then there's one of the greatest places to see materials science in action—the dentist. Fillings for your teeth have different thermal expansion coefficients than your natural tooth enamel, which can cause pain or cracks when eating foods with low or high temperatures. That's why, nowadays, its much more common to get porcelain fillings, as opposed to gold or silver, as they coefficients of expansion much closer to that of your teeth.

So, whether we're talking about your house, your car, your teeth, or your hip replacement, I hope you've got a nice appreciation for how materials expand when heated. It's an essential component of life, and a key ingredient in thermodynamics.

Phases of Matter—Gas, Liquid, Solid

Lecture 14

Matter can exist in a variety of phases, which are uniquely identifiable forms of a material separated by other forms by an interface. The phases we encounter most often are gases, liquids, or solids of differing crystal structures. Thermodynamic properties can vary dramatically from one phase to another of the same material, and at those transition points between phases, some properties are not even well defined. In this lecture, you will learn that the phase maps of materials represent one of the most important outcomes of thermodynamics. These maps, also called phase diagrams, provide enormous amounts of information about a given material, or a given composition of different materials.

The Phases of Water

- Water is one of the very few materials that we are intimately familiar with in all three of its phases—ice (solid), water (liquid), and steam (gas). Most materials only exhibit a single phase in our normal living conditions. In fact, water has not just one but, rather, a whopping 15 different phases of solid ice, each one with a different crystal structure and density.

- One of the simplest ways to change a material from one phase to another is by changing its temperature. This leads us to our very first phase maps, where we examine the dependence of the phase of a material on its temperature.

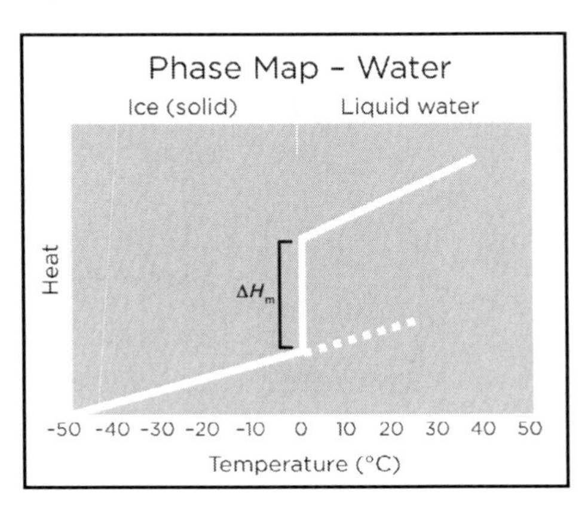

- At the scale of the atom, there are one oxygen atom and two hydrogen atoms for each water molecule. The oxygen atom grabs some of the charge away from the hydrogen atoms, making the oxygen have a slight negative charge, while the hydrogen atoms have a slight positive charge.

- Opposite charges attract, so this charge redistribution in the molecule is what drives the bonding behavior in the solid: The negatively charged oxygen from one molecule wants to find and get close to the positively charged hydrogen atoms from another molecule. These types of bonds are called hydrogen bonds, and they are the key type of bonding in water.

- In the solid phase of water, when we start to add heat, the energy goes into the vibrational degrees of freedom, so the molecules start shaking more and more vigorously—that is, until the material reaches the temperature of its phase change.

- Up until that point, the hydrogen bonding between molecules keeps them all together, even though they are vibrating with more and more energy. But at the phase-change temperature, while the shaking doesn't stop, the temperature is enough that the hydrogen bonding itself starts to break apart.

- Because that is the bonding that holds the positions of those molecules fixed on average, it means that the molecules can start to slip and slide out of their fixed crystalline positions—resulting in a liquid.

- The heat energy we added to break up the hydrogen bonds did not go into anything related to the kinetic energy of the system. It did not go into motion.

- Recall that the three types of degrees of freedom are translation, rotation, and vibration. At the phase transition, the heat goes into none of those degrees of freedom; rather, the heat energy goes entirely into raising the potential energy of the system. That's what happens when we break apart those hydrogen bonds. That's why the temperature during the phase transition doesn't change, even though we're adding heat.

- To go from solid to liquid at zero degrees Celsius—so, without changing the temperature by a single degree but just changing the phase—we need to add an amount of heat equivalent to that needed to raise the temperature of water as a liquid by 80 degrees Celsius. It's enough energy to get water from the tap all the way up to its boiling point.

- And from liquid to vapor, even more thermal energy is needed to change the phase. With the temperature fixed at 100 degrees Celsius, we need to add an amount of heat equivalent to raising the temperature by 540 degrees, just to change from liquid to vapor.

- This type of heat that gets a material to change phase is referred to as latent heat. For water, it's 80 calories for the solid-to-liquid transition and 540 calories for the liquid-to-vapor transition.

- Sometimes, other terms are used for these phase transition energies, such as the heat of fusion for going from solid to liquid or heat of vaporization for going from liquid to vapor.

- In the case of supercooling water, when we reach the phase boundary temperature, instead of stopping the temperature drop and changing the phase, we simply continue along this same line going below the solidification transition temperature. These boundaries separate the equilibrium phases of the material.

- For the transition of water into ice, we can have metastable phases, and that's exactly what we have with water that remains as a liquid below zero degrees Celsius. It wants to be a solid, but turning into a solid from a liquid takes time, and it is helped by having what are called nucleation sites. These are tiny places in the system where the liquid can get stuck, forming a tiny piece of solid, which then coaxes more liquid to form more solid, and so on.

- As we make phase maps of materials, the boundaries we draw between phases come from thermodynamics, and therefore, they are always referring to the equilibrium properties of the system. It does not mean that a material cannot exist in a different phase than is indicated in our phase map, because it can be kinetically trapped, but it does tell us what the preferred equilibrium phase will be. And that's the starting point for any study of any material.

- If we can supercool a material, we can also superheat it. In fact, water can be superheated and often is when it is put in a microwave. The temperature goes above the boiling point, but we grab it before vapor has time to form, and there's little around to help the transition along, so it stays a liquid.

- Superheating and supercooling is possible for all materials. For the case of water, it can be cooled down to nearly −50 degrees Celsius before it is simply no longer possible for it to stay a liquid at atmospheric pressure, which is the temperature at which the material no longer needs any external help from nucleation sites but will simply nucleate on its own. So, you can never supercool below that temperature.

Involving Other Thermodynamic Variables

- Measuring heat transfer for a process at constant pressure gives us directly the enthalpy of the process, and this includes the phase transitions. In fact, at constant pressure, the latent heat of a phase transition is exactly equal to the enthalpy change during that phase transition, which is often called the enthalpy of melting or the enthalpy of vaporization.

- But what about the changes in other thermodynamic variables, such as entropy, occurring in a system at these phase transitions? From the reversible process definition of the entropy, we know that δQ is equal to temperature (T) times the change in entropy (dS).

- Suppose that we perform a constant-pressure experiment, heating a sample through its melting point. The heat released upon melting is the enthalpy of melting, so dH is equal to TdS, and we can integrate both sides of this equation over the heat required to make the phase transition occur.

- Because the temperature does not change during a phase transition, this gives us the following: The change in enthalpy at a phase transition equals the temperature of that phase transition times the change in entropy for the phase transition. Just by measuring the heat added at constant pressure, we can get the entropy change as a material is heated through its different phases.

- The entropy for water as a function of temperature is always increasing with temperature. In addition, just as with the heat or enthalpy change, we have jumps in the entropy as well at the phase transition. These are often called the entropy of fusion or the entropy of vaporization.

Controlling Pressure

- Other than temperature, what properties of a material can we use to control what phase a material exhibits? There are many ways to push on materials with different thermodynamic driving forces—chemically, electrically, magnetically, optically, and so on—but apart from temperature, one of the most important variables we can control to change phases is pressure.

- In the temperature-versus-pressure phase map, we examine a plot of the different phases of water with temperature on one axis, and in this case, instead of heat, we have pressure on the other.

- The phase a material exhibits depends on the state of the material, and a state is a unique set of values for the variables that describe a material on the macroscopic level. These are the state variables for the system. Because we're dealing with just water by itself, we have only one component in our system; we're not mixing water with anything else, at least not yet.

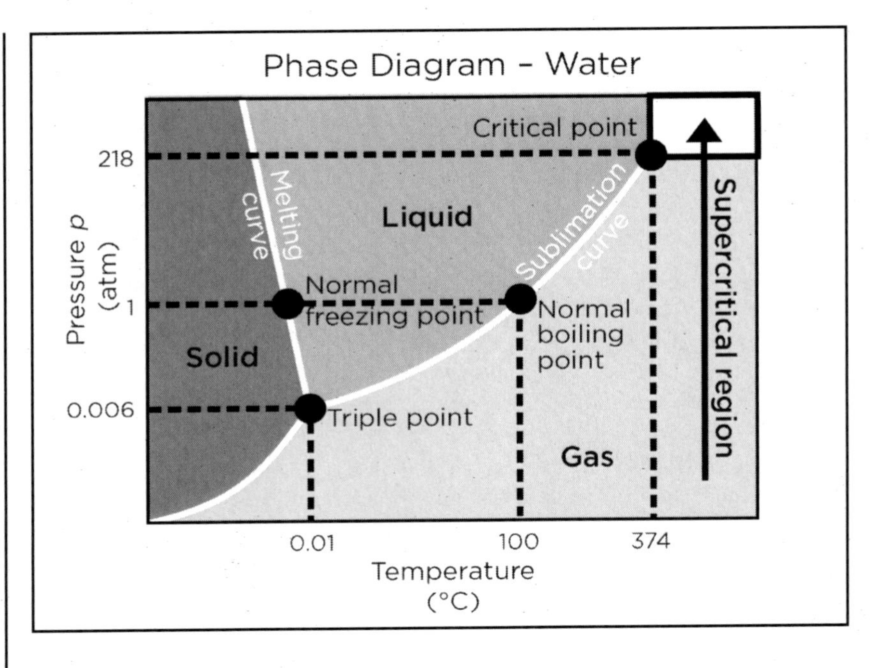

- So, that means that knowing three thermodynamic variables uniquely defines the state of a given closed system. Usually, these variables are pressure, temperature, and volume. Assuming that there are a fixed number of water molecules, then one of the most useful plots is that of the equilibrium phases of water versus the temperature and pressure.

- These phase maps only tell us about the equilibrium properties of matter. And as we've learned, in a closed system with independent thermodynamic variables pressure and temperature, it is the Gibbs free energy that determines what phase is in equilibrium.

- Each phase has a free-energy value at a given pressure and temperature, and the one that is the minimum energy at a particular temperature and pressure will be the equilibrium phase. A plot of the different phases a given system can exhibit at different thermodynamic variable states is known as a phase diagram.

- The pressure-versus-temperature phase diagram for water shows that as we lower the pressure, the freezing point of water remains about the same, but the boiling point gets significantly lower.

- If we go to extremely low pressures, such as 0.001 atmospheres, then there's not even the possibility for a liquid phase of water, at any temperature. That whole phase just disappears, and as temperature is increased, we go straight from solid to gas—a phase transformation called sublimation.

- In the case of water, increasing the pressure will lead to much higher boiling points. For example, at 100 atmospheres, the boiling point is 300 degrees Celsius.

Suggested Reading

Callen, *Thermodynamics and an Introduction to Thermostatistics*, Chaps. 2–3.

Callister and Rethwisch, *Materials Science and Engineering*, Chap. 9.

DeHoff, *Thermodynamics in Materials Science*, Chaps. 4–5 and 10.

Questions to Consider

1. What is a material that exhibits all three phases of matter at standard living conditions?

2. How can it possibly be energetically favorable for water to be below its freezing point and not freeze?

3. Where does the heat come from to rapidly freeze the water when it transforms from being supercooled into a solid?

4. What properties of a material can we use to control what phase the material exhibits?

5. What is latent heat?

Phases of Matter—Gas, Liquid, Solid
Lecture 14—Transcript

We've already mentioned the word "phase" a number of times in this course. Matter can exist in a variety of phases. To be precise, a phase is a uniquely identifiable form of a material separated by other forms by an interface. The phases we encounter most often are gases, liquids, or solids of differing crystal structures.

I've talked about how thermodynamic properties can vary dramatically from one phase to another of the same material, and how at those transition points between phases, some properties are not even well defined. But so far, the phases of materials have been discussed in terms of some other important thermodynamic concept, like say, heat capacity or thermal expansion.

Today, I'll make the phase itself the topic of discussion. Folks, this is big time, because ultimately, as I talked about way back in the introduction to the course, it is the phase maps of materials that represent one of the most important outcomes of thermodynamics. These maps, also called phased diagrams, they provide enormous amounts of information about a given material, or a given composition of different materials.

Let's begin with a question. What is one of the very few materials that we are all intimately familiar with in all three of its phases? Most materials only exhibit a single phase in our normal living conditions. However, there is one material familiar to everyone that exhibits multiple phases at standard living conditions. Quite bluntly, if you weren't familiar with it, you would be dead. The material is dihydrogen monoxide, or more commonly, water.

We can't live without knowing and drinking it in its liquid phase, and most of us have also held ice cubes and felt steam. And we also know how to change water from one phase to another. Lower its temperature to zero degrees Celsius and it becomes ice, or, raise it above 100 degrees Celsius to make it steam.

But how rigid are these temperature boundaries around the phases? Come take a look at how I've cheated the freezing point boundary and made something called super-cooled water. This is a really fun demo and something you can all try at home.

Okay, so in this demo we're going to talk about what happens when you super cool water. First of all, here, I have just some regular water. When I pour it, well, what happens? What happens is you get regular water in the bowl. But, here is water that's been cooled below its freezing point. When I pour this water into the bowl, what happens is something very different. It, as you can see, as I pour it and it hits the bottom of the bowl, it actually turns into ice.

How is that possible? How did that happen? Well, it's because I was able to bring the temperature of the water down below the actual freezing temperature, the phase transition point, of that liquid. It needed a trigger to get out of that metastable state. That's why it didn't just convert into ice while it was inside the bottle.

The trigger in this case was just pouring it into the bowl and having it strike the bottom of the bowl. And you could see that as it did so, it froze instantly. And another thing I could do instead of pouring the water and watching it turn into ice as it pours, is I could just trigger it by shaking it or hitting the bottle and then all the water inside of the bottle will turn into ice instantly. Let's take a look at that.

So I'm just going to hit it, hit the bottom of the bottle to trigger it. And then you can see that the whole bottle of water just, essentially, froze instantly, and it went back to the phase that it wants to be in, which is, solid ice.

So there you saw an example of how we can push materials into being metastable, that means, it's not exactly in its happy place; and any small nudge could send it out of its current phase and into another, happier, or, in more technical terms, lower energy state.

By the way, did you know that there isn't just one form of ice? Remember, that for solids, we can have different phases, since the atoms or molecules can arrange in different crystal structures. Diamond and graphite are a well-known example of this—two different materials made up of exactly the same atom, carbon, but just arranged differently in its repeating lattice structure. And what a difference that makes! We don't usually spend thousands of dollars on a graphite ring for someone we love; that would be kind of strange, since graphite is the core ingredient to most pencils. But in the case of water, we have not just one, but rather a whopping 15 different phases of solid ice, each one with a different crystal structure and density.

Now I'm going to spend most of today talking about one of the simplest ways to change a material from one phase to another using a single control knob, namely, by changing its temperature. This leads us to our very first phase map, where we examine the dependence of the phase of a material on its temperature.

In our lecture about heat capacity we looked at where energy from heat actually goes in a material, how it gets distributed among the various degrees of freedom. And we talked about how heat capacity is defined as the change in heat added, divided by the change in the resulting temperature of a material. That's just the slope of a plot of heat added vs. temperature. And while we did talk about how the heat capacity varies considerably from one phase to another, we did not examine closely what happens for this plot across a range of phases. Let's do that now for the case of water.

Here's a plot of heat vs. temperature for a fixed pressure of one atmosphere. I'll start at a low temperature, say minus 50°C, where we know water is a solid. As we add heat to the solid ice, its temperature increases, and it goes fairly linearly with a slope equal to the heat capacity; in this part of the diagram, note that it will be the heat capacity of the solid phase.

Now, as we approach the melting point of water, at 0°C, we find something very interesting occurs. Namely, I keep adding heat to the system, but the temperature does not change. In fact, I need to add a whole lot of heat before the temperature starts changing again.

On the plot, this results in a vertical line at the temperature corresponding to the phase boundary. Once we add enough heat, the temperature, again, starts to increase linearly with heat added, just as before, although now with a different slope corresponding to the heat capacity of the liquid phase. The same type of transition then occurs at 100°C, where water undergoes another phase transition.

So what's going on here? We can see from the plot that in order to make the material go from one phase to another, a lot of heat needs to be added. This heat is not raising the temperature of the material, but rather, it is doing something else to make it change its phase. What is it doing?

Well, to answer this, let's do something we like to do when it's time to get all fundamental and gain some understanding. Let's cruise down to the scale of the atom. At this scale, we can see those water molecules up nice and close. We can see the one oxygen and two hydrogen atoms for each molecule, and if we look closely, we might be able to see that the oxygen atom grabs some of the charge away from the hydrogen atoms. This makes oxygen have a slight negative charge, while the hydrogen atoms have a slight positive charge.

Now, as we know, opposite charges attract, so this charge redistribution in the molecule is what drives the bonding behavior in the solid; the negatively charged oxygen from one molecule wants to find and get close to the positively charged hydrogen atoms from another molecule. Those types of bonds, by the way, are called hydrogen bonds, and they are the key type of bonding in water.

So, here we are in the solid phase of water, and we start to add heat. Remember from our discussions of heat capacity and thermal expansion, that as I add heat to a solid, what happens is that the energy goes into the vibrational degrees of freedom. So for this case, those molecules start grooving and shaking more and more vigorously. That is, until the material reaches the temperature of its phase change.

Up until that point, the hydrogen bonding between molecules keeps them all together, even though they are vibrating with more and more energy. But, at the phase-change temperature, while this grooving and shaking doesn't stop, something else happens. That temperature is enough so that the hydrogen bonding itself starts to break apart. And since that is the bonding that holds the positions of those molecules fixed on average, it means that the molecules can start to slip and slide out of their fixed crystalline positions. And so we have a liquid.

Notice that the heat energy we added to break up the hydrogen bonds did not go into anything related to the kinetic energy of the system. It did not go into motion. Remember, the three types of degrees of freedom—translation, rotation, and vibration? At the phase transition the heat goes into none of those degrees of freedom. Rather, the heat energy went entirely into raising the potential energy of the system. That's what happens when we break apart those hydrogen bonds. That's why the temperature during the phase transition doesn't change, even though we're adding heat.

And what a lot of heat we needed to add! To go from solid to liquid at 0°C, so, without changing the temperature by a single degree, but just changing the phase, we need to add an amount of heat equivalent to that needed to raise the temperature of water as a liquid by 80°C! It's literally enough energy to get water from the tap all the way up to its boiling point.

And for the next transition, from liquid to vapor, even more thermal energy is needed to change the phase. With the temperature fixed at 100°C, we need to add an amount of heat equivalent to raising the temperature by 540°, just to change from liquid to vapor. This type of heat that gets a material to change phase is referred to as latent heat. For water, it's 80 calories for the solid-to-liquid transition and 540 calories for the liquid-to-vapor transition. Now, sometimes you might hear other terms used for these phase-transition energies, like the heat of fusion for going from solid to liquid, or, the heat of vaporization for going from liquid to vapor.

The discovery of latent heat is actually quite interesting. As with so many scientific discoveries, it was discovered in the context of trying to make sense out of nature and of the world in general. It goes back to the 1700s when it was thought that once a material was uniformly heated to its melting point, the addition of only a tiny additional amount of heat would melt the entire sample. In other words, as soon as heat was added to raise the temperature of the material above the melting point, the phase transformation spontaneously occurred with no extra input. In that case, a graph of heat added vs. temperature for the melting ice would have looked something like this, with the solid and liquid lines each having a slope still equal to their corresponding heat capacities, but now connected at 0° continuously, with no latent heat. This was the view of how materials changed their phase back in the 1700s.

But, for Joseph Black, who was a professor of medicine at the University of Glasgow, this did not make any sense, and he wrote the following observation:

> "If we attend to the manner in which ice and snow melt when exposed to the air of a warm room, or when a thaw succeeds to frost, we can easily perceive that they soon warm up to their melting point and begin to melt at their surfaces.
>
> If the complete change of them into water required only the further addition of a very small quantity of heat, the mass, though of a considerable size, ought all to be melted within a very few minutes or seconds by the heat from the surrounding air. Were this really the case, the consequences of it would be dreadful in many cases; for, the melting of large amounts of snow and ice occasions violent torrents and great inundations in the cold countries or in the rivers that come from them. They would tear up and sweep away everything, and this so suddenly that mankind would have great difficulty in escaping their ravages. This sudden liquefaction does not actually happen…"

And so, latent heat was discovered!

In the context of what we know to be the correct heat vs. temperature plots, ones that include latent heat, let us now return to our demo of super-cooled water. I'll start over here and start to cool down the water, so that means that I'm taking heat out of the material. Now, what happens in the case of super cooling, is that, when I reach the phase boundary temperature, instead of stopping the temperature drop and changing the phase, we simply continue along this same line going below the solidification temperature.

How is this possible? Well, it turns out that these boundaries separate the equilibrium phases of the material. Remember, way back when, we talked about how thermodynamics is concerned with equilibrium properties of materials? That means, we take a kind of, "eventually it will happen" approach to materials. Thermo tells us all about what a material would like to be like if it could have its way—that's what I call equilibrium.

But what thermodynamics does not go into is the kinetics of different processes. As we talked about in our lecture on equilibrium, coffee would like to spill out of a cup, but the boundaries of a cup prevent it from doing so. And those expensive diamond rings would like to turn into graphite rings, since graphite is thermodynamically favored at atmospheric pressure, but it doesn't, because there are barriers that prevent such a transformation, well, at least on short time scales. Eventually, that is, after perhaps hundreds of thousands or millions of years, those diamond rings will, in fact, become good, very expensive pencils, since again, "eventually it will happen." Diamond is, therefore, a metastable phase of carbon.

So, for the transition of water into ice, we can also have metastable phases, and that's exactly what we have with water that remains as a liquid blow 0°C. It wants to be a solid, but turning into a solid from a liquid takes time, and it gets helped by having what are called nucleation sites. These are tiny little places in the system where the liquid can get stuck, forming a tiny little piece of solid, which then coaxes more liquid to form more solid, and so on.

That's why, in order to make super-cooled water, we needed to use distilled water, with no impurities, like trace amounts of other elements that are usually in water and help solidification along by serving as nucleation sites.

We also used ultra-smooth plastic containers so that there was less chance of nucleation at the container walls. By doing this, and by very, very carefully not shaking the water while it cools, we were able to keep it from transitioning to a solid, even though we measured a temperature of −10°C. I had to use a cooler with salted ice water in order to cool it, instead of a freezer, because the vibrations that come from the compressor turning on and off are enough to provide the nucleation of water to ice and the phase transition to go. Now again, I want to emphasize that, eventually, this super-cooled water would have turned into a solid, since that is its thermodynamic equilibrium, but we were able to trick it for a bit by taking away some of things that help the phase change occur.

The reason I wanted to spend time on this point, is that I want you to keep in mind that as we make phase maps of materials, the boundaries we draw between phases come from thermodynamics, and therefore, they are always referring to the equilibrium properties of the system. It does not mean that a material cannot exist in a different phase than is indicated in our phase map, since it can be kinetically trapped, but, it does tell us what the preferred equilibrium phase will be. And that's the starting point for any study of any material.

Now, let's go back to where the water wound up on this heat vs. temperature plot in our demo, since I'd like to make one last point about it. When it's up here, as I just said, the material is only metastable. It wants to get down to here, where it's in its happy equilibrium place. So, how does it get back?

Well, you can see that it can basically do one of two things; it could drop straight down, keeping its temperature constant and releasing a whole bunch of heat; that would be called an isothermal transition, since the temperature is constant. Or, the super-cooled water could increase its temperature in another way until it reaches the freezing point, and then be back on this equilibrium curve and drop down to ice in the usual path. In that second case, it would be called an adiabatic transformation back to equilibrium, since it occurs without any heat transfer. Either way, the water winds up freezing into a solid, but the two paths are different in nature, and which path is actually taken will depend on the conditions under which the transformation takes place.

And some of you may be wondering at this point, if I can super cool a material, couldn't I also superheat it? Absolutely. It just goes the other way on this same plot, as you can see here. In fact, water can be superheated and often is when it's put in a microwave. The temperature goes above the boiling point, but we grab it before vapor has time to form, and there's little around to help the transition along, so it stays a liquid. Place a tea bag into superheated water, and a massive rush of water will boil instantly, but watch out, since that can actually be quite dangerous!

Superheating and supercooling are possible for all materials. For the case of water, it can be cooled down to nearly −50°C before it's simply no longer possible for it to stay a liquid at atmospheric pressure. That's the temperature at which the material no long needs any external help from nucleation sites but will simply nucleate on its own; so you can never super-cool below that temperature. Apart from water, take a look at just how far below the solidification temperature we can push these different materials. In the case of the element niobium, we can cool the liquid down to 525°C below its solidification temperature!

Okay, I've been talking about these simplest forms of phase maps, plotting heat against temperature. In the backs of your minds, you might have been thinking, but I know that heat is also the same as enthalpy for a constant pressure process. And you'd definitely be on to something here. Measuring heat transfer for a process at constant pressure gives us directly the enthalpy of the process. And this includes the phase transitions. In fact, at constant pressure, the latent heat of a phase transition is exactly equal to the enthalpy change during that phase transition, which is often called the enthalpy of melting, or the enthalpy of vaporization.

But what about the changes in other thermodynamic variables occurring in a system at these phases transitions? Well, in fact, I can use the relations I've been drawing to work my way over to other important thermodynamic variables, like entropy! More precisely, we know from the reversible process definition of the entropy that δQ is equal to temperature, T, times the change in entropy, dS.

Suppose we perform a constant-pressure experiment, heating a sample through its melting point. The heat released upon melting is the enthalpy of melting, so dH is equal to TdS, and I can integrate both sides of this equation over the heat required to make the phase transition occur. Since the temperature does not change during a phase transition, this gives me the following expression: the change in enthalpy at a phase transition equals the temperature of that phase transition times the change in entropy for the phase transition. And isn't that a glorious moment? Just by measuring the heat added at constant pressure, in a simple open-to-air experiment, we can get the entropy change as a material is heated through its different phases.

Take a look at the entropy for water as a function of temperature. Notice first that it's always increasing with temperature, which we know must be true. And also notice that just as with the heat or enthalpy change, we have jumps in the entropy as well at the phase transitions. These are often called the entropy of fusion, or the entropy of vaporization.

So, here's our next question, What properties of a material can we use to control what phase the material exhibits? I've been focused on just one single property, and that would be the temperature. Put water in the freezer to get ice or on the stove to get steam. But other than temperature, how can we change the phase of a material? Well, remember that there are a lot of ways to push on materials with different thermodynamic driving forces—chemically, electrically, magnetically, optically, and so on. But apart from temperature, one of the most important variables we can control to change phases is pressure.

In the temperature vs. pressure phase map, we examine a plot of the different phases of water with temperature on one axis, and in this case, instead of heat, we have pressure on the other. Backing up for just a moment, remember that the phase of a material depends upon the state of the material. And, as I'm sure by now you happily aware, a state is a unique set of values for the variables that describe a material on the macroscopic level.

These are the state variables for the system. Now, since we're talking about just water here by itself, we have only one component in our system; we're not mixing water with anything else, at least not yet. So, that means that knowing three thermodynamic variables uniquely defines the state of a given closed system. Usually these variables are pressure, temperature, and volume. Assuming there are a fixed number of water molecules, then one of the most helpful and useful plots is that of the equilibrium phases of water versus the temperature and pressure.

Now, let's think back to our lecture on equilibrium, since remember, these phase maps only tell us about the equilibrium properties of matter. And of course, as soon as I say the word equilibrium, I'm sure your hearts and minds immediately turn to one thing, the Gibbs free energy. As we've learned, in a closed system with independent thermodynamic variables—pressure and temperature, it is the Gibbs free energy that determines what phase is in equilibrium. Each phase has a free energy value at a given pressure and temperature, and the one that is the minimum energy at a particular temperature and pressure will be the equilibrium phase.

A plot of the different phases a given material can exhibit at different thermodynamic state variables is known as a phase diagram. So here's the pressure vs. temperature phase diagram for water. Now, what we're all pretty used to is the single line at a pressure of one atmosphere; that goes across like this—from solid to liquid to gas. That gives us the freezing and boiling temperatures of 0°C and 100°C that we all know and love, and it's the line we've been talking about for this whole lecture in all of our pressure vs. temperature graphs. Remember, we have assumed a constant pressure in all of our analysis thus far. But that line is just for a single pressure.

Notice how much the properties and phase boundaries change as I change the pressure. For example, as I lower the pressure, the freezing point of water remains about the same, but the boiling point gets significantly lower. If you live in Denver, Colorado you're up about one mile above sea level, and the pressure in the air is about 0.83 atmospheres. According to our phase diagram, this means that, in Denver, water boils at around 95°C, so plan to boil those potatoes a little while longer, because they'll need the extra time. But if you go even higher, like, say, to the top of Mount Everest, you'll be at about $^{1}/_{3}$ the atmospheric pressure of sea level, and again, from this phase diagram, we can see that water will boil, in that case, at around 70°C. I wouldn't even try to boil the potatoes up there; just eat them raw.

Now, check this out. If I go to even lower pressures, like say, 0.001 atmospheres, then there's not even the possibility for a liquid phase of water—at any temperature. That whole phase just disappears, and as temperature is increased, we go straight from solid to gas. That kind of phase transformation, from solid to gas, is called sublimation, as opposed to, say, freezing, melting, or boiling.

By the way, have any of you seen dry ice? That's actually got nothing to do with water, so, sometimes people get confused because they associate the word ice with water. Actually, dry ice is the solid phase of carbon dioxide, or

CO_2. How does it work? Well, let's take a look at the phase diagram of CO_2 to find out. See, in this case, at a pressure of one atmosphere, that's back to our standard sea level pressure, as we increase the temperature, we see that we have the same behavior as for water at very low pressures, namely, there is no liquid phase, and the material goes from solid straight to gas. That's why dry ice is so useful, for example, in shipping something that has to be kept cold. It can keep things cold—quite cold in fact, since its sublimation temperature is negative 78°C at one atmosphere—without becoming a liquid as water would, which would, in turn, ruin the packaging or container.

And notice from this phase diagram that for CO_2, if I were to increase the pressure to, say, 10 atmospheres, well, now the liquid phase becomes possible for certain temperature ranges. For the case of water, increasing the pressure will lead to much higher boiling points. At 100 atmospheres, the boiling point is now 300°C! And you can see that even more strange things happen at higher pressures, like the fact that this entire phase boundary line disappears altogether. What is that all about? Well, to find out you'll need to wait few lectures.

But for now, take in the fact of just how important this type of diagram is and how much information it contains. We can go even higher in pressure and temperature too, and the information keeps coming. There are many materials with lots of phases and various boundaries between phases, and we'll be learning a lot more about them in the coming lectures. But to end this lecture, let's think about how water is at the heart of all life, and just look at how lucky we are. Check out where Earth's standard operating conditions are in this phase diagram. Unlike almost any other material, we sit right in the perfect spot, where the solid, liquid, and gas are all nearby. And that, my friends, is a beautiful thing.

Phase Diagrams—Ultimate Materials Maps

Lecture 15

In the last lecture, you learned about the phases of matter and what happens to the phase of a material as heat is added to it. You also were exposed to maps of phases where both temperature and pressure can vary. In this lecture, you are going to travel more deeply into the thermodynamics that is under the hood of such phase diagrams. What you learn in this lecture will prepare you to consider the properties of mixtures of materials in multicomponent phase diagrams, which are presented later in the course.

Vapor Pressure

- Suppose that we're almost at the point where a liquid is going to turn into a gas. In order to become a gas, the water molecules in the liquid need to gain enough kinetic energy so that they can break free from their liquid bonds to become a molecule in the gas phase. But that act of jumping away from the liquid will be hindered if there are a lot of atmospheric gas molecules bumping them back down, pushing on them into the liquid, applying a pressure.

- The reason water boils at lower temperatures when the pressure is lower is that there are fewer gas molecules to counter the desire of the liquid water molecule to hop out. In the most extreme case—for example, in outer space, where the pressure is essentially zero—nothing is there to hold the water back, and the liquid water would immediately vaporize as a result (at any temperature, even in the cold darkness of deep space).

- At higher pressures, on the other hand—for example, higher than the one atmosphere at sea level on Earth—there's more to push down on the liquid, hindering the boiling and making it so the liquid molecules need more kinetic energy and, therefore, a higher temperature to escape.

- Let's focus on the phase boundary between liquid and gas. For any given temperature, there is a range of velocities that the atoms and molecules have. In the case of water at room temperature, which is around 25 degrees Celsius, the average velocity is about 600 meters per second, or roughly 1,300 miles per hour. Indeed, just because the molecules are in a liquid, it doesn't mean they don't move around quite fast.

- It's not that every water molecule moves at exactly this speed; in fact, there is a very broad distribution of speeds. Some molecules are moving much, much faster, even two or three times as fast, while others are barely moving.

- Even at room temperature, there will always be some molecules with enough kinetic energy to escape the bonding in the liquid and fly out into the air. Whenever a water molecule leaves the liquid, it becomes a water molecule in the gas phase, which we also call water vapor.

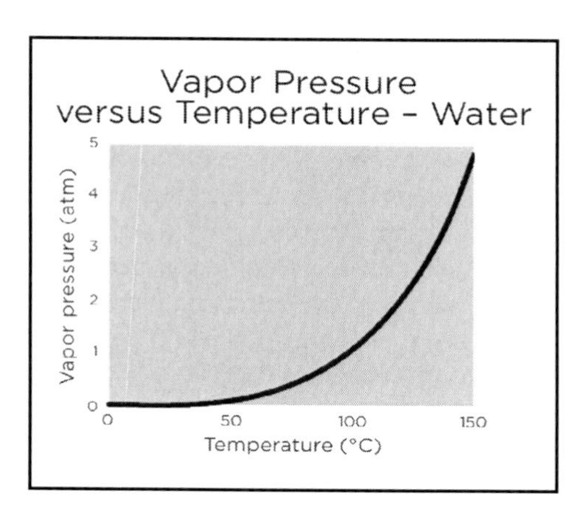

- At any given point, there will be a bunch of these vapor molecules hanging out above the liquid surface. In fact, just as some of the liquid molecules can gain enough kinetic energy to break free of the liquid, some of those vapor molecules may have just the right collision to make them lose enough kinetic energy so that they really want to fall back into the liquid. This is the real picture at the phase boundary: molecules constantly going back and forth between the liquid and vapor phases.

- The vapor pressure is the pressure at which the vapor and the liquid are exactly in equilibrium with one another. So, the gas molecules are going back into the liquid at exactly the same rate as the liquid molecules are going into the gas. The vapor pressure of a material is highly dependent on the temperature, the type of bonding that takes place in the liquid, and how heavy the atoms or molecules are.

- In a vapor-pressure-versus-temperature phase map for water, the pressure increases—in fact, exponentially—with temperature. At the boiling point of water, the vapor pressure increases to exactly one atmosphere.

- And this is the definition of boiling: When the vapor pressure is larger than the atmospheric pressure, then the atmosphere no longer has enough strength to hold that vapor near the surface. The vapor is able to win a tug-of-war with the atmosphere, and it can freely break out and go off to see the world. At that point, the vapor near the surface can leave that vicinity, which leaves room now for the liquid to send more molecules out into the vapor phase.

- The way in which pressure from the atmosphere influences the liquid is not direct. Rather, it pushes on the vapor phase of the liquid, which is sitting right above the surface. The atmospheric pressure is holding those vapor molecules near the surface of the liquid. For large atmospheric pressures, there's a lot of force to hold the vapor down, and with low pressures, there isn't much force to push on the vapor.

- Boiling is not the same thing as evaporation. Evaporation is a process that only occurs on the very surface of the liquid. Unlike evaporation, boiling is a phase transformation that involves the entire liquid, from the top to the bottom of the container.

- Furthermore, once water is boiling, if we turn the heat up, it does not get any hotter while it's boiling; instead, the water boils more rapidly. Because the material is undergoing a phase transition, its temperature will remain constant—more bubbles more quickly, same temperature.

The Triple Point

- The triple point is a very special point in a phase diagram: It is the single point at which all three phases of a substance (gas, liquid, and solid) meet. We know that we are near the triple point in the phase diagram when the liquid is both boiling and freezing simultaneously.

- The triple point happens at one point and nowhere else. It's different than, for example, a phase change, where the boundary can be moved around by changing pressure or temperature. There's no moving around of the triple point.

- The process of boiling is one that releases heat from the material. The higher kinetic energy molecules leave, so the average temperature gets just a tad lower each time a molecule from the higher end of the kinetic energy spectrum leaves.

- With no heat source feeding thermal energy into the system, this means that boiling lowers the temperature of the liquid. A little pressure drop leads to more boiling, in turn leading to a temperature drop, until we hit that singular point where all three phases coexist in one happy place.

The Gibbs Phase Rule

- When a closed system (that is, one with fixed mass and composition) is considered and there is no mass transport allowed across the boundaries, the system is in stable equilibrium when it has the lowest possible value of the Gibbs free energy. The Gibbs free energy is connected to the equilibrium of processes occurring under laboratory conditions, such as constant temperature and pressure. A change in G, dG, is equal to volume times dP minus entropy times dT: $dG = VdP - SdT$. Because both T and P can control the equilibrium by changing G, changing T and P can control equilibrium.

- The Gibbs free energy of a closed system with independent thermodynamic variables pressure and temperature determines what phase is in equilibrium. Each phase has a free-energy value at a given pressure and temperature, and the one that is the minimum energy at a particular temperature and pressure will be the equilibrium phase. This is how we construct pressure-versus-temperature phase diagrams.

- Many materials exist in multiple forms as a function of the environmental conditions. Solids melt, liquids boil, and solids can sublimate directly to vapor. But materials also exhibit different crystalline forms in the solid state, depending on the conditions of temperature and pressure.

- Solid water, or ice, has 15 distinct crystal structures. These crystal structural variants are known as polymorphs or allotropes, where the word "allotrope" is typically reserved for different structural forms of pure elements, and "polymorph" more generally covers different structures of any substance.

- One of the most important aspects of a phase diagram is to know how many degrees of freedom there are at any point in the plot. The rule for this is called the Gibbs phase rule, which states that the number of degrees of freedom, F, is related to the number of components, C, and the number of phases we have in equilibrium, P, by the following relation: $F = C - P + 2$. (The 2 comes from the fact that the phase diagram has 2 independent variables: pressure and temperature.)

- The reason it's useful to know what F is anywhere in the plot is that the number of degrees of freedom tell us how many different control knobs we have available for changing the material into a different phase. The Gibbs phase rule determines how many phases can be in equilibrium simultaneously and whether those phases are stable along a field, a line, or a point in the phase diagram.

- Because it is the Gibbs free energy that determines the equilibrium phase, it is extremely useful to be able to go back and forth between a phase diagram and the free-energy curves for the various phases.

- Phase diagrams contain a lot of information. These materials maps capture all of the Gibbs free energy behavior for the different phases, as a function of any line at constant pressure or temperature.

Suggested Reading

Callen, *Thermodynamics and an Introduction to Thermostatistics*, Chap. 9.

DeHoff, *Thermodynamics in Materials Science*, Chaps. 4–5, 7, and 10.

Questions to Consider

1. What thermodynamic property governs the phase of a material?

2. What is the microscopic picture of water molecules at the surface of a pot of boiling water? How would you describe pressure and temperature for this system?

3. Explain the process of freeze-drying, using the pressure-versus-temperature phase diagram of water.

4. Why is the triple point of a material of technological importance?

Phase Diagrams—Ultimate Materials Maps
Lecture 15—Transcript

Hi there, and welcome to our lecture on phase diagrams. In the last lecture, we talked about the phases of matter, and we worked our way through what happens to the phase of a material as we add heat to it. In the end, we looked at a few maps of phases where both temperature and pressure can vary. In this lecture, we're going to travel more deeply into the thermodynamics that is under the hood of such phase diagrams. We're going to stick with a single component, as we did last time, but the work we do today will prepare us to go even further later on, as we consider the properties of mixtures of materials in multi-component phase diagrams.

So, today, we're going to ask the question why. Why does ice become water above 0°C? And if we continue heating, why does water become vapor above 100°C? What quantity governs their equilibrium at different temperatures?

First, let's go with an intuitive explanation. We talked about the pressure vs. temperature phase diagram for water at the end of the last lecture. And we saw how the phase can change with either temperature or pressure changes. Remember how water boils at a much lower temperature at the top of Mount Everest? We chatted about that last time, and I said it was because it's at a lower pressure up there, and as you can see from the phase diagram, at lower pressures the transition between liquid and gas for water moves to lower temperatures.

But why does it do that? Well, let's get some intuition using this particular phase transition as an example. Let's imagine the water molecules in their liquid phase, say, sitting in a pot on a stove. Now, above the surface of the liquid, sits air, which really just means mostly a bunch of nitrogen and oxygen dimers in their gas phase. Yes, there's also a touch of Argon atoms, and some CO_2, and trace amounts of other gases too, but nitrogen and oxygen certainly are the 800-pound gorillas of our atmosphere.

So, these gas molecules hanging out above the water are moving around very fast, bouncing around in all sorts of ways. And when they bounce against the water molecules in the liquid, well, that exerts a force onto the water, a force we know and love, which we call pressure. So, that's what we're talking about when we say there's a pressure of 1 atmosphere, or 0.01 atmospheres, or 10 or 100 atmospheres. We're talking about how much of this atmospheric gas there is bumping into the top of that liquid, how fast it's bumping, and overall, what sort of force that exerts per area onto the liquid's surface.

Now, think about that for a moment. Let's suppose we're almost at the point where the liquid is going to turn into a gas. For water at a pressure of 1 atmosphere, that would mean a temperature just shy of a 100°C. In order to become a gas, the water molecules in the liquid need to gain enough kinetic energy so that they can break free from their liquid bonds and go off and become a happy gas-phase molecule. But that act of jumping away from the liquid will be hindered if there are a lot of atmospheric gas molecules bumping them back down, pushing on them into the liquid, applying a pressure.

That's the reason why water boils at lower temperatures when the pressure is lower; there are fewer gas molecules to counter the desire of the liquid water molecule to hop out. In the most extreme case, say, in outer space, where the pressure is essentially zero, nothing is there to hold the water back at all, and liquid water would immediately vaporize as a result—at any temperature, even in the cold darkness of deep space. At higher pressures on the other hand, say, higher than our one atmosphere at sea level back here on Earth, there's more to push down on the liquid, hindering the boiling and making it so the liquid molecules need more kinetic energy, and therefore, a higher temperature to escape.

Now, there's a pretty important point here, and that would be that the real story is a little bit more complicated than what I just described. I need to introduce the concept of vapor pressure in order to get our intuitive picture completely correct.

So, let's zoom in to that phase boundary between liquid and gas. In our temperature lecture, we discussed how, for any given temperature, there is a range of velocities that the atoms and molecules have. So, for water at, say, room temperature, which is around 25°C, the average velocity is about 600 meters per second; that's roughly 1300 miles per hour. Indeed, just because the molecules are in a liquid, it doesn't mean they don't move around really fast!

But the point I want to make is that it's not like every water molecule moves at exactly this speed. In fact, there is a very broad distribution of speeds. Some molecules are moving much, much faster, even two or three times as fast, while others are barely moving at all. Notice that even at room temperature, there will always be some molecules with enough kinetic energy to escape the bonding in the liquid and fly out into the air. Whenever a water molecule leaves the liquid, it becomes a water molecule in the gas phase, which we also call water vapor.

Now, you may have noticed that I've used gas and vapor interchangeably. The thing is, that at any given point, there will be a bunch of these vapor molecules hanging out above the liquid surface. In fact, just like some of the liquid molecules can gain enough kinetic energy to break free of the liquid, likewise, some of those vapor molecules may have just the right collision to make them lose enough kinetic energy so that they really want to fall back into the liquid. This is the real picture at the phase boundary—molecules constantly going back and forth between the liquid and vapor phases.

And this is where the concept of vapor pressure comes in. The vapor pressure is the pressure at which the vapor and the liquid are exactly in equilibrium with one another. So, the gas molecules are going back into the liquid at exactly the same rate as the liquid molecules are going into the gas. And you can imagine that the vapor pressure of a material will be highly dependent on the temperature, on the type of bondings that takes place in the liquid, and, on how heavy the atoms or molecules are.

For example, at room temperature, the vapor pressure of water is around 0.03 atmospheres. This means that the atmospheric pressure at sea level of 1 atmosphere will be more than 30 times greater than the vapor pressure at room temperature. So the atmosphere is really able to hold that vapor down close to the liquid surface. And that's a critical point; if that water vapor cannot go anywhere because it's being held close to the surface, then it kind of sits around, in equilibrium with the liquid, not allowing more liquid to come out, except in this constant exchange with the vapor.

Now, take a look at the vapor pressure vs. temperature for water. You can see that the pressure increases, in fact, exponentially, with temperature. At the boiling point of water, the vapor pressure has increased to exactly 1 atmosphere. And this is, indeed, the definition of boiling. When the vapor pressure is larger than the atmospheric pressure, then the atmosphere no longer has enough strength to hold that vapor near the surface. In that case, the vapor is able to win against a tug of war with the atmosphere, and it can freely break out and go off and see the world. At that point, the vapor near the surface can leave that vicinity, which leaves room now for the liquid to send more molecules out into the vapor phase, that's boiling.

So, you can see that all of what I said before is true, regarding the effects of atmospheric pressure and how it impacts the liquid. It's just that the way in which pressure from the atmosphere influences the liquid is not direct. Rather, it pushes on the vapor phase of the liquid, which is sitting right above the surface. The atmospheric pressure is holding those vapor molecules near the surface of the liquid. So, for large atmospheric pressures, there's a lot of force to hold the vapor down, and with low pressures, there isn't much force to push on the vapor.

So that's our intuitive picture of how pressure and temperature are tied to one another at a phase transition. But since I've been talking about boiling and vapor pressure, I cannot leave the subject without mentioning that boiling is not the same thing as evaporation. Water will leave the surface at any temperature, even far below the phase transition temperature. That's something we're all pretty familiar with. If you leave a jar of water out in the open air, it will slowly evaporate until there's nothing left. So what's going on? I just mentioned that at lower temperatures the pressure from the atmosphere holds the vapor molecules around, so the liquid guys don't boil out until those vapor molecules can go where they want by overcoming the atmospheric pressure. And that is true in general, but, the atmosphere does not stay still.

Evaporation happens when the vapor molecules near the surface are brushed aside, maybe by a nice, fresh breeze, or simply because over time they find ways to escape. Either way, if some vapor molecules are taking away from the vicinity of the surface, then the liquid molecules, once again, will send some more up into the gas state, until the vapor pressure is reached. That is, once again, until the pressure of the gas phase is just right to balance the rate of molecules able to leave the liquid, something that, again, depends heavily on the temperature.

We can see the dependence of evaporation on the ability of the vapor molecules to find a way out quite clearly by putting a lid on that jar of water. In that case, water will initially evaporate, but eventually, when enough water vapor forms, it cannot float off freely away, and no breeze will carry it off. So, the vapor pressure remains the same, and the molecules condense back into the liquid at the same rate as the liquid evaporates them off into the vapor. In a sense, you could say that evaporation is still constantly happening, it's just that it's perfectly countered by condensation, so, no liquid actually disappears in this case.

One difference between evaporation and boiling is that evaporation is a process that only occurs on the very surface of the liquid. Even if a molecule deep down in the liquid did have enough kinetic energy to escape, if it's not near the surface, it will never make its way out, since along the way, it will bump into other liquid molecules, dissipating and trading its kinetic energy as it goes.

But unlike evaporation, boiling is a phase transformation that involves the entire liquid, from the top to the bottom of the container. In this case, water molecules anywhere in the system are ready to try to jump into the gas phase. That's why gas bubbles, the formation of which is actually quite complicated, and I won't go into the details here, those gas bubbles can and do form at the bottom of the pot, where heat energy is flowing into the system. These bubbles, however, must still be able to beat the pressure exerted by the atmosphere above the liquid in order for the gas to be able to leave the system.

And one last point on boiling water; once the water is boiling, if we turn up the heat, does it get any hotter while it's boiling? The answer is no, not at all. What does happen is that the water boils more rapidly. But I hope you know, by this point, that, since the material is undergoing a phase transition, its temperature will remain constant. More bubbles, more quickly, same temperature.

Okay, so that's our intuitive picture of why both pressure and temperature push and pull on different phases of matter in different ways. The P-T phase diagram helps us find our way around the interplay between these thermodynamic forces on the equilibrium phase. Take a look again at the phase diagram for water. We just talked about walking around one of these lines, the one that separates the liquid and gas phases shown here. Last time, we also talked about how, at low enough pressures, you wouldn't even have a liquid phase. The transition across this line here is called sublimation, as the solid phase transforms directly into the gas phase. I hope with our intuitive picture, you can now understand how this is possible.

At one atmosphere of pressure, as solid ice absorbs heat, its temperature increases, and at the phase transition, while the temperature is fixed, the internal energy increases. This is because the heat energy is flowing into breaking apart all those hydrogen bonds as we discussed last time. But, at low enough pressures, it becomes so easy for water molecules in the solid to leave the surface, that extra heat energy simply prefers to go into that process, as opposed to the process of breaking apart hydrogen bonds internal to the material. It's quite amazing to think about; just by changing the pressure, we've completely eliminated even the possibility for one of the phases to exist!

Actually, there's a very important technology that many of you probably know about and have experienced, that relies on just this process. It's called freeze drying. We know that what makes up most of the stuff we eat is composed of water, mostly, at least, of water. By taking the water out of foods, they can be made to last much longer, since bacteria and mold don't like to grow without water. Now, ideally, we want to dry out the food with minimal loss in nutritional content, as well as taste. Let's say I want to dry out an apple starting at room temperature. One way to do this would be to heat it up. But unfortunately, this leads to the destruction of the rest of the materials in the apple; we're simply cooking it, Eventually all the water will leave as gas, but the apple will also not be anywhere close to its original form. So that was going from our starting point here, over to the right on the phase diagram.

But, what if instead, we head to the left first in the phase diagram, and then down on the phase diagram? Now, we've frozen the water inside the apple first, which changes its volume a little bit, but not usually enough to cause much damage. Then, when I go down in pressure and perhaps increase the temperature a little bit, the ice will sublimate and turn directly into gas. Because the sublimation occurs only from the surface of the ice, it's a much gentler process, and the original apple structure, as well as its flavor, is preserved much better, I mean, apart from the fact that it's now all dried out, but that was the whole point in the first place.

Now, I wanted to bring back this phase diagram not just to mention sublimation and freeze drying, but also to talk about a very special point on the plot. It's something you've probably noticed already, and you may have noticed that I've been kind of ignoring it so far. Fear not, we're not only going to discuss it now, but I'm going to bring it to life in our demo. It's called the triple point, and it's where all three phases meet at a single point in the phase diagram. Let's take a look.

In this demo, we're going to get right up close to a really cool part of a phase diagram. It's called the triple point. And at that point, all three phases of a material—the gas, the liquid, and the solid—coexist at the same time.

So how are we going to do this? Well, what I have here in this bowl is tert-butyl alcohol. And that's an alcohol that actually has a phase transition from liquid to solid that's a little higher than water, so that means that it's easier to reach that transition temperature under normal, standard conditions.

What I'm going to do, is I'm going to lower the pressure, I'm going to put it in the vacuum chamber and lower the pressure. And what that does, is it removes force from the surface of the liquid, removes air molecules that are causing a force downward on the liquid. And that allows the liquid to start to boil. Let's see what happens when that happens. I'm going to put it in here and put this back on and turn on the pump. Let's take a look.

So the liquid is now boiling, but, the thing is, remember when we talked about boiling, that what we said was that it's the highest kinetic energy molecules in the liquid that are leaving it. And because of that, the temperature of the liquid, when it starts to boil, goes down. Let's see what happens when it keeps on boiling and the temperature also starts going down at the same time.

So there what you saw, is that the liquid for a little bit there was both boiling and freezing at exactly the same time. That is the triple point.

So that was totally cool. With the liquid both boiling and freezing simultaneously, we know we were near that triple point in the phase diagram. Actually, it's interesting to examine how the system got to the triple point.

So that was totally cool. With the liquid both boiling and freezing simultaneously, we know we were near that triple point in the phase diagram. Actually, it's interesting to examine how the system got to the triple point. First of all, why did I use something called tert-butyl alcohol instead of water? Well, take a look at the phase diagrams of these two materials. Both of them have a triple point, but notice that the triple point for water occurs at an exact temperature and pressure of 0.01°C and 0.006 atmospheres. Now, it's not hard to get down to that temperature, but a pressure of 0.006 atmospheres is pretty low. In fact, the pump I'm using is only rated down to just above that, or around 0.01 atmospheres. And because this is a single point in the phase diagram, there's absolutely no room for negotiation. The triple point happens here, and nowhere else. That's different than, say, a phase change, where we saw that the boundary can be moved around by changing pressure or temperature. There's no moving around of the triple point.

And you can see from the phase diagram that for tert butyl alcohol, that triple point is at a different place. In that case, the triple point pressure is much higher, at around 0.05 atmospheres, so, well within the range of our pump. The triple point temperature is a bit higher than that of water, at about 25°C, so very accessible in our experiment.

And speaking of temperature, you may have noticed in the experiment that I did not have any way to control temperature. In fact, I started the liquid at a slightly heated temperature, just to make sure it was well within its happy liquid state. So it started around here in its phase diagram. I then lowered the pressure. And a we saw, what happened is that, first, the liquid began to boil.

That's because we went like this in the phase diagram, traveling down across the liquid-gas phase boundary line.

But the process of boiling in itself is one that releases heat from the material. Remember, it's those higher kinetic energy molecules that leave, so the average temperature gets just a tad lower each time a molecule from the higher end of the kinetic energy spectrum leaves. With no heat source feeding thermal energy into the system, this means that boiling lowers the temperature of the liquid. That means we travel this way, to the left, on our phase diagram. And so the process continued, slowly but surely heading over to the triple point from where I started. A little pressure drop from the vacuum pump sucking out the gas molecules that just got boiled off, leading to more boiling, in turn, leading to a temperature drop, until we hit that fantastic singular point where all three phases coexist in one happy place.

Okay, I hope that by now you're starting to really feel your oneness with a phase diagram such as this, with a good sense of what these phase boundary lines imply and what it means for them to meet in a single place. As I've flashed up various phase diagrams for different materials, you'll have noticed that they are different, in fact, quite different from one another, depending on the material. In order to come back to more of our why questioning, to dig deeper into what makes a material tick and give it a curve here, or a slope there on its phase diagram, let's now turn to our wonderful thermodynamic variables and the framework of thermo we have built up in this course.

As I hope you might be thinking, especially if you watched our lecture on equilibrium, whenever there's a why question and we're talking about equilibrium, we can probably find the answer in the Gibbs free energy. When a closed system, that is, one with fixed mass and composition, is considered, and there is no mass transport allowed across the boundaries, the system is in stable equilibrium when it has the lowest possible value of the Gibbs free energy. And also remember that the whole reason we brought the Gibbs free energy into our thermodynamic framework was because it's connected to the equilibrium of processes occurring under laboratory conditions, like constant temperature and constant pressure. A change in G, dG, is equal to volume times dP minus entropy times dT. So, what does this imply? Since both T

and P can control the equilibrium by changing G, it implies that changing T and P can control equilibrium.

As an example, consider what is happening with the Gibbs free energy at a phase transition. Let's bring back our plot from the last lecture of the enthalpy, or heat added at constant pressure, as a function of the temperature of the system. As more heat is added, the temperature rises, until it reaches a phase transition, say, melting from solid to liquid, or vaporizing from liquid to gas. At these phase transitions, remember, that we have a jump in the enthalpy, since a lot of heat is needed to make the transition occur, even though the temperature does not increase. So, what's happening to the free energy?

Well, take a look at the melting transition as an example. The change in enthalpy, delta-H, at that point is equal to the amount that the curve jumps up from the solid to the liquid side. That's called the enthalpy of melting, or delta-H sub m. And as we saw before, the change in entropy for the phase transition is equal to delta-H sub m at the melting temperature divided by the melting temperature, T sub m. Remember that since G equals H minus TS, we can write a change in the free energy as delta-G equals delta-H minus T times delta-S for a constant temperature, which is the case during a phase transition. So putting this all together, for the change in free energy during the melting transition, we have delta-G for solid to liquid equals delta-H sub m minus T sub m times delta-H sub m over T sub m. And voila, everything cancels, and we get that delta-G for the solid-to-liquid transition is zero.

In fact, we knew that this had to be the case. If we plot the free-energy curve as a function of temperature, we'll have a separate curve for each phase, and in each case, the slope is negative, which makes sense, since the minus TS part of G makes it lower and lower as temperature goes up.

But the really key part of this G vs. T plot is that the curves for each phase intersect at certain points. They intersect at exactly the points that correspond to phase transitions. Since it's always the lowest free energy that dictates the equilibrium phase, once one of the curves goes below the other, like when the liquid-free energy curve crosses the solid-free energy curve at the melting point, then, it is that lower one that tells us which phase is stable.

We just worked out with our thermodynamic relationships that delta-G at the crossing point is zero, but now you can see this quite clearly and visually from the graph. If two curves cross, then at the point they cross the difference in their values is, of course, equal to zero, since they have the same value.

So, to summarize, the Gibbs free energy of a closed system with independent thermodynamic variables—pressure and temperature—determines what phase is in equilibrium. Each phase has a free-energy value at a given pressure and temperature, and the one that is the minimum energy at a particular temperature and pressure will be the equilibrium phase. This is how we construct our pressure vs. temperature phase diagrams.

Now, we know from day-to-day experience that many materials exist in multiple forms as a function of the environmental conditions. Solids melt, liquids boil, and solids can sublimate directly to vapor. But I'd also like to remind you that materials also exhibit different crystalline forms in the solid state, depending on the conditions of temperature and pressure. Remember from the last lecture that solid water, or ice, as we like to call it, has 15 distinct crystal structures. These structural variants are known as polymorphs, or allotropes, where the word allotrope is typically reserved for different structural forms of pure elements, and polymorph more generally covers different structures of any substance.

Okay, back to Gibbs. One of the most important aspects of a phase diagram is to know how many degrees of freedom there are at any point in the plot. It turns out that we have a rule for this, and it's called the Gibbs phase rule. This states that the number of degrees of freedom, call it F, is related to the number of components, call that C, and the number of phases we have in equilibrium, call that P, by the following relation. F equals C minus P plus 2.

Now the two here comes from the fact that I am looking at a phase diagram with two independent variables—pressure and temperature. The reason it's useful to know what F is anywhere in this plot, is that the number of degrees of freedom tell me how many different control knobs I have available to me for changing the material into a different phase. The Gibbs phase rule determines how many phases can be in equilibrium simultaneously, and whether those phases are stable along a field, a line, or a point in space on the phase diagram.

Well, rules are good and all, but let's go back to the Gibbs free energy. Because it's this quantity that determines the equilibrium phase; it is extremely useful to be able to go back and forth between a phase diagram and the free-energy curves for the various phases. So, here's the phase diagram for carbon dioxide. I just showed you some free energy plots as we examined what happens at the transition point. But more generally, how do we go from this phase diagram to a set of free energy curves? In order to do that, the first thing we need to do is to make a decision. Do we want to walk along a line of constant pressure, or constant temperature? As an example, here's a walk along a constant pressure line shown on the phase diagram. I can then, staying at this pressure, plot the free energy as a function of temperature.

And as you can see, at low temperatures, the free energy of the solid phase is lowest, then, crossing with the liquid free-energy curve at the phase boundary, and the liquid curve is crossed by the vapor phase free-energy curve at the subsequent-phase boundary. Now I know I've shown you these types of free-energy plots before. But I really want you to feel comfortable with the idea of being able to go back and forth between this plot and the phase diagram.

Take a look at what happens if I take a different walk along a different pressure. Down at this lower pressure, the free energy curves are completely different. Here, the liquid free-energy curve is higher than both the solid and gas free-energy curves at all temperatures. The liquid free-energy curve still exists; it's just always higher in energy. We'll be doing a whole lot more of this back and forth, especially as we move on to another type of phase diagram, one that gives us a materials map for what happens when two components are combined.

I hope that by looking at it this way you can see just how much information is contained in phase diagrams. These materials maps capture all of the Gibbs free energy behavior for the different phases as a function any line at constant pressure or temperature. And that is what I call packing a serious thermodynamic punch!

Properties of Phases

Lecture 16

In this lecture, you will continue your adventure into the phases of materials, the diagrams that tell us about them, and the properties they have by digging even more into the properties of phases and phase diagrams. You will learn about the thermodynamics behind why a curve in a phase diagram goes a certain way and why it has the slope that it has. In addition, you will examine your first multicomponent phase diagram.

The Clausius-Clapeyron Equation

- A given phase has a set of properties that defines its state. Depending on how many phases are in equilibrium with each other, the number of these properties that can vary decreases as the number of phases in equilibrium increases.

- When, for example, water is at its triple point, its temperature and pressure are fixed for all three phases. Knowing what the properties of state variables are for a given system allows for the ability to predict what a given system will do when subjected to a new set of conditions.

- In general, the properties of a system change drastically at a phase transition. The enthalpy is characterized by the latent heat of the phase transformation. The entropy is directly related to the enthalpy; more ordered systems, such as solids, are much lower in entropy than their fluid forms.

- For the heat capacity, the local atomic arrangement of species in a system drastically affects how the system absorbs and releases heat. For the volume, more ordered systems tend to have a higher density and thus lower volume than their fluid forms. One very notable exception to this last trend is water.

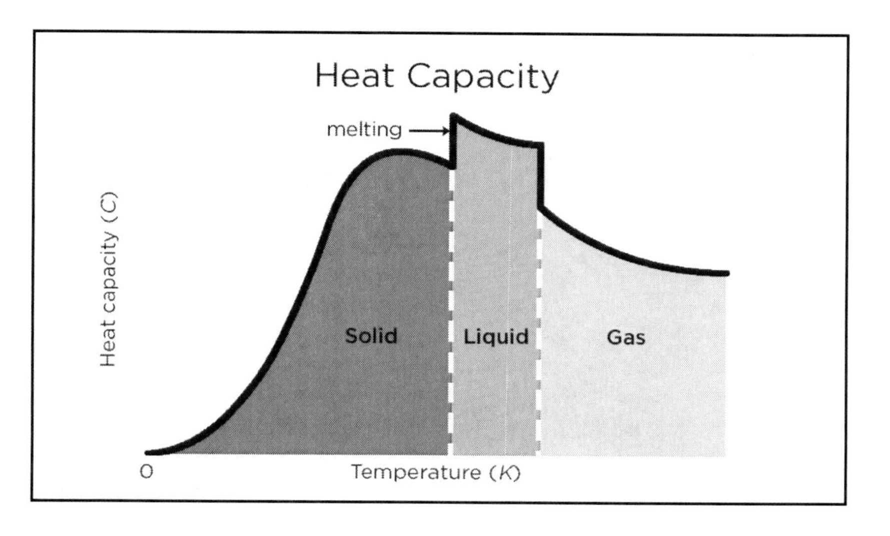

- The fact that ice is less dense than water right around zero degrees is at the heart of why so much aquatic life can survive. But why is the behavior of water so different than other materials? More generally, what is it that governs the slope of a phase boundary on the phase diagram?

- For water, the solid-liquid phase boundary—called a phase coexistence curve—is negative. But for most other materials, that solid-liquid coexistence curve has a positive slope.

- This negative slope is precisely the reason why you wouldn't want to leave pipes filled with water during the winter. If the water isn't moving, then eventually it freezes, and you have a massive amount of pressure from the phase change, which can burst the pipe.

- If you want to calculate the volume change that occurs upon freezing, it's pretty straightforward. When the water freezes, the initially liquid water is at a density of 1 gram per centimeter cubed, which decreases to a density of about 0.92 grams per centimeter cubed once it becomes ice at zero degrees Celsius.

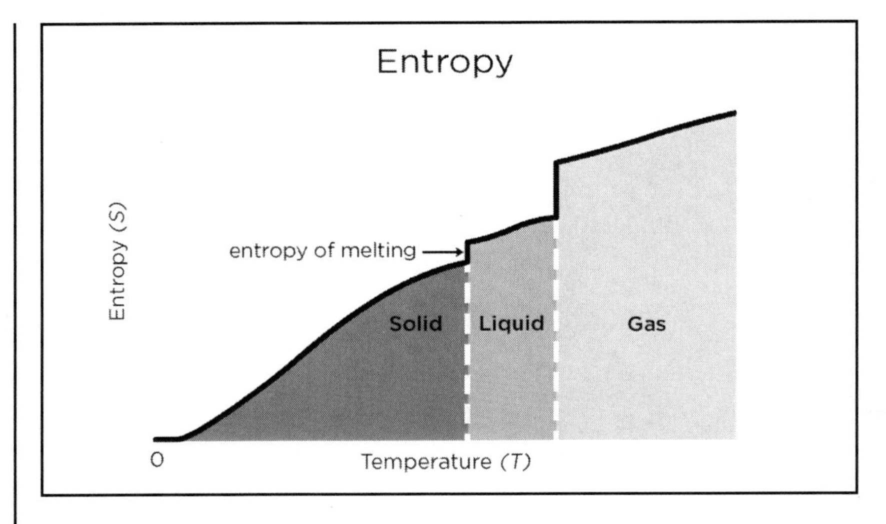

- If you start with a certain number of grams of water and that amount does not change (no mass transfer at the boundaries of your system), then the volume increases upon the phase change.

- The relationship between thermodynamic variables dictates the shape of free-energy curves. Similarly, equilibrium relationships have something to say about the shape of coexistence curves on a phase diagram. An important relationship in thermodynamics, known as the Clausius-Clapeyron equation, provides a link between the conditions for two-phase equilibrium and the slope of the two-phase line on the pressure-versus-temperature diagram. It was derived independently by Émile Clapeyron in 1834 in France and in 1850 by Rudolf Clausius in Germany.

- With the Gibbs phase rule, when we are sitting on a two-phase coexistence curve, there is only one degree of freedom, which means effectively that the derivative of pressure with respect to temperature must equal some function of pressure and temperature. The Clausius-Clapeyron equation gives us that function: dP/dT equals the latent heat of the phase transition divided by the temperature times the change in volume of the phase transition: $dP/dT = L/(T\Delta V)$. This equation gives us a way to get the slope of any two-phase coexistence curve on a phase diagram.

- The Clausius-Clapeyron equation helps give us some explanation for the properties of phase diagrams by directly relating the coexistence curves to the temperature and changes in heat and volume. However, at the atomic scale, why does water do what it does in terms of anomalous solid-to-liquid density change and therefore negative dP/dT slope?

- That has to do with the very particular type of bonding network that forms between water molecules. They interact with one another mainly through the hydrogen bonds that are formed between the hydrogen atoms from one water molecule and the oxygen atoms from another.

- It turns out that this is not a superstrong bond, but it's also not superweak; rather, it is somewhere kind of in between. These hydrogen bonds lock the crystal structure into place when it solidifies in a way that leaves a lot of open space—so much that it's actually less dense than the liquid form.

- Even though the bonds are weaker, and, therefore, the average distance between molecules is larger in the liquid compared to the solid, the liquid does not leave as much open space, resulting in overall higher packing. This is a pretty unusual phenomenon, and it has to do with the particular moderate strength of the bonds and also the very directional nature of these bonds—that is, the fact that they prefer to have quite distinct angles with respect to one another.

- Keep in mind that we're talking about expansion (or lowering of the density) through the phase change. This is not the same thing as expansion by heating. Expansion by heating is caused by the asymmetry in the potential energy curve for the atoms in a material.

- On a phase diagram, there is a point at which the two-phase coexistence curve between the vapor and the liquid simply stops, and the line just ends in a point. This is called the critical point, and beyond this point, we say that the material is supercritical. That's a fancy way to say that the material is no longer one phase or another, and it's not even a coexistence between two phases. It really is a new phase of the material that combines the properties of both the vapor and liquid phases together.

- For a supercritical fluid, the solid phase simply disappears. But the material does not just all of a sudden vaporize to become all gas or condense to become entirely a liquid. Instead, it becomes a new phase of matter—one that has about half the density of the liquid and can effuse through solids just like a gas but also still dissolves materials just like a liquid.

Multicomponent Phase Diagrams

- Next, we want to understand what happens to the properties—and phases—of a material when the material itself is comprised of more than one component. As an example, let's examine what happens when salt (NaCl) is added to water (H_2O).

- First, let's examine the interface between liquid and solid water. On the liquid side, water is moving more quickly and is able to move, while in the solid phase, the molecules can sort of jitter around but cannot translate or rotate freely. So, for a water molecule to freeze from the liquid to the solid phase, it has to get organized into this regular crystalline lattice.

- Instead of having just water molecules, suppose that we take a few of those in the liquid phase and replace them with a salt ion. Because the sodium or chlorine atoms are quite different in their size and bonding compared to a water molecule, they just aren't going to fit as easily into this solid matrix. So, they tend to prevent water molecules from crystallizing easily. Every time the solidification runs into one of these big solute molecules, it frustrates the whole process because the solid can't accommodate them like the liquid can.

- In thermodynamic terms, what we're doing is lowering the free energy of the liquid with respect to the solid. In fact, by mixing the solvent—that is, water in this example—with a solute particle, we always lower the free energy to some extent.

- The entropy always increases when we mix a solute into a pure solvent. Simply put, the system gets more degrees of freedom to play with, and the number of available microstates for any given macrostate is greater, leading to positive entropy of mixing.

- In addition, the change in chemical potential for a given process, which is equal to the change in the Gibbs free energy per mole at constant temperature and pressure, dictates the direction of the process.

- Equilibrium is reached for each component in the system when the chemical potentials of that component are equal across all phases. Molecules will tend to move from higher chemical potential to lower chemical potential until this equality is reached. So, with our entropy-only model, we deduce that the change in free energy must be negative as a function of the solute concentration.

- At a certain point, we cannot add more salt to the water at a particular temperature and pressure. Well, we can, but it won't dissolve in the liquid anymore; rather, it will fall to the bottom of the container as salt crystals. The point at which we cannot dissolve the solute into the solvent is known as saturation, and for salt and water at standard conditions, it occurs at a concentration of around 23 percent by weight salt.

- When we add some salt to the liquid, the free energy of the liquid phase goes down. And when we push that curve down, the crossing point between the liquid and solid curves moves over to lower temperatures. The farther down we push the liquid free-energy curve, the lower the crossing point temperature. That is why salt melts ice.

- In fact, this same liquid phase curve also changes its crossing point with the gas phase curve. By lowering the free energy of the liquid phase with respect to the gas phase, we increase the boiling point of the liquid.

- The intuitive picture on that end is similar in a way: The water molecules are a bit more "stuck" in the liquid phase because of the presence of solutes at the surface, so the vapor pressure is in effect lowered, which in turn raises the boiling point. And all of this comes from our simple model for the molar free energy of mixing.

Suggested Reading

Callen, *Thermodynamics and an Introduction to Thermostatistics*, Chap. 11.

Callister and Rethwisch, *Materials Science and Engineering*, Chap. 9.

DeHoff, *Thermodynamics in Materials Science*, Chaps. 8–9.

Questions to Consider

1. Why do pipes sometimes crack in the wintertime? Explain this using the phase diagram of water.

2. Why do most things contract when frozen, but water expands?

3. How does the entropy of water change as it freezes?

4. What equation relates the conditions for two phases to be in equilibrium and the slope of that two-phase coexistence line on the pressure-versus-temperature phase diagram?

Properties of Phases

Lecture 16—Transcript

Hello and welcome to our continued adventure into the phases of materials, the diagrams that tell us about them, and the properties they have. Today I'm excited about digging in even more into the properties of phases and phase diagrams. But first, let's go have some fun with the phase change of water. In the last lecture, we did a demo where I made a liquid boil and cool at the same time, simply by changing its pressure. Here, I'll show that even without the ability to control pressure, I can make water boil. And in fact, I'll do so by cooling it as opposed to heating it! Let's take a look.

So we know that water boils when we heat it up, but did you also know that it can boil when we cool it down? In this flask I have water that's hot, but it's clearly not hot enough to be boiling, since it's just sitting there, happy to be in the liquid phase. In this container, I have ice-cold water, and what I'm going to do is pour it on top. Watch what happens to the water inside when I do that.

Now you can see the water inside starts to boil. You see it will boil a little bit, and it will slow down, and if I add some more cold water on top, it will boil some more. So what's happening in here. Well, and you can see the water is still boiling. How is this possible? Well, it all has to do with pressure. I've lowered the temperature of the water vapor in side. So some of it condenses back down into the liquid. Less water vapor means lower vapor pressure, which lowers the boiling point of the liquid. This is a beautiful example of how the phase of materials can be controlled in different ways, depending on where we are in the phase diagram.

So that was yet another fun example of moving around in a phase diagram. By pouring cold water over the flask, the water vapor condenses, which leads to a drop in the pressure of the vapor. Feeling the lower pressure above it, the water begins to boil.

So we've certainly seen some cool ways and learned how to push across the phase boundary lines of a phase diagram. But today, we'll talk about the thermodynamics behind why a curve in a phase diagram goes a certain way in the first place, and why it has the slope that it has. Also, for the past two lectures we've stayed in the realm of single-component phase diagrams. Today, we'll begin with a single-component phase diagram, but by the end of the lecture, we'll be examining our first multi-component phase diagram, one that explains why ice melts when we sprinkle salt on it.

So, a given phase has a set of properties that define its state. As we learned in the last lecture, depending on how many phases are in equilibrium with each other, the number of these properties that can vary decreases as the number of phases in equilibrium increases, that's the Gibbs phase rule. We saw that when, for example, water is at its triple point, its temperature and pressure are fixed for all three phases. Knowing what the properties of state variables are for a given system allows for the ability to predict what a given system will do when subjected to a new set of conditions.

In general, the properties of a system change drastically at a phase transition. For example, the enthalpy is characterized by the latent heat of the phase transformation and often looks something like this. The entropy is directly related to the enthalpy; more ordered systems, like solids, are much lower in entropy than their fluid forms, as you can see from this plot. And for the heat capacity, as we've seen, the local atomic arrangement of species in a system drastically affects how the system absorbs and releases heat. And for the volume, more ordered systems tend to have a higher density, and thus, lower volume than their fluid forms. But, one very notable exception to this last trend is water.

I already talked about how the fact that ice is less dense than water right around zero degrees is literally at the heart of why so much aquatic life can survive. But I didn't tell you why the behavior of water is so different than other materials. So, the question we need to answer is, what is it that governs the slope of a phase boundary on the phase diagram? For water, the solid-liquid phase boundary—remember, we call that a phase coexistence curve—that slope of that boundary is negative. But for most other materials, and I really mean almost every other single component phase diagram, that solid-liquid coexistence curve has a positive slope. And it turns out that this negative slope is precisely the reason why you wouldn't want to leave pipes filled with water during the winter. If the water isn't moving, then eventually it freezes, and you have a massive amount of pressure from the phase change, which can burst the pipe.

If you wanted to calculate the volume change that occurs upon freezing, it's pretty straightforward. When the water freezes, the initially liquid water is at a density of around 1 gram per centimeter cubed, which decreases to a density of about 0.92 grams per cm^3 once it becomes ice at 0°C. If you start with a certain number of grams of water, and that amount does not change, so, no mass transfer at the boundaries of your system, then the volume increases upon the phase change.

But let's now tie this back to the pressure-temperature phase diagram. We've learned that the relationship between thermodynamic variables dictates the shape of free energy curves. Similarly, equilibrium relationships have something to say about the shape of coexistence curves on a phase diagram. An important relationship in thermodynamics that I'm about to explain provides a link between the conditions for two-phase equilibrium and the slope of the two-phase line on the pressure vs. temperature diagram. This relationship is known as the Clausius-Clapeyron equation. It was derived independently by Emile Clapeyron in 1834 in France and a little bit later in 1850 by Rudolf Clausius in Germany.

So, remember the Gibbs phase rule from our last lecture? How, when we are sitting on a two-phase coexistence curve, there is only one degree of freedom, which means, effectively, that the derivative of pressure with respect to temperature must equal some function of pressure and temperature. Well, the Clausius-Clapyron equation gives us that function.

Now, as you know by now, I don't always go around deriving stuff in this course. But I'd like to do so here, since we have a nice, simple derivation, and besides, it's always good once in a while to see what's under the thermo hood. So let's begin by considering the differential Gibbs free energy for ice and water at 0°C. We'll refer to these phases more properly as solid and liquid. So, with no chemical driving forces, and since we're looking at an equation for the change in Gibbs free energy for each phase separately, we won't have any change in number of moles, so no chemical potential terms. So in that case, we write dG for the solid equals minus the entropy S of the solid times dT plus volume V of the solid times dP. We can write the same dependence of the change in free energy similarly for the liquid phase.

And now, since we know the two phases are in equilibrium at the phase transition, then at the transition this means that their Gibbs free energies must be equal. So G for the solid phase equals G for the liquid phase, and dG of the solid equals dG of liquid. So equating our two expressions for dG, we have the following. Minus entropy of the solid time dT plus volume of the solid times dP equals minus entropy of the liquid time dT plus volume of the liquid times dP. And this gives us a way to solve for dP divided by dT; just group the terms together and divide. And what we arrive at is the expression: dP/dT at a phase transition. So one of those coexistence curves in the phase diagram, is equal to the change in entropy divided by the change in volume between the two phases.

By the way, keep in mind that we can always rewrite such expressions in terms of the molar versions of these variables. Here, if I divide both the top and bottom by the number of moles, I get that dP/dT equals the change in specific entropy, or molar entropy, divided by the change in specific volume between the two phases. I wanted to remind you of that, since often it's this intensive version of the thermodynamic variables we choose to work with, or that you may find in tables.

Okay, so the volume change doesn't seem too bad, but how do we obtain the change in entropy during the phase transformation? It's a hard one to measure directly, but you may remember that something we can measure quite easily is the latent heat of the phase transformation. This is the heat energy required to make the phase transition happen, or the enthalpy change over the phase transformation at constant pressure.

We also know that since delta-G equals delta-H minus T times delta-S, then at the phase transition, since delta-G equals zero, delta-H must equal T times delta-S. So in the previous equation for the derivative of pressure with respect to temperature, we can now substitute delta-H divided by temperature in for that delta-S term. This gives us, finally, a really cool expression, known as the Clausius-Clapyron equation. It gives us that the derivative of pressure with respect to temperature, so dP/dT, equals the latent heat of the phase transition, divided by the temperature times the change in volume of the phase transition. And, there we have it, a way to get the slope of any two-phase coexistence curve on a phase diagram, brought to you by basic thermodynamics!

So, we see now from this relation how the volume change is connected to the *P*-*T* diagram. Let's take a look back at the phase diagram of water. See, if I'm going from the solid to the liquid phase across this boundary, then we know that the heat of melting must be a positive value, since we are adding heat to the system to go from the solid to the liquid. But, we also know that ice is more dense at zero degrees as a liquid than a solid, so the volume change is negative. Clausius-Clapyron tells us that this means that the slope of this *P*-*T* two-phase coexistence curve has to be negative.

It's also interesting to point out the verticality of this line. Why does the solid-liquid boundary line go nearly straight up and down, while the liquid-gas boundary line is much more flat? Well, we can understand this by thinking about the change in volume. While the volume change between solid and liquid for water goes in a funny direction compared to most materials, it's still a small change. This is also true for most other materials, regardless of the direction of volume change. It's usually quite small for the solid to liquid transformation. On the other hand, it's often enormous in going from liquid to gas. Since that goes in the denominator of the expression, that volume change, the larger the volume change, the smaller the slope will be of the phase boundary line. Of course, there's also an interplay of this volume change with the change in enthalpy and the temperature of the phase transition as you see from the equation.

So, let's take a look at what happens when we expand our phase diagram to include higher pressures. By the way, the graph now goes up to pressures that you have inside the core of a nuclear bomb, so this is, indeed, pretty high! And it's up here where we see all these other solid phases of water start to appear. Notice that the slopes of the phase boundaries are in general pretty flat, indicating that there's likely to be either a large volume change between the phases, or, not a lot of heat energy needed to covert from one phase to another, or, a combination of these two attributes. And we can see by looking at the different densities of these phases that, in fact, there is quite a large change in volume among them, with, for example, ice 10 and 11 having densities two and half times larger than the good, old ice we know and love. The one that we have atmospheric pressure, which I suppose you could call ice 1, if you wanted, or hexagonal ice, or well, maybe just plain old ice.

But notice that certain phase boundaries, like this one here between what is known as ice 7 and ice 8, are not flat, but rather possess a very high slope, at least at these pressures. We know now that this means either that the volume change between these phases is very small, or, that the amount of heat energy it takes to undergo the phase change is very large. As we look at all of these different phases and phase boundaries, armed with the Clausius-Clapyron equation, which we just derived, we can admire with renewed appreciation the complexity that's built into such materials' maps.

Think about all of the different pushes and pulls going on at the atomic scale, about how heat flow in and out, the changes in crystal structure that result in changes in volume, and how sensitive this is to pressure and temperature. In some regions, one phase becomes two with temperature, while in other regions, a phase stops existing altogether. And with all these phases, we also have new triple points, where three different phases coexist together.

Seeing such a triple point in action, as we did for tert-Butyl alcohol in the last lecture, might not be as exciting in these cases, since two or even all three of the phases are solid, but still, the material at such singular points in the phase diagram possesses all three phases at once, which, on the scale of the atom, can mean just as drastic differences as we saw with the liquid that freezes and boils at the same time.

And by the way, I told you that Clausius-Clapyron would help give us some why for the properties of phase diagrams. And it does, by directly relating the coexistence curves to the temperature and changes in heat and volume. But what about the question of why at the atomic scale? Why does water do what it does in terms of anomalous solid-to-liquid density change, and therefore, negative dP/dT slope?

Well, that has to do with the very particular type of bonding network that forms between water molecules. They interact with one another mainly through what are called hydrogen bonds that are formed between the hydrogen atoms from one water molecule and the oxygen atoms from another. So it turns out that this is not a super-strong bond, but it's also not super weak. It's rather sort of somewhere in between. These hydrogen bonds, they lock the crystal structure into place when it solidifies in a way that leaves a lot of open space, so much so, that it's actually less dense than the liquid form.

So even though the bonds are weaker, and therefore, the average distance between molecules is larger in the liquid compared to the solid, the liquid does not leave as much open space, resulting in overall higher packing. This is a pretty unusual phenomenon, and it has to do with the particular moderate strength of these hydrogen bonds, and also, the very directional nature of these bonds, that is, the fact that they prefer to have quite distinct angles with respect to one another.

By the way, I keep saying how unique water is in this regard, but there are other materials that expand upon solidification for physical reasons similar to what I just described for water. For example, bismuth and antimony are other examples of materials that expand when they solidify. Now, keep in mind that I'm talking about expansion, or lowering of the density, through the phase change. This is not the same thing as expansion by heating. Remember from our lecture on thermal expansion, that expansion by heating is caused by the asymmetry in the potential energy curve for the atoms in a material.

Before moving on to our first glimpse into what happens when we mix materials together, there's just one more thing I want to point out on this phase diagram. It's something you may have noticed before, and it's this point all the way over here where the two-phase coexistence curve between the vapor and the liquid simply stops. The line just ends in a point. This is called the critical point, and beyond this point, out here in this region, we say that the material is supercritical. That's a fancy and, admit it, pretty cool-sounding way to say that the material is no longer one phase or another, and it's not even a coexistence between two phases. It really is a new phase of the material that combines the properties of both the vapor and liquid phases together. For a supercritical fluid, the solid phase simply disappears! But the material does not just all of a sudden vaporize to become all gas or condense to become entirely a liquid. Instead, it becomes a new phase of matter, one that has about half the density of the liquid and can effuse through solids, just like a gas, but also still dissolves materials just like a liquid.

One of the most commonly used supercritical materials is carbon dioxide. This is because it's a completely non-toxic material, so it can be used as a very efficient solvent in both medical and food industry applications. For example, did you know that when you drink decaffeinated coffee, it's very likely because supercritical CO_2 was flowed through the originally caffeinated beans? Because it acts like a gas, it can go right through the bean, but because it's also like a liquid, it's able to grab onto those caffeine molecules and extract them out. So the next time you switch to decaf, since you're already a bit jittery from those first three regular tall lattes of the morning, just think about the fact it's yet another reason why knowing where we are in a phase diagram is extremely important and useful.

Okay, now, for the remainder of this lecture, I'd like to move away from just a single component. Note that when I say single component, I'm not specifying a single atom, rather, I'm specifying a species, which could, of course, be just an atom, or, as in the focus of last few lectures, it could be a molecule, such as water. But now we want to understand what happens to the properties and the phases, of a material when the material itself is comprised of more than one component. As an example, we'll discuss what happens when salt is added to water. One component, therefore, is H_2O, and another is NaCl, or just plain, old salt.

First, conceptually, let's look up close at the interface between liquid and solid water. On the liquid side, water is moving more quickly and is able to move, while, in the solid phase, the molecules can sort of jitter around but cannot translate or rotate freely. So for a water molecule to freeze from the liquid to the solid phase, it has to get organized into this regular crystalline lattice.

Now, instead of having just water molecules, let's suppose we take a few of those in the liquid phase and replace them with a salt ion. What do you think will happen, intuitively? The issue is that, because the sodium or chlorine atoms are quite different in their size and bonding compared to a water molecule, they just aren't going to fit as easily into this solid matrix. So, they tend to prevent water molecules from crystallizing easily. Every time the solidification runs into one of these big solute molecules, it frustrates the whole process, since the solid can't accommodate them like the liquid can.

In thermodynamic terms, what we're doing is lowering the free energy of the liquid with respect to the solid. In fact, by mixing the solvent, that is, water, in this example, with a solute particle, we always lower the free energy to some extent. Remember our lecture on mixing and osmosis? In that lecture, we went through a simple lattice model to derive a simple expression for the change in entropy upon mixing. And we found that the entropy always increases when we mix a solute into a pure solvent. Simply put, the system gets more degrees of freedom to play with, and the number of available microstates for any given macrostate is greater, leading to positive entropy of mixing.

And remember that we also talked about how the change in chemical potential for a given process, which is equal to the change in the Gibbs free energy per mole at constant temperature and pressure, dictates the direction of the process. So, equilibrium is reached for each component in the system when the chemical potentials of that component are equal across all phases. Molecules will tend to move from high chemical potential to lower chemical potential until this equality is reached. So, with our entropy-only model, we can deduce that the change in free energy must be negative as a function of the solute concentration.

Let's plot that here. The molar free energy is on the y-axis and the concentration of the solute, in this case, salt, is on the x-axis. We'll start it here at some value which would correspond to the chemical potential of pure water in its standard state—that's a pressure of 1 atmosphere and a temperature of 25°C, or 298 Kelvin. Now, as I add salt to the mixture, so *X*-sub salt goes from zero up to some fraction, the molar-free energy goes down. As long as we continue to ignore enthalpy changes, then the entropy of mixing is a maximum at *X*-sub-salt equals one half, so the chemical potential will have a minimum at that point and then start to turn back up. Finally, on the far right hand side of this graph, we'd have *X*-sub salt equals

1, so the chemical potential should arrive at a value equal to that of pure salt at standard state conditions.

Now, we're going to be seeing a whole lot of these plots in the next lecture, which is all about phase diagrams of mixtures, so I wanted to give you a feel for it here. But since our question is about what happens to the freezing point of water with just a little salt added, we'll focus on the low salt concentration part of the graph. In fact, at a certain point, we can't add more salt at this temperature and pressure. I mean, we can, but it won't dissolve in the liquid anymore, rather, it will just fall to the bottom of the container as salt crystals. This is quite easy to see, and you can try it at home. That point at which we cannot dissolve the solute into the solvent is known as saturation, and for salt and water at standard conditions, it occurs at a concentration of around 23% by weight of salt.

Okay, so, back to our free energy; notice that this is the free-energy curve for just one phase of the salt-water mixture. We could plot the free-energy curves for other phases as well; each plot we make on this graph corresponds to a concentration dependence at a given temperature and pressure. If we go back to the graphs we made before, of the free energy curves vs. temperature for each phase, then remember it looks like this.

So, what happens when we add some salt to the liquid? As you can see from the concentration plot, the free energy of the liquid phase goes down. So on this plot over here, we can push that curve down with respect to the solid and gas free energy curves. And now notice what magical thing happens when I push that curve down: You see, the crossing point between the liquid and solid curves moves over to lower temperatures. The further down we push the liquid free-energy curve, the lower the crossing-point temperature. That is why salt melts ice!

Actually, you can see that further over, this same liquid-phase curve also changes its crossing point with the gas-phase curve. By lowering the free energy of the liquid phase with respect to the gas phase, we increase the boiling point of the liquid. The intuitive picture on that end is similar in a way: The water molecules are a bit more "stuck" in the liquid phase because of the presence of solutes at the surface, so the vapor pressure is in effect lowered, which in turn raises the boiling point. And all of this comes from our very simple model for the molar free energy of mixing!

We'll be building on this type of multi-component phase diagram in the next lecture, where as you may have guessed we'll also need to stop ignoring enthalpy. For now, I want to make one more cool point about this water-salt phase diagram: Did you notice that since we ignored interactions, our model is only based on entropy and therefore only based on counting states? That means it doesn't even care what the solute is in the first place! It tells us that, regardless of what we add to water, if we add anything its freezing point will go down. And it turns out to be quite true.

Try adding sugar or alcohol or any other kind of salt to your ice next time you need to do some de-icing. You'll get the same effect. Where there's solute added, the ice will melt (that is of course, if we stay above the new phase transition temperature, dictated by how much the free-energy curve lowered). So why is it that we use salt to melt ice? You probably guessed: simply because it's cheap. Try sprinkling sugar on your driveway instead: You'd get the same results but have to pay two to five times as much.

And with all this talk of de-icing, I cannot help but end with a story involving MIT and the mayor of Boston back in 1948. During the winter of that year, there was an unrelenting amount of snowfall, and the city had a lot of trouble staying out from under it all. The mayor at the time, a man by the name of James Curley, began pleading for help from the MIT president at the time, Karl Compton. The mayor wrote in a letter, "I am very desirous that [MIT] have a competent group of engineers make an immediate study as to ways and means of removing the huge accumulation … be it by the use of flame throwers or chemicals or otherwise." The mayor was clearly a bit desperate.

And Compton's response? He recommended that flamethrowers, "would be neither practicable nor efficient," but would be "hazardous." And instead, he suggested that the city try salt: "In many sections of the country salt, usually calcium chloride, is used quite effectively." Well, what can I say, he knew his thermo!

To Mix, or Not to Mix?

Lecture 17

What is it that dictates how much of one thing can mix with another thing in a given phase? In this lecture, you will learn about the thermodynamic conditions under which mixing is more or less likely. This is crucial to know for any situation in which the material you're making is composed of more than one component. Along the way, you will discover that there is something called a miscibility gap, and depending on the particular mixture and application you are going for, such a gap could either offer opportunities or present difficulties.

Mixing Components

- We are comfortable with the idea that some things mix—such as salt and water, at least for low concentrations of salt. The reason is shown by a simple entropy-of-mixing argument that leads to a lowering of the free energy upon mixing.

- We are also comfortable with the idea that some liquids don't mix—such as oil and water. You can shake a combination of oil and water enough so that it looks like it's mixed, but given a little time, it separates again. This can also be understood from an entropic argument, although for oil and water, it's not a simple entropy-of-mixing argument—which would have implied spontaneous mixing—but, rather, has to do with the specific bonding structure of the water molecules and the lower entropy shape they would have to take in order to work their way around the oil molecules. The mixing here is still dominated by entropy, but in this case, it goes the other way, driving the mix not to take place.

- Consider two generic components: component A and component B. Suppose that you make powders out of these pure components, and each powder has its own molar free energy, μ^o_A and μ^o_B. These two components are massively large compared to the atoms and molecules that make them.

- In this case, if we assume that the A and B atoms do not "feel" each other through some kind of interatomic forces between them, then the average free energy of such a mixture is simply the fraction of component A times its chemical potential plus the fraction of component B times its chemical potential.

- We use the variable X to represent these fractions, so if X is how much we have of B, then $1 - X$ is how much we have of A. Then, the free energy is $(1 - X)\mu^o_A + X\mu^o_B$. This is called a mechanical mixture.

- A solution is different from a mechanical mixture; it is taken to describe a mixture of atoms or molecules. There are two properties we need to account for in a solution: enthalpy change, which is due to the change in the strength and type of interactions between components, and entropy change, which is due to a higher number of arrangements of the two different components.

- The change in enthalpy upon mixing could be zero, especially if what is being mixed is similar in size, shape, and bonding type. But enthalpy change could also be positive or negative, which means it can either increase the free energy of mixing (making mixing less favorable) or decrease the free energy (making the mixing more favorable). On the other hand, the simple entropy contribution, the one based on counting states, always ensures a reduction in free energy of mixing. It is the relative contributions between these two quantities—the enthalpy and the entropy—that determine whether mixing is favorable or not.

- For any given concentration, the difference between the true free energy of the solution and the free energy of the mechanical mixture is known as the free energy of mixing. This quantity is equal to the enthalpy of mixing minus temperature times the entropy of mixing.

- The entropy of mixing for an ideal solution can be approximated by R (the ideal gas constant) times T (temperature) times the sum over the two components of the mole fraction times the natural logarithm of that mole fraction: $RT\sum_i X_i \ln X_i$. Also, if the solution is ideal, that means, just as for an ideal gas, we do not consider interactions, which in turn means that the enthalpy of mixing must be zero.

The Regular-Solution Model

- Now we want to see what happens when we do consider interactions, which for most materials and systems are pretty important. One of the simplest ways to include enthalpy in the game is to think about the two components randomly mixed together on a lattice.

- For the enthalpy, what we're interested in is how the *A* and *B* components bond together differently compared to how *A* components bond with *A*'s and how *B* components bond with *B*'s. This difference is the mixing energy, and it can be approximated by the following: The molar energy of mixing component *A* with component *B* is equal to the energy change per *A*–*B* contact relative to the pure *A*–*A* or *B*–*B* contacts times the number of *A*–*B* contacts.

- This is a "nearest neighbor" model in the sense that it only considers the change in bonding due to the immediate surroundings of each component. What's nice about this model is that it once again just comes down to counting. In this case, what we need to know is how often an *A* will be next to a *B*: The probability of an *A* being in contact with a *B* is proportional to the fraction of component *A* times the fraction of component *B*.

- Taken together, this simple picture leads to the following expression for the molar free energy of mixing: The molar ΔG of mixing equals a constant times the fraction of *A*, or X_A, times the fraction of *B*, or X_B, plus $RT(X_A \ln X_A + X_B \ln X_B)$: $\Delta G_M = \omega X_A X_B + RT(X_A \ln X_A + X_B \ln X_B)$. The second term is the entropy of mixing term.

- With the interaction term set to zero, this is the ideal-solution model. Including that interaction term in this simple framework, we call this the regular-solution model. The approximation that the atoms are randomly distributed is retained (so the entropy of mixing is the same as the ideal solution), but enthalpy of mixing is not zero.

- Once we have the behavior of the free energy, then we can get to a phase diagram, which has regions where the solution is mostly one phase or the other—because, for example, at low concentration of component *B* it's dominated by *A*, and at high concentrations it's dominated by *B*. The system in these regions is always unmixable, at any concentration. Because the region is unmixable, we call this part of the phase diagram a miscibility gap (the term for "mixability" in thermodynamics is "miscibility").

- So, to mix, or not to mix? That is the ultimate question. The answer lies in the free energy, in a competition between enthalpy and entropy, with temperature playing a key role. Temperature moderates the relative importance of the two mixing terms.

- The regular-solution model is a very simple way to incorporate interactions. It was introduced in 1927 by a chemist named Joel Hildebrand, and it's an easy way to understand the behavior of mixtures that show an ideal entropy of mixing but have a nonzero interaction energy. And it helps us understand a whole lot about the behavior of mixtures.

- However, it's not so easy to obtain all of those interaction parameters for a wide range of different materials. Also, this is a pretty simple model of interaction because it only accounts for nearest neighbors. And finally, we're still using that same ideal-solution model for the entropy. In some cases, the regular-solution model breaks down.

Oil and Water

- The mixture of oil and water is a mixture that cannot be described by the regular-solution model. Oil is a hydrophobic substance that does not bond to water because it doesn't form hydrogen bonds, which are the bonds that form between the hydrogen atoms of one molecule and the oxygen atom from another.

- A mixture of oil and water is a clear-cut case where component *A* does not like component *B* as much as *A* and *B* like themselves. However, because the oil molecules occupy space in the water

Oil and water do not mix because of the entropy that drives them to separate.

solution, the water molecules have to rearrange their hydrogen bonds around the oil. What ends up happening is that the water molecules, which are trying to hydrogen bond with each other, form open cage-like structures around the non-hydrogen-bonding oil molecules. This cage-like structure is quite ordered in nature, because the water molecules at the oil-water interface are much more restricted in their movement.

- In this mixture, what we have is a system that wants to minimize the amount of oil-water interface because it means anywhere we have that kind of interface, we take away microstates, which reduces entropy, which as we know from the second law is not the direction systems like to go. And the best way for the system to minimize the surface area of the oil-water interface is to keep all of the oil together so that the volume-to-surface-area ratio is maximized.

- So, in the oil-water mixture, we do not have the typical entropy-of-mixing argument but, rather, something more complicated. The entropy increase associated with mixing the two components together is completely sacrificed in favor of gaining more entropy by not allowing those open cage-like structures to form.

- The whole thing happens in a way because of the bonding nature of hydrogen bonds and the networks they like to form. However, this has led to a common misconception among many people and even some quite well-known scientists. The reason oil and water do not mix is not because of the enthalpy term; instead, it is because of the entropy that drives these two fluids to separate.

Suggested Reading

Anderson, *Thermodynamics of Natural Systems*, Chap. 7.

Callister and Rethwisch, *Materials Science and Engineering*, Chaps. 8–9.

DeHoff, *Thermodynamics in Materials Science*, Chaps. 9–10.

Porter and Easterling, *Phase Transformations in Metals and Alloys*, Chaps. 1.3–1.5.

Questions to Consider

1. What is the difference between a mechanical mixture and a solution?

2. Describe the ideal solution model: Explain in particular what part of the free energy expression is considered and what part is ignored.

3. What is the role of enthalpy in mixing?

4. Can materials always be mixed at any concentration?

To Mix, or Not to Mix?

Lecture 17—Transcript

So by now, we've nailed the single component pressure-temperature phase diagram stuff pretty well. In the last lecture, I introduced our first binary phase diagram, that would be a multi-component phase diagram with two components. In this lecture, we're continuing our journey into what happens to phases of materials as they are mixed together. When I talked about mixing salt with water, I mentioned that at a certain point, if we keep adding salt to water, it will saturate and no longer form a mixed solution with the water. Instead, after we've already added about 23% by weight salt, any additional salt will just sink to the bottom as pure salt crystals.

We've all heard and seen that oil and water don't mix, but is that always true? And do, say, ethanol and water always mix? What is it that dictates how much of one thing can mix with another thing in a given phase? In this lecture, we're going to look at the thermodynamic conditions under which mixing is more or less likely. And that's crucial to know, not just for saltwater, mixed drinks, and salad dressing, but for any situation in which the material you're making is composed of more than one component. Along the way, we'll discover there's something called a miscibility gap, and depending upon the particular mixture and application we're going for, such a gap could either offer opportunities or present difficulties. So, to mix or not to mix, that is our question!

I'm going to start by bringing to life a little thermo mixology. Check it out.

In this demo, we're going to talk about how different liquids mix or don't mix, and sometimes, a little bit of both. Now, we're comfortable with the idea that some liquids don't mix. What's an example? Well, how about oil and water. Let's take a look at that right now.

Here I have some oil, and I'm going to pour it into this beaker. Okay? And I think most of us have a good sense that if I pour the water in on top, it's going to want to separate, okay. So you can see it kind of mixed a little bit there for just a brief minute, a brief second, really, and then it phase separated. So you really got a layer of water, followed by a layer of oil on top.

Now, oil and water is an example of two different liquids that always separate, at least at room temperature. No matter how much we have of either one in the mixture, and no matter how much I try to shake it; it's always going to go back to that phase-separated state. But in other cases, as we'll learn in this lecture, liquids are mixable, but only to a certain concentration. Let's take a look at that case.

So over here, I have acetic acid, fixed with a little bit of water, and I've dyed it blue so that you can see it very clearly, okay. And that's because I want to contrast it with this, which is ethyl acetate. And I'm going to mix these two together. Now, notice that when I pour a little bit of ethyl acetate in to the acetic acid, you can see that it's still completely mixable. So we see that it's just a blue liquid, and I can take a look at it here and see that it's completely mixed.

But now I'm going to add a little bit more ethyl acetate to the mixture. And what you see is that as I added more ethyl acetate, so I changed to concentration of the ethyl acetate, now, something interesting is starting to happen. Now, if I look at it, I can see that the two liquids are starting to separate. Let me give a little more ethyl acetate so you can see it really clearly. And you can see that now there's a really clear separation between the acetic acid, which is blue, and the ethyl acetate, which is clear.

So you see, for this mixture, unlike oil and water, there is a clear range of concentrations where the two liquids are completely mixable. And then, there's also a clear miscibility gap in the phase diagram, where, when I add a little bit more concentration, they completely phase separate and become unmixable.

So there we had some very different examples of mixing at room temperature and atmospheric pressure. How can we explain these different behaviors? We are comfortable with the idea that some things mix, like salt and water, at least for low concentrations of salt. We showed the reason for that in the last lecture with a simple entropy of mixing argument that leads to a lowering of the free energy upon mixing.

We are also comfortable with the idea that some liquids don't mix, like oil and water. I can shake it up enough so that it looks like it's mixed, but then, given a little time, it re-separates. This can also be understood from an entropic argument, although, for oil and water, it's not a simple entropy-of-mixing argument as we learned earlier, which would have implied spontaneous mixing. But rather, has to do with the specific bonding structure of the water molecules and the lower entropy shape they would have to take in order to work their way around the oil molecules. The mixing here is still dominated by entropy, but in this case, it goes the other way, driving the mix not to take place. I'll come back to this example a bit later, since it's a bit complicated, but it's an important point.

So, what was going on in that demo, where we seemed to have a bit of both—mixing, and not mixing? Two liquids mixing together that are in between these two cases, mixable, but only to a certain concentration. Starting with the colored acid/water mixture, I added ethyl acetate and it was fully miscible. That means it can be mixed.

But as I added more, it eventually it became unmixable. Now, here's the cool part about this; it's not like the salt-water mixture. It's not that I kept adding ethyl acetate, and initially it mixed fully, but then after a certain concentration the ethyl acetate saturated and just started separating. No, that's not actually what's going on here. Instead, what's happening is that the solution went from being fully mixable to being fully un-mixable with complete phase separation. What's strange about it is that the unmixability occurs just by varying the concentration. The more technically correct term in thermodynamics for mixability, by the way, is miscibility.

So, how can we explain that case? Yes, you know what I'm going to say. You can feel it! Of course, it's all understandable through the Gibbs free energy. Let's turn to our thermo now to help us figure out what's happening. Consider two generic components, call them component A and component B. Suppose I make powders out of these pure components, and each powder has its own molar free energy, μ^0_A and μ^0_B. Now, these powders may seem pretty finely ground, like your morning coffee grounds, but, at that size, they would still be very large chunks of pure material.

Just for reference, the typical size of a ground-up particle of coffee is somewhere between about a quarter of a millimeter in diameter for espresso, to a bit over one millimeter for the French press. And just for fun, a cubic millimeter of coffee weighs about one half of a milligram. And how much of that is caffeine? Well, it varies, but somewhere around one and a half percent of a coffee bean is caffeine by weight. That means that each little tiny ground-up particle of coffee contains about 0.0075 milligrams. That sounds like just a tiny bit, but that still leaves us with a whopping ten to the 16th number of caffeine molecules in each grain of coffee! Think about that next time you make a nice fresh cup

So, the point is, I'm mixing these two components the size of coffee grounds, which means massively large compared to the atoms and molecules that make them. In this case, if we assume that the A and B atoms don't "feel" each other through some kind of interatomic forces between them, then the average free energy of such a mixture is simply the fraction of component A times its chemical potential, plus the fraction of B, times its chemical potential.

And remember we use the variable *X* to represent these fractions, so, if *X* is how much we have of B then 1-*X* is how much we have of A. Then we write the free energy as one minus *X* times mu0A, plus *X* times mu0B. This is called a mechanical mixture, and the molar-free energy plot for such a mixture would look like the following. As a function of the concentration of component B, on the left we have just pure A, and the curve starts at the chemical potential of pure A. And on the right, we have pure B and the chemical potential of pure component B. And in between, for a mechanical mixture, we simply have a straight line. The free energy at any point is just a weighted mean of the components.

But a solution is different from a mechanical mixture; it's taken to describe a mixture of atoms or molecules. Of course, that's what we had in our demo, but I wanted to build up our free energy plots from that simplest case of the mechanical mixture. You can see that there are two properties we need to account for in a solution, first, enthalpy change, which is due to the change in the strength and type of interactions between components. We have completely ignored it thus far. And second, there is entropy change, which is due to a higher number of arrangements of the two different components.

That is the simplest type of entropy of mixing change we covered last time, and it always increases upon mixing.

So, the change in enthalpy upon mixing could be zero, especially if what is being mixed has similar size, shape, and bonding type. But enthalpy change could also be positive or negative, which means it can either increase the free energy of mixing, making mixing less favorable, or it can decrease the free energy, making the mixing more favorable.

On the other hand, the simple entropy contribution, the one based on counting states, always ensures a reduction in free energy of mixing, as we discussed last lecture. But the key point here is that, as you can see, it is the relative contributions between these two quantities—the enthalpy and the entropy—that determines whether mixing is favorable or not. For any given concentration, the difference between the true free energy of the solution and the free energy of the mechanical mixture is known as the free energy of mixing. This quantity is equal to the enthalpy of mixing, minus temperature, times the entropy of mixing.

Now, we know already that the molar entropy of mixing for an ideal solution can be approximated by $-R$ times the sum over the two components of the mole fraction times the natural log of that mole fraction. Remember that R is the ideal gas constant. Also, if the solution is ideal, that means, just as for an ideal gas, we do not consider interactions. But now we want to see what happens when we do consider interactions, which for most materials and systems, are pretty important.

One of the simplest ways to include enthalpy into the game is to think about the two components randomly mixed together on a lattice, like you see here. Starting with a simple lattice like this is the same way we built a model for the entropy of mixing. For the enthalpy, what we're interested in is how the A and B components bond together differently compared to how A components bond with A and B components bond with B. This difference is the mixing energy, and it can be approximated by the following, The molar energy of mixing component A with component B is equal to the energy change per A-B contact relative to the pure A-A or B-B contacts times the number of A-B contacts.

This is called a nearest neighbor model in the sense that it only considers the change in bonding due to the immediate surroundings of each component. What's nice about this model is that it once again just comes down to counting. In this case, what we need to know is how often an A will be next to a B. Now, I won't go into probability theory here or anything like that, but I'll just tell you the answer; the probability of A being in contact with a B is proportional to the fraction of component A times the fraction of component B. So, taken together, this simple picture leads to the following expression for the molar-free energy of mixing. The molar delta-G of mixing equals some constant times the fraction of A, or X_A, times the fraction of B, or X_B, plus R times T times the sum of $X_A \log X_A$ plus $X_B \log X_B$. This second term is the entropy of mixing term we derived earlier.

With the interaction term set to zero, this is the ideal solution model. Including that interaction term in this simple framework, we call this the regular solution model. The approximation that the atoms are randomly distributed is retained, so the entropy of mixing is same as ideal solution. But enthalpy of mixing is not necessarily zero.

Let's take a look at what this does to the free energy of mixing as a function of concentration. So here's that plot we had before, showing the mechanical free energy of mixing curve, or really I should say, straight line, as well as the lower free energy of mixing from entropy in the ideal solution. Now, when we add this enthalpy term to get the regular solution model, we have several possibilities. First, suppose the mixing interaction parameter is a negative number. That means that the two components really like each other, even more than they like themselves. This will lower the free energy curve even more upon mixing. Actually, it also means that even for very low temperatures, or even in the extreme case of T=0, where the entropy contribution is minimal, the free energy will still favor mixing, in this case driven by interactions.

Next, suppose that the mixing interaction parameter is a positive number. In that case, it means that the two components don't bond together as strongly as they bond with themselves, and it will counter the entropy term in this model. If the interaction term is positive but small, then, the free energy of mixing will still go down but may soften a bit as you can see here. If on the other hand the interaction parameter is a large positive number, then this can completely change the nature of the free energy of mixing, leading to a net increase overall as a function of concentration, as you see here.

Let's take a closer look at this case of strong interactions where the interaction parameter is strongly positive. So that means the components either really like themselves, or, they really don't like each other, or both. If we look closely, the enthalpy term from the regular solution model takes on a maximum value right at a concentration of 50/50, which is the same concentration where entropy of mixing takes on a minimum value. However, the shapes of these curves are different, since one is based on a logarithm while the other is a simple multiplication.

This means that under certain conditions the free energy of mixing may actually start out by going negative for, say, low concentrations, but then, turn up and become positive for intermediate concentrations. And then, at even higher concentrations, it goes back down to become negative again, as you see here. Now that's quite an interesting result, since it seems related to that miscibility experiment I did in the demo. Notice that, in general temperature plays a very important role here. It moderates the relative importance of the two mixing terms.

So, here's a plot showing the molar free energy of mixing vs. concentration for the case of a large positive interaction parameter. Each of these curves corresponds to a different temperature. At lower temperatures, the enthalpy contribution to mixing plays a more important role regardless of its sign, although in this case, since it's positive, no mixing is possible, while at higher temperature entropy tends to dominate, and mixing is always favored. So there is this range in the middle where neither term dominates, and that's where we can have both a net decrease and increase of the free energy of mixing as a function of concentration.

So, how does that explain the miscibility behavior we observed in the demo? Well, once we have the behavior of the free energy, then we know we can get to a phase diagram, so let's do that now. Here, on this side, I'm showing you that same molar energy of mixing vs. temperature for the case of positive interaction energy, so the components don't like each other as much as they like themselves. This is the high temperature plot where entropy dominates.

On the other side, over here, I'll construct our phase diagram, where I'm plotting the phases on a temperature vs. concentration plot. By the way, this is a standard way to look at multi-component phase diagrams. That is, by plotting the temperature vs. concentration, while holding pressure fixed to some value, often, standard atmospheric pressure. Okay, so on our phase diagram so far, we have that this higher temperature region corresponds to a fully mixed phase of A and B at any concentration of B, so we just write A + B in solution.

But now, back to the free energy plot. As we lower the temperature, the molar free energy curve gets more and more shallow until eventually, at some critical temperature, it no longer has a distinct minimum point. That means, any little bit cooler, and the free energy is going to start going back up, so it seems like an important moment in temperature space. Well, I also called it a critical temperature, so that may have given it away. But let's draw a line on our phase diagram at that temperature and then mark a point at the concentration of B where the free energy curve went flat.

Now, as we lower the temperature further, the free energy tends up in the middle here, just like we saw before, since, for those concentrations, the interaction starts to dominate over the entropy. That means that there are two points at which the molar free energy takes on a minimum value, and it goes up in between, as well as on each side of these minima. On the phase diagram, that means that there will be two special concentration values, one over here, and the other over there. And as you can see, if we continue lowering the temperature and plotting these concentrations that correspond to minimum points in the free energy diagram, then we get a very nice curve on the phase diagram plot.

This curve marks a critical boundary between phases. Above it, as we mentioned, components A and B are fully miscible. To either side of this boundary, at lower temperatures, you can see that the free energy of mixing goes down, so the components are also miscible. Although, pretty quickly, we find ourselves in regions where the solution is mostly one phase or the other, since at low concentration of component B, it's dominated by A and at high concentrations, it's dominated by B.

So, what's going on inside of this boundary? That corresponds to this region here on the free energy plot. Now, here's the key. You would think that the free energy is simply equal to whatever this curve gives us. Ah, you would think that, but, nature is much smarter than that! Because there are two minima in this plot. So, we can draw a line in between them. This is called a tangent line, because, well, because it's a tangent to both of those points. But the point here is that this line, which is reminiscent of the mechanical mixture straight line, gives the system a way to lower its overall free energy. And that's exactly what it chooses to do, and we know from the second law that it has no choice in the matter; it will always lower its free energy if it can.

And here's why it can. If we look to the regions outside of these minima in the free energy curve, mixing is favored because unmixing would take us back up to that mechanical mixing line, which is much higher in free energy. On the other hand, in between the minima, if the liquid un-mixes in a way that gets it over to these two minima, then it can lower its free energy.

You can see this just visually by noting the following; at any concentration point in between the minima, the free energy of the solution, where the two components are fully mixed together at the given concentration, is given by the curve drawn. However, there's another possibility. The free energy of a mechanical mixture comprised of two separate solutions is lower than the free energy corresponding to a fully mixed state. If we take one solution with the concentration of one of those minimum points and another solution with a concentration of the other minimum point, then the free energy corresponding to a mechanical mixture of the two distinct solutions will be on the line connecting those two minima points.

And this is indeed what nature does. In this in-between region, the system can always find its lowest possible free energy by being comprised of exclusively the two distinct solutions, one at each of the concentration points corresponding to those two minima. That means that the system in this region is always unmixable. We cannot ask the material to mix together into a solution at any concentration, since the system will always go to just these two concentration points and nowhere else in between. Since the region here is unmixable, we call this part of the phase diagram a miscibility gap.

Here, inside this boundary, we have both components A and B, but they are not able to mix freely. I can pick any point in here, and if I draw a horizontal line out to the boundary lines, then those points where the line intersects the boundary give us the fixed concentration values for that temperature. This is pretty profound if you think about it. And I know it's also a bit strange and unintuitive. But it's really important in thermodynamics, so, let's be sure we understand this whole thing as best we can. Let me give a few examples using this phase diagram I've just constructed.

Suppose I'm at a temperature of 300°C. Well, then I'm up here in the phase diagram. And, as you can see, anywhere I travel from left to right along the 300°C line shows that the two components, A and B, are fully mixed in a solution. If I add together equal amounts, by mole fraction remember, since that's how we're plotting the horizontal axis, if I add together equal amounts of A and B, then I'm going to get a solution of 50/50 A and B.

But, now, let's come down in temperature to 100°C. Down here, when I draw a line across, it's a different picture. At this temperature, up until a mole fraction of 10% component B, I can specify any concentration of B, and the two components will mix into a solution. But, if I specify that same 50/50 concentration, I will not get it. Instead, I get a mixture of two different materials; one is comprised of 10% mole fraction of B mixed with 90% A, and the other is made of 90% of B with 10% of A. These two solutions are the only two materials I can have inside of this region at this temperature. As I move along on the concentration axis, I change the relative amounts of the two solutions, but I do not change their composition. This is exactly what happened in our demo. We started miscible, then I added enough of one of the components to go into the miscibility gap, where the mixture immediately phase separated into two distinct solutions. Then, if I go to very high concentrations of component B, higher than 90% mole fraction, well, then I'm back to being fully mixable and will get a solution comprised of exactly the concentrations that I mix together.

So, to mix or not to mix, that is the ultimate question, And now you know; the answer lies in the free energy, in a competition between enthalpy and entropy, with temperature playing a key role. Now, the regular solution model is a very simple way to incorporate interactions. It was introduced in 1927 by a chemist named Joel Hildebrand, and it's a very nice, simple way to understand the behavior of mixtures that show an ideal entropy of mixing but have a non-zero interaction energy. And it helped us understand a whole lot about the behavior of mixtures. However, it's not so easy to obtain all of those interaction parameters for a wide range of different materials. Also, this is a pretty simple model of interaction, since it only accounts for nearest neighbors. And finally, we're still using that same ideal solution model for the entropy. In some cases, the regular solution model does break down.

And this reminds me that I had wanted to return to the example of oil and water. This is a mixture we all know quite well, and it is one that cannot be described by the regular solution model. You see, oil is a hydrophobic substance that does not bond to water, since it doesn't form hydrogen bonds. Remember, those are the bonds that form between the hydrogen atoms of one molecule and the oxygen atoms from another.

A mixture of oil and water is a clear-cut case where component A does not like component B as much as A and B like themselves. Yet, because the oil molecules occupy space in the water solution, the water molecules have to rearrange their hydrogen bonds around the oil. And what ends up happing is that the water molecules, which are trying to hydrogen bond with each other, form open cage-like structures around the non-hydrogen-bonding oil molecules. This cage-like structure is quite ordered in nature, since the water molecules at the oil-water interface are much more restricted in their movement.

In this mixture, what we have is a system that wants to minimize the amount of oil-water interface, because it means anywhere we have that kind of interface, we take away microstates, which reduces entropy, which, as we know from the 2nd law, is not the direction systems like to go. And the best way for the system to minimize the surface area of the oil-water interface, is to keep all of the oil together so that the volume-to-surface-area ratio is maximized.

So, in the oil-water mixture, we do not have the typical entropy of mixing argument, but rather, something more complicated. The entropy increase associated with mixing the two components together is completely sacrificed in favor of gaining more entropy by not allowing those open cage-like structures to form. The whole thing happens, in a way, because of the bonding nature of hydrogen bonds and the networks they like to form. However, this has led to a common misconception among many students and even quite well-known scientists. The reason oil and water do not mix is not because of the enthalpy term, As we have just seen, it's because the entropy that drives these two fluids to separate.

So, there we have it. By adding interactions, we have been able to see how complicated mixing can be and how to go back and forth between the molar free energy curves and the phase diagram. In the next lecture, we're going to take this further as we explore even more complicated types of mixing, ones that involve the mixing of different phases of materials. The miscibility gap here was just a warm-up. Wait until you see what a eutectic does to a phase diagram!

Melting and Freezing of Mixtures

Lecture 18

You have been learning about what happens when two components are mixed together in a binary phase diagram. The last lecture focused on whether a material will form a solution or not, which is done by examining the free energy of the system. This lecture continues with binary mixtures and the free energy. You will also continue to focus on the properties of these mixtures. In particular, you will learn what happens for the phase transitions of mixtures. Does a mixture of two different components melt in the same way a pure component does?

Molar-Free-Energy-versus-Concentration Curves

- In a molar-free-energy-versus-concentration curve, two components are mixed together, A and B, with the mole fraction of component B on the x-axis. The mole fraction is just a measure of the concentration of B in the mixture, and it is written in terms of the moles of B divided by the total number of moles.

- That means that we also have the concentration of A at that point, because $X_A = 1 - X_B$. On the left, the curve joins up at the molar free energy of pure A, while on the right is the molar free energy of pure B. The molar free energy is also equal to the chemical potential for constant temperature and pressure, so the values on this plot at $X_B = 0$ or 1 are equal to the chemical potentials of the pure, unmixed components.

- In between, at mixed concentrations, the free-energy curve may go down. If this were an ideal solution, then you know that the curve would have to go down, because the only contribution we have to the free energy of mixing in an ideal solution is from the entropy of mixing. There are simply more possible ways to arrange the components when they mix together—the number of microstates increases, and in turn, so does the entropy.

- But in this case, the solution is not ideal, because while it does go down, it has a little bit of a shape to it, and the minimum is not at 50 percent but, rather, at 25 percent. That means that interactions are playing a role, but it doesn't stop the two components from mixing at all concentrations.

- In this mixed region, the chemical potentials of the two components are changing. They are not the same as the values they have for the pure components. However, we can still obtain the chemical potentials in the mixed region in the following manner: We draw a tangent line to the free-energy curve, and the values at which that line intersects the y-axis at $X_B = 0$ and $X_B = 1$ are equal to the chemical potentials of the components for that concentration.

Examining Different Phases

- What happens when we check out different phases of this mixture? The liquid phase of these components is lower in free energy at every concentration than the solid phase. That means that we must be at a temperature such that the material is always liquid.

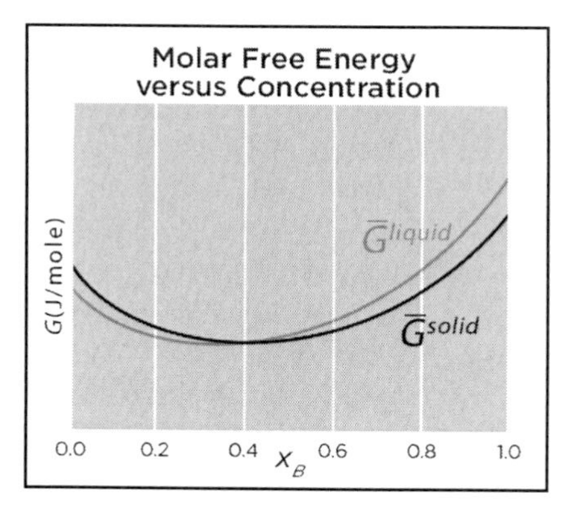

- If we were to lower the temperature by a lot, then the curves for the liquid phase and the solid phase might switch places: The solid-phase free-energy curve would be lower than the liquid-phase free-energy curve at all concentration values. This indicates that at the lower temperature, the material is always a solid at any mixture.

- At intermediate temperatures, the two free-energy curves will intersect at some point along the concentration axis. This crossing point could be anywhere and will depend on temperature, but for example, let's say that these two curves cross at X_B = 50 percent, so a 50-50 mix in terms of the mole fraction of component *A* and component *B*.

- When the free energy of the system can find a way to go lower, it does. So, we would think that the free energy of the system is simply a trace along the lowest-possible free-energy path. In this case, for concentrations of X_B less than 50 percent, we're on the liquid curve, and at concentrations greater than X_B = 50 percent, we move over onto the solid curve because it's lower in energy.

- It would mean that while maintaining this temperature, as we increase the concentration of component *B*, the mixture stays a liquid until it gets to 50 percent concentration, at which point it undergoes a phase change and becomes a solid.

- However, this is not what actually happens. We can understand this by noting what happens when the free-energy curve shows two separate minima along the concentration axis. In that case, in between those minima, something very special happens. As in the case of immiscibility, the actual free energy of the system will not follow one of these two curves. That's because the system has a means to go even lower, by phase separating into a mixture of what lies at those two minima.

- In the last lecture, those two minima corresponded to two different solutions made up of different ratios of components *A* and *B*. In this case, the two minima actually correspond to different phases of the system. But the same thing happens. Namely, in between these two points in concentration, we get a combination of two different phases of the material: Some of the material is a liquid, and some of it is a solid.

- Once again, we can find the free energy by drawing a tangent line between these two minima. For concentrations in between them, the free energy of the system will lie on this straight line.

- This type of graphical approach, drawing a tangent line on the free-energy curve to figure out where there is phase separation and by how much, is so important that it gets its own name: the common tangent construction.

Using Curves to Draw a Phase Diagram

- For the phase diagram, we plot the temperature on the *y*-axis versus the concentration, still in terms of the mole fraction of component *B*. That means that we need to examine what happens with these free-energy curves as we change the temperature.

- We've already encountered the extremes, where one free-energy curve is always above the other. Working our way down in temperature, starting up at the top, the liquid free-energy curve is always lower than the solid free-energy curve. As we lower the temperature, the solid curve goes down because it's more and more favorable. Let's stop at the temperature for which the two curves first begin to cross. For this case, that's going to be the point at which $X_B = 1$. So, there is a point on the phase diagram at that temperature and concentration.

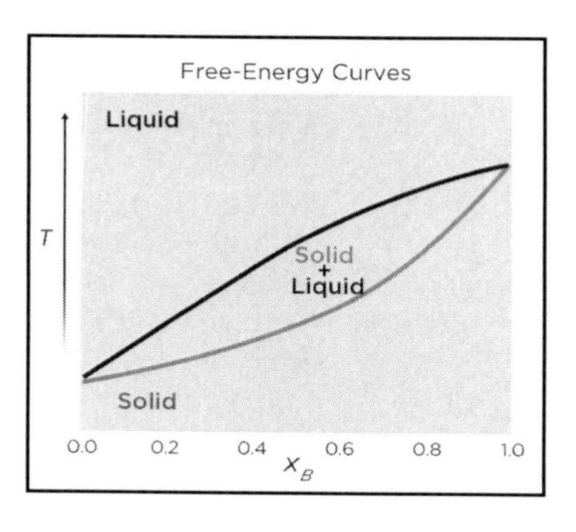

- As we lower the temperature a little more, the two curves cross at a slightly lower concentration. What we have now is a region in the free-energy curve where we can draw a line tangent to the two free-energy curves, and in between those two tangent points, the free energy is along the line, not on either the liquid or solid curve. The tangent is not between two minima as before; regardless, the same exact behavior occurs and for all the same reasons, as long as we can draw a tangent line between the two free-energy curves.

- As we lower the temperature further, the solid free-energy curve continues to go down in energy (that is, it's becoming more favorable to be a solid), and the tangent points occur at different concentrations. Finally, when we get to low enough temperatures, the free-energy curves once again intersect only at a single point, this time at $X_B = 0$. For temperatures lower than that, it's all solid, all the time.

- All we need to do is connect the dots together with curves, resulting in the completed phase diagram. This is one of the simplest types of phase diagrams, but it's packed with information. This type of phase diagram is known as a lens phase diagram. Inside the lens shape, both the liquid and solid phases coexist together.

- At any point in the lens, we can determine how much of each phase we have and what their compositions are. The liquidus phase boundary separates the two-phase region from the liquid; the solidus phase boundary separates the two-phase region from the solid.

- In the lens region of the phase diagram, as in the case of the miscibility gap, we find that the material will not take on any arbitrary composition and remain fully mixed. Instead, it will phase separate into liquid and solid—but at fixed compositions dictated by the phase boundary lines at the given temperature.

The Eutectic Point

- Many materials are highly miscible in the liquid state but have very limited miscibility in the solid state. This means that much of the phase diagram at low temperatures is dominated by a two-phase region of two different solid structures: one that is highly enriched in component *A* and one that is highly enriched in component *B*.

These binary systems, with unlimited liquid-state miscibility and low or negligible solid-state miscibility, are referred to as eutectic systems.

- There's a point on such phase diagrams where the system melts at a certain composition, called the eutectic point. For example, soldering involves passing right along through the eutectic point of a phase diagram.

- For example, tin melts at around 232 degrees Celsius, while bismuth melts at about 271 degrees Celsius. Naively, we might think that if we mix tin and bismuth together, the melting temperature will be somewhere in between those two values. But that's not at all the case.

- We can lower the melting point considerably below that of either of the two pure elements simply by mixing them together. The lowest melting point in this case occurs at 139 degrees Celsius—nearly half of the average of tin and bismuth's melting temperatures and lower than tin's melting temperature by almost 100 degrees.

- The eutectic point offers the lowest possible melting temperature, so it is extremely useful in soldering and other industrial applications.

Suggested Reading

Callister and Rethwisch, *Materials Science and Engineering*, Chap. 9.

DeHoff, Thermodynamics in Materials Science, Chaps. 9–10.

Questions to Consider

1. List two graphical approaches to getting useful thermodynamic data from a phase diagram.

2. When solutions are not able to mix, can you still choose any concentration of the mixture you desire?

3. How can the melting point of a solution be lower than the melting point of either of the components that have been mixed together?

4. What is a eutectic point?

Melting and Freezing of Mixtures
Lecture 18—Transcript

So we've been talking about what happens when two components are mixed together in a binary phase diagram. Last time we really focused on whether a material will form a solution or not, and we did that by examining the free energy of the system. Here, we're going to continue our obsession with binary mixtures and the free energy, after all, it is the thermodynamic property that tells us when a system is in its own happy place, otherwise, known as equilibrium. But we'll also continue to focus on the properties of these mixtures. In particular, we're going to take a look at what happens for the phase transitions of mixtures. Does a mixture of two different components melt in the same way a pure component does?

For example, let's take tin and bismuth. Tin melts at around 230°C, while bismuth melts at about 271°C. Now, naively, we might think that if we mix tin and bismuth together, the melting temperature will be somewhere in between those two values. But that's not at all the case. Let's cook up some of these mixtures now to see what really happens…

So let's say I go to the store and I buy a piece of metal. Well, I might get something like this; this is one single element, one pure metal. But, I could take a piece of this and combine it with another metal, and that would be called an alloy. What I want to show you today is what happens when you mix two pieces of metals together to make an alloy, what happens to the melting point.

So, here I have a piece of metal, I have a piece of pure tin. And here, I have another metal. This is pure bismuth. And I'm going to put them on a hot plate, okay. And what you can see when I put them down is that not much happens. In fact, that's because the temperature of this griddle is about 200°C, and the temperature of melting for tin is around 230°C, and the temperature for melting of bismuth, is around 270°C. So you can see that if I try to move them around, they're not melting at all; they're not leaving any liquid tracks.

But now, here, I have a piece of a mixture, I have two pieces in here, that are different mixtures of these two metals. This is 40% tin and 60% bismuth.

And this is 40% bismuth and 60% tin. So when I put them down on the hot plate, let's take a look at what happens. So it's the same temperature, and you can see that these guys on the end, the pure metals, have still not melted. But in the mixture, you actually have the metal completely melting, going through a phase transition, and becoming a liquid.

These are the same two pure elements that are not melting at all, but you can see that at just the right composition, at just the right composition, you have a much lower melting point. And the lowest melting point for any mixture of two different metals, that's the eutectic point.

So there you saw that we can lower the melting point considerably below that of either of the two pure elements simply by mixing them together. The lowest melting point in this case occurs at a lowly 139°C. That's nearly half of the average of tin and bismuth melting temperatures on their own, and lower than tin's melting temperature by almost 100 degrees! How can we understand this surprising result?

Well, you know what I'm about to say; it will all become clear as we look at some free energy curves. Those curves, in turn, allow us to construct the phase diagram for this mixture. And that special point with the lowest melting temperature? That has a special name as well; it's called a eutectic point.

But, I'm getting a little ahead of myself here, because first, I want to go back to the free energy curves for a mixture of two different materials. Now, in the last lecture, we were concerned with whether or not two components would mix, and we said that in an ideal solution, they always would because of entropy, but, if we start including the effects of interactions, they may mix sometimes, and the may not other times.

But there's something we have left out thus far, whether in the case of ideal, regular, or general solution models. And that would be what happens to the phase change in a mixture upon mixing. Let's take a look at that right now. So, we return to our phase diagram happy place. That would be, of course, a molar free energy vs. concentration curve. We're mixing two components together, A and B, with the x-axis corresponding to the mole fraction of component B. Don't get confused by the term mole fraction; that's still just a measure of the concentration of B in the mixture. It's just written in terms of the moles of B divided by the total number of moles.

And by the way, that means, of course, that we also have the concentration of A at that point, since X_A equals one minus X_B. So on the left, the curve joins up at the molar free energy of pure A, while on the right, we have the molar free energy of pure B. As a reminder, the molar free energy is also equal to the chemical potential for constant temperature and pressure, so these values on this plot at X_B equals 0 or 1 are equal to the chemical potentials of the pure, unmixed components.

Now, in between, at mixed concentrations, the free energy curve may go down like this. If this had been an ideal solution, then you know that the curve would have to go down, since the only contribution we have to the free energy of mixing in an ideal solution is from entropy. There are simply more possible ways to arrange the components when they mix together; the number of microstates increases, and so in turn does the entropy.

But in this case, the solution is non ideal, since you can see that, while it does go down, it has a little bit of a shape to it, and the minimum is not at 50% but rather over here at 25%. That means that interactions are playing a role here, but it doesn't stop the two components from mixing at all concentrations.

In this mixed region, the chemical potentials of the two components are changing. They are not the same as the values they have for the pure components. But don't despair! We can still obtain the chemical potentials in the mixed region in the following manner. We draw a tangent line to the free energy curve, and the values at which that line intersects the y axis at $X_B = 0$ and $X_B = 1$, those values are equal to the chemical potentials of the components for that concentration.

Let's walk through that carefully. Suppose my molar free energy plot looks like this. Over here and here are the chemical potentials of the pure components. Now, say I pick a concentration on this graph of X_B equals 25%. At the point, I draw a line tangent to the molar free energy curve. Now, that tangent line hits the y axis up here at $X_B = 0$, so, I can read off the value of the molar free energy there, and that will give me the chemical potential of component A at this concentration.

Similarly, the tangent line intersects the y axis at X_B equals 1 over here, so the molar free energy value at that point is exactly equal to the chemical potential of component B at concentration of $X\_B = 25\%$. So, if you want to know the chemical potentials of mixtures, you now have a nice, simple, and fun way to get them!

Okay, now, let's see what happens when we check out different phases of this mixture. Suppose the free energy curve I just showed you is for the liquid phase of these components. Let's not worry about mixing or not right now, but just assume that the system mixes these components in a nice solution at all concentrations. Let's now consider a free energy curve corresponding to the solid phase of these mixtures. I've decided that this one has a minimum at a different concentration than the liquid, at X_B equals 75%. So the curve starts up here.

Notice the way these curves are drawn; the liquid phase is lower in free energy at every concentration than the solid phase. What does this mean? Well, it means that we must be at a temperature such that the material is always a liquid. If I were to lower the temperature by a lot, then these two curves may switch places; we'd have the solid phase free energy curve lower than the liquid phase at all concentration values. This indicates that at the lower temperature, the material would always a solid at any mixture.

Now, here's where it gets interesting. What happens at intermediate temperatures? In that case, the two free energy curves will intersect at some point along the concentration axis, like you see here. This crossing point could be anywhere, and it will depend on temperature, but, for this example I've designed these two curves such that they cross at X_B equals 50%, so a 50/50 mix in terms of the mole fraction of component A and component B.

But here's the thing. Just as we learned in the lecture for the miscibility gap, when the free energy of the system can find a way to go lower, it does. So, we would think that the free energy of the system is simply a trace along the lowest possible free energy path. In this case, that trace would look like you see here, where for concentrations of X_B less than 50% we're nice and happy on the liquid curve, and at concentrations greater than X_B equals 50% we move over onto this solid curve, since it's lower in energy.

It would mean that while maintaining this temperature as I increase concentration of component B, the mixture stays a liquid until I get to 50% concentration, at which point it undergoes a phase change and becomes a solid. But, as you may have picked up on, this is not what actually happens. We can understand this by remembering from the last lecture what happens when the free energy curve shows two separate minima along the concentration axis. In that case, in between those minima, something very special happens. As in the case of immiscibility, which we examined last time, here again, the actual free energy of the system will not follow one of these two curves. That's because the system has a means to go even lower in its free energy, by separating into a mixture of what lies at those two minima.

So last time, those two minima corresponded to two different solutions made up of different ratios of components A and B. In this case, the two minima actually correspond to different phases of the system, but the same thing happens, namely, in between these two points in concentration, we get a combination of two different phases of the material—some of the material is a liquid, and some of it is a solid. And the free energy? Well, once again, we can find that by drawing a tangent line between these two minima. For concentrations in between them, the free energy of the system will lie on this straight line.

Have you noticed that I like to draw tangent lines on such plots? That's because, being based on derivatives and all, there's some good thermo packed into them, and I always like it when I can learn something graphically. And speaking of learning something graphically, there is something else I can get from this tangent line, and that is the answer to how much.

You can imagine that, as we come across the x-axis here, as a happy liquid-only phase first, and then reach the minimum of the liquid free energy curve, that at that point, we're still pretty much all liquid. And that's true. Same thing goes for the solid side. At the beginning of the two-phase region coming over from the right, it's essentially all in the solid phase. As I move in from these boundaries, the system takes on both phases. The amount is simply given by the amount along this line we've travelled, divided by the full length of the line. So if I'm, say, here at X_B equals 30%, then if I want to know how much of the material is a liquid, I can take the distance from there to the minimum of the solid phase curve, and divide by the total length. By the way, this type of graphical approach, drawing a tangent line on the free energy curve to figure out where there is phase separation and by how much, that's so important, it gets it's own name in thermodynamics; it's called the common tangent construction.

Okay, so, I know that was a whole lot of fun. I mean, honestly when is it not fun to think about free energy curves? But, as you know, the fun gets even greater once we use such curves to draw a phase diagram. So let's do that now for the example I just showed you. For the phase diagram, we plot the temperature on the y-axis vs. the concentration on the x-axis, still in terms of the mole fraction of component B. That means that we need to examine what happens with these free energy curves as I change the temperature.

Well, I already kind of discussed the extremes, where one free energy curve is always either above or below the other. Let's work our way down in temperature, starting up here.

In that case, the liquid free energy curve is always lower than the solid free energy curve, so, I can write up here the word liquid on my phase diagram. Now, as I lower the temperature, the solid curve goes down, since it's more and more favorable. By the way, it can also change shape as I change the temperature, but I'm ignoring that for now. Let's stop at the temperature for which the two curves first begin to cross. As you can see, for this case, that's going to be all the way over here for X_B equals 1. So we'll put a point on the phase diagram at that temperature and that concentration.

Now, as I lower the temperature a little more, the two curves cross at a slightly lower concentration. What we have now is a region in the free energy curve where I can draw a line tangent to the two free energy curves, and in between those two tangent points, the free energy is along the line—not on either the liquid or solid curve. Notice here that the tangent is not between two minima, as I showed before just to help make it easier to see.

But that's fine; the same, exact behavior occurs and for all the same reasons, as long as I can draw a tangent line between the two free energy curves. And here, as you can see, I can.

For our phase diagram, we're going to track these tangent points. Let's use red for the first tangent, where it comes to the liquid phase curve, and blue for the second tangent point, where it comes to the solid phase curve. Okay, I have three points on my phase diagram so far—Isn't this fun? Let's keep going! As I lower the temperature further, the solid free energy curve continues to go down in energy, that is, it's becoming more favorable to be a solid, and the tangent points occur at different concentrations. Here, I'm continuing to track those tangent points with the same red and blue dots. And finally, when I get to low enough temperatures, the free energy curves once again intersect only at a single point, this time over here at $X\_B = 0$. For temperatures lower than that, it's all solid, all the time, so, I'll write the word solid over down here on phase diagram.

And all we need to do is connect these dots together with curves, and here we have our completed phase diagram. This is one of the simplest types of phase diagrams we can get, but it's packed with information. Now, the naming of this was not exactly done in a moment of originality, since this type of phase diagram is known as a lens phase diagram. In any case, as you can see, inside the lens shape I have both the liquid and solid phases coexisting together, so I'll write solid plus liquid inside.

And at any point in this lens, I can determine how much of each phase I have and what their compositions are. For example, if I'm at this point here, then I look for the phase boundaries at that same temperature by simply drawing a horizontal line. Over on this phase boundary, which is called the liquidus phase boundary, since it separates the two-phase region from the liquid, the point where that line intersects gives me the composition of the liquid. Over on the other side, we intersect the solidus phase boundary at a point representing the composition of the solid. In this lens region of the phase diagram, once again, as in the case of the miscibility gap, we find that the material will not take on any arbitrary composition and remain fully mixed. Instead, it will phase separate into liquid and solid, but at fixed compositions dictated by the phase boundary lines at the given temperature.

I know this can seem a little bit confusing, especially if it's your first time seeing it. So let me take a stroll through this lens phase diagram by picking a concentration, starting at high temperature and virtually cooling down the material. Let's start with a concentration that's right in the middle of the phase diagram, at 50%, and I'll start up here at high temperatures.

Now, at this temperature, I know that the system can fully mix at any concentration, so I have a liquid made up of 0.5 mole fraction of component B and 0.5 mole fraction of component A. As I cool it, I hit this top part of the lens in the phase diagram. Immediately what happens is that the system can no longer have the 50/50 composition. That's because it can lower its free energy by phase separating, which is exactly what it does.

Part of the system becomes a solid—a small part, since I'm just entering the two phase region from the liquid side. And that solid phase will not have a 50/50 mix of components A and B. Instead, it will have a mix dictated by this point over here on the lens boundary, which looks more like 95% B and 5% A. Meanwhile, the liquid phase, which remember, just a degree ago used to have a 50/50 mixture of A and B, now has a different composition, one with less of component B in it, say, a mix of 45% B and 55% A.

As I lower the temperature to continue freezing this mixture, more and more of the mixture becomes a solid, although, each time I lower the temperature, the composition of that solid phase changes. As you can see, it gets more and more of component A into it, although it's still dominated by component B for this example. The liquid phase, which is getting less and less in terms of amount, is also changing its composition with each degree change. For the liquid, as I lower the temperature, it takes a composition with more and more of component A in it and less of component B. And this continues on in this way until I reach the bottom of the lens. At that point, my solid has come back to the original composition that I started with, namely, 50/50. My liquid is essentially gone, although, if there were a minuscule amount left, it would have a composition given by the intersection all the way over here.

So there you have a completely different way of undergoing a phase transition than for a pure component. Notice that the phase transition for the pure components is actually on this same plot. It's simply the points on each end where the lens connects to the y-axis. Here is the melting temperature of pure A, and here is the melting temperature of pure B. But you see, these are just points. There is, of course, no 2-phase region, since there are not two phases, and the material simply melts, or freezes, or whatever. But with a mixture of components, we have the more intricate behavior I just described.

And why would we want to know all of this? Well, it turns out that most of the time, we have mixtures of materials. Steel is not pure iron, but rather, a mixture of iron and a little bit of carbon. How the fact that we have a mixture impacts the phase diagram is of enormous importance.

Now, I mentioned that the lens is one of the simplest things that can happen between phases in a two-component phase diagram. There are all sorts of other fun things that happen, leading to really complicated phase boundary lines and behavior of materials. Just take a look at some of these binary phase diagrams. Look at all of the wonderful lines between phases that occur, and while you're thinking about that, I hope you're also thinking about other stuff, like, how some of these lines separate liquid and solid, while others separate two different solids; about how different the different boundaries are, some going vertical, while others, with small slopes or even horizontal; and about in each region on this diagram, there's a corresponding set of free energy curves, with some of them lower than others, leading to the identification of equilibrium.

For the rest of this lecture, I'm going to tell you about one of the most common other occurrences in a phase diagram, something that also has a whole lot of important applications. Many materials are highly miscible in the liquid state but have very limited miscibility in the solid state. This means that much of the phase diagram at low temperatures is dominated by a two-phase region of two different solid structures—one that's highly enriched in, say, component A, and one that is highly enriched in component B. These binary systems, with unlimited liquid state miscibility and low or negligible solid state miscibility, are referred to as eutectic systems.

There's a point on such phase diagrams where the system melts at a certain composition, called the eutectic point, and in fact, I'll bet that a number of you have already done or at least seen some eutectic melting in your lives.

That is, if you've done any soldering at all, you'd be passing right along through the eutectic point of a phase diagram.

And this returns us to the very beginning, where I talked about tin and bismuth, and where in the demo I showed you how the melting temperature of a mixture of the two can be much lower than that of either one of these by itself. So here we have the phase diagram for these two components. On the left is 100% tin, while on the right, is 100% bismuth, with the melting temperatures of each in their pure phases indicated on each axis.

Up here in this region, we're at temperatures where it's a fully miscible liquid. Let's freeze a solid from this liquid from two different concentrations. First at 5% mole fraction of bismuth mixed in with tin, we're going to be cooling down along this vertical line here. When I cross this liquidus phase boundary, the material phase separates, just as we saw in the lens phase diagram before.

Once I've cooled enough to be in this two-phase region, I'll have a combination of liquid and solid phases. The liquid will have a composition of bismuth and tin, as given from this point on the liquidus curve, while the solid will have a composition given by this boundary over here. Now, as I continue to cool, I enter into the solid-only phase. It's a single phase on this phase diagram, which means that it can take on the concentration that is given, which in this case is 5%. But as we continue to cool down, we pass below this next boundary line, and enter yet another two-phase region.

Unlike our previous two-phase regions, this one is comprised of two solid phases, a solid phase of mostly tin and a solid phase of mostly bismuth. But that doesn't change anything that we discussed before. When we cool down into this two-phase region on a phase diagram, the system separates into two different solid phases, with concentrations dictated by the phase boundaries. As I cool further, this doesn't really change, although the amount of the tin-rich and bismuth-rich phases varies, as does their precise composition.

But let's now do another experiment; let's cool down at a concentration of $X\_B = 20\%$. In this case, the first phase boundary I cross is just as before, although notice that now, at this composition, the solidification begins at a lower temperature. But now the system does not enter the solid tin-rich phase as before. Instead, the next boundary we cross upon further cooling is this horizontal line here. Once I cross that line, all of the material that is in the liquid phase becomes a combination of these two solids, one, tin rich, and the other, nearly pure bismuth. The system is then comprised of these solids that come from the liquid freezing, plus, all the solid particles that had formed up in this lens-type region here, which have a composition of this point over here.

And finally, we'll cool down through the eutectic point. For tin and bismuth, this point occurs at a mixture of around 57% by weight of bismuth. Notice that as we cool down through this point, we do not enter a lens-like shape by crossing a liquidus curve. Rather, we go directly from liquid to solid at a single temperature. That's just like the freezing behavior of a pure component!

But, at the eutectic point, things are a little different. We do, in fact, freeze completely as we cool from one side of this point to the other, but when frozen, the solid consists of two different phases, one tin-rich and one bismuth-rich, as you see here by drawing the line across. That actually leads to some very interesting structure, since the solidification of the two phases happens all at once.

Take a look at this image of the eutectic structure of this material; you can see that it consists of alternating very thin layers of each of the two solid phases that form when solidification occurs. As you can imagine, a material with this type of structure has all sorts of differences in terms of its properties, compared to a pure material, or, for that matter, one with a homogenous mixture of the two components. The eutectic point offers, also, the lowest possible melting temperature, so it is extremely useful in soldering and other industrial applications. Pretty cool example of materials mixology!

In the next lecture we're going to turn our attention to non-phase diagram-related thermo stuff. We're going to be talking about heat engines. But, over the last five lectures, I do hope that you've gained a solid appreciation, along with some good understanding, of what phase diagrams mean, how they can be used, and how important they are to the study of thermodynamics.

The Carnot Engine and Limits of Efficiency

Lecture 19

In this lecture, you will learn about heat engines and the ultimate limit of efficiency in making heat do work. In an earlier lecture, you learned that one way to state the second law of thermodynamics is that heat can never fully be converted into work. In this lecture, you will learn about the Carnot limit, the fact that the efficiency limit of this conversion is set not by temperature of the heat we put into the engine but, rather, by the temperature difference between the heat we put in and the cold side somewhere else. In other words, heat alone cannot do any work; it has to be in the context of a temperature difference.

Heat Engines

- The term "heat engine" is pretty generic, and there are a wide variety of heat engines, from the ideal case of the Carnot engine, to the Stirling engine, to the Otto cycle commonly used in your car engine.

- In general, the term "heat engine" refers simply to the conversion of thermal energy into useful work. The heat could come from a wide variety of sources, such as gasoline, electricity, or the Sun. And the work is done by what is often called a working fluid. This could be an ideal gas, but it could be many other materials as well, including water.

- In general, for a heat engine we assume that there is an operating cycle, which means that the operation of the engine can be sustained and that the working substance (such as gas) must be able to follow a cycle that can be repeated over and over again.

- In a real engine, which covers all the ones ever made in this world that do useful work for us, the processes are nonideal—meaning that they are irreversible. This means that for any engine, we always expect dissipative forces to be present.

- In the very nonreal case of an ideal engine, on the other hand, we assume that all processes are reversible, so there are no dissipative forces whatsoever. An ideal engine is called a Carnot engine.

- The basic cycle of a Carnot engine involves some source of heat, which is known as the hot reservoir. You can think of this as some vast resource of thermal energy—in a car, it comes from making fire by burning gasoline. This hot reservoir is held at some temperature (T_h), and it is in contact with the working substance. As a result of being in contact with the thermal reservoir, the working substance absorbs some amount of heat.

- We will be using an ideal gas as our working substance. Let's call the amount of heat energy absorbed by the gas in a cycle Q_h because it comes from the hot reservoir. Then, that gas converts a part of the heat absorbed into useful work; let's call that amount of work W. Any unused heat—that is, the heat that did not get converted into work—goes into the cold reservoir. Let's call that unused, or rejected, heat Q_c because it goes to a thermal reservoir at a lower temperature. And let the temperature of the cold reservoir be T_c.

- We know from the second law of thermodynamics that heat can only flow from a hot body to a cold body, and not the other way around, so we know that heat will flow from the hot reservoir to the working fluid and that any unused heat will flow to the cold reservoir.

- The thermal efficiency of a heat engine is the ratio of the net work done by the engine during one cycle to the energy input at the higher temperature during the cycle. So, efficiency equals W/Q_h. We already know that Q_c is whatever heat was not used for work, so that means that, by definition, $W = Q_h - Q_c$. That means that the efficiency is equal to $(Q_h - Q_c)/Q_h$, which can be rewritten as $1 - Q_c/Q_h$.

- The efficiency of a heat engine is simply the ratio of what we gain (which is the work done by the engine) to what we give (which is the energy transfer at the higher temperature). It turns out that this ratio is typically a lot less than 1, so most heat engines operate at much, much lower efficiencies than 100 percent.

- Your car engine only operates at an efficiency of about 25 percent. That means that from all of that precious ancient sunlight that we are extracting from the ground to put into our cars and trucks, we are only using a measly 1/4 to make them move—the other 3/4 is simply wasted as extra heat that's not used for anything.

- Notice from our efficiency equation that in order to truly get to an efficiency of 100 percent, we would need for Q_c to be zero. This implies that a perfect engine cannot expel any energy whatsoever to the cold side, and all of the input energy would be converted to work. However, this is not at all possible.

The Carnot Engine

- The Carnot engine is the absolute best we can ever do. It's an ideal heat engine that starts and ends in the same place, is based on reversible processes, and gives us a total amount of work equal to the area enclosed by the path from the first process to the second process to the third process to the fourth process (the four key processes of the Carnot cycle). The change in internal energy is zero for the full cycle.

- Internal energy is a state function, so if we return to the same place in pressure-volume space, it must have the same value. And because the internal energy change is zero, the net work done by the gas over one cycle must equal the net energy transferred into the system, which is the difference: $Q_h - Q_c$.

- The beauty of the Carnot cycle is that because there is a set of concrete processes, we can compute these Q's for each of the processes. And as a result, we can arrive at a very important result: It just comes down to integrating the P-dV state functions over each process to find the work term.

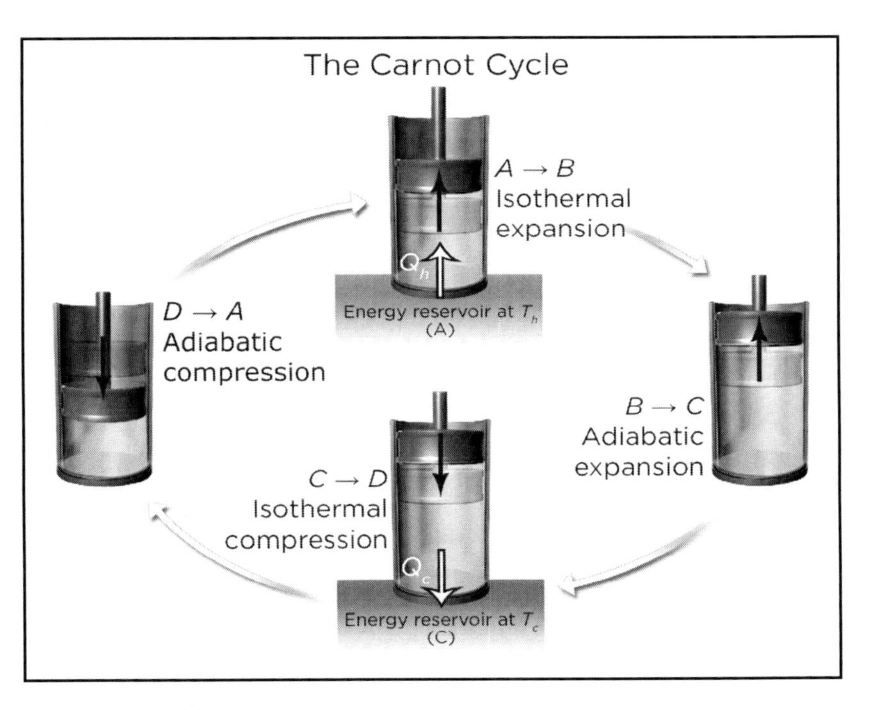

- By computing explicitly the work and heat terms for each process, we find that for this cycle, $Q_h/Q_c = T_h/T_c$. And that means that the efficiency of a heat engine, which is $1 - Q_c/Q_h$, can be rewritten for a Carnot engine as $1 - T_c/T_h$.

- The efficiency of the ultimate, ideal-case heat engine is only dependent on the temperature difference between hot and cold reservoirs. If the temperature difference is zero, then we have 1 − 1, which is an efficiency of zero. We can understand this intuitively: No temperature difference means there cannot be heat flow, and if there's no heat flow, then basically nothing can happen.

- On the other hand, we can approach an efficiency of exactly 1—or, in other words, a perfect engine—if the temperature of the hot side tends toward infinity or the cold side tends toward zero, but neither of these limits is actually possible.

- The Carnot engine tells us the absolute theoretical limit to the efficiency of making heat do work. An actual engine doesn't operate anywhere near this efficiency, because it's in no way an ideal case.

- In most cases, such as in a car or in a power plant, T_c is near room temperature—it would be highly impractical to maintain a cold reservoir of ice in real applications because that alone would require an enormous amount of energy. So, because the cold side is stuck around 300 kelvin for most cases, it's usually the goal in such engines to increase the efficiency by increasing T_h.

- There are many, many losses in a real engine, so the Carnot limit, based only on temperature difference, is often not the actual operating efficiency of a heat engine. For example, in a real engine—for example, the one in your car—there is a very short time interval for each cycle.

- At 1,000 revolutions per minute, the pistons in your car's engine are firing more than 16 times per second. The result is that the gas inside of them (which for a car engine is simply air) doesn't quite make it all the way up to that high temperature of the hot reservoir, and similarly, there's not enough time for it to get all the way back down to the low temperature of the cold reservoir. That's just one example of the complications that arise when we move from the ideal case of the Carnot engine to a real working engine.

Heat and Work

- Heat pumps are different than heat engines. A heat pump effectively works in the reverse of a heat engine: Instead of putting in heat to do work, we put in work to move heat.

- Do not confuse the first and second laws and how they apply to heat engines. The first law is a statement of conservation of energy, $\Delta U = Q + W$: The change in the internal energy of a system occurs because of energy transferred to or from the system by heat or work, or by both.

- Notice that the first law does not make any distinction between whatever happens because of heat or whatever happens because of work—either heat or work can cause a change in internal energy.

- However, in this lecture, which is devoted to one of the most practical applications of thermodynamics, we find that there is a crucial distinction between heat and work, one that is not at all evident from the first law. In order to understand the distinction, we need the second law of thermodynamics.

- From the point of view of engineering and practical applications, perhaps the single most important implication of the second law is the limited efficiency of heat engines. The second law states that a machine that operates in a cycle, taking in energy by heat and putting out an equal amount of energy by work, cannot be constructed.

Suggested Reading

Callen, *Thermodynamics and an Introduction to Thermostatistics*, Chap. 4.

Halliday, Resnick, and Walker, *Fundamentals of Physics*, Chap. 20.

Serway and Jewett, *Physics for Scientists and Engineers*, Chap. 22.

Questions to Consider

1. What are the essential characteristics of a heat engine?

2. How do steam engines convert heat into mechanical work? List all of the processes.

3. Can mechanical work be done with only a single temperature—for example, a really high temperature?

4. How does the choice of working fluid affect the efficiency of an engine?

The Carnot Engine and Limits of Efficiency
Lecture 19—Transcript

Hello and welcome to our lecture on heat engines and the ultimate limit of efficiency in making heat do work. This will bring us back to the earlier lectures, especially Lectures 5, 6 and 7, where we learned about the ideal gas, the first law, and the equivalence of work and heat. We also learned in our lecture on the second law that one way to state that second law is that we can never fully convert heat into work. In this lecture, we'll dig deeper into this statement and show, by way of what is known as the Carnot limit, the fact that the efficiency limit of this conversion is set not by temperature of the heat we put into the engine, but rather, by the temperature difference between the heat we put in and the cold side somewhere else. It's a pretty cool concept when you think about it, and it's a bit counterintuitive; heat alone cannot do any work; it has to be in the context of a temperature difference. We'll learn all about why that's the case in this lecture.

Now, you may be wondering, why didn't I do this heat engine stuff back when I covered the first law and work heat equivalence? Well, I wanted to cover engines this late in the course because this lecture is focused on heat engines, but the next lecture will be all about how we can get work out of materials in other ways. And, you may have guessed it, if we want to put materials to work, we really need to know what their properties are. Which, in case you've skipped to this point in the course, has been the topic of the last five lectures.

So why is the heat engine so important? Well, if you ever been in a car, a train, or a plane, or for that matter, plugged anything into an outlet to use electricity, or turned on a light, just to name a few examples, well then you've used a heat engine. In fact, heat engines are the foundation of the industrial revolution. And because they are so important, they are also a big reason for the existence of the entire field of thermodynamics itself. I mean, the heat engine is about as literal as we can get to "heat in motion," the very definition of thermodynamics.

Humans mastered fire a million years ago, give or take a half million, so we've known how to make heat for a very long time. But, it wasn't until just 500 years ago that we really began to understand how to harness that heat to do mechanical work in an engine. Explosive heat, such as that from gunpowder, was used in weapons much earlier, but for the purpose of today's discussions I'm talking here about continuous, repeatable mechanical work, as opposed to an explosion. So, trying to figure out how to get the most value out of heat, to do the most work, has been a major topic of enormous importance for centuries. Let's take a moment to check out our very own heat engine!

So in this demo we're going to talk about the Stirling engine. This is a good example of different thermodynamic processes acting on a material. In this case, the material is simply air. With a temperature gradient, work can be extracted. Now, here is a Stirling engine, and the working fluid inside is the air inside of this container. And what I'm going to do, is I'm going to put the Stirling engine on top of a cup of hot water. And now, what's happening is that the steam from the water is touching the bottom of this container, so we have the bottom side that's getting how, while, the top side stays at roughly room temperature. So the bottom side is getting close to 100°C, whereas the top side is room temperature, which is about 25°C. That's a temperature difference of 75°C.

So let's see if it's gotten hot enough, and I'm going to give it a little start. And you can see that it keeps going. This is an engine that's turning heat, or really, I should say, a temperature difference, into mechanical motion.

Now, what if we wanted this engine to go faster. What would we do? Well, one thing we could do is we could make the hot part hotter. Because again, it's all about how big of a difference in temperature we have. But in this case, this Stirling engine would actually, some of the parts inside would start to break if I got the hot part above 100°C. So instead, I can make it go faster by making the cold side colder.

Let's take a look at what happens when I put a little bit of ice on top of the engine.

Notice that with just a little bit of ice on top, it's actually spinning a lot faster. So I made it colder, and I got more work out of it. Why? Because I just made one side colder. And so the temperature difference is closer to 100°C, since ice is 0°C, then 75°C, I've increased the temperature difference, and that's what matters in terms of how much work I can get out of a heat engine.

Now, the Stirling engine is an example of a thermodynamic cycle, where the pressure and volume and temperature are all changing in the phase diagram. There are different ways to do this, though, in order to extract work—mechanical work—from heat. There are many different kinds of thermodynamic cycles that can be used to do that.

So, here I have an example of two different fuels that are used in two very different cycles to make your car go. First I have ethanol, and I'm going to take a little bit of ethanol and put it on top of this piece of metal. And I'll show you that if I light this now on fire, you see the ethanol lights on fire. You can see the flame there that's burning the ethanol.

That's what's happening in gasoline car engines; that's how the heat is coming into the system. However, if I take a different kind of fuel, so, let's say, biodiesel, and I'm going to put some drops of biodiesel onto this piece of metal. Now you see if I try to light that on fire, it simply doesn't light. It's not flammable.

However, If I take this piece of metal with the biodiesel on it, and I raise its temperature, okay, I'm going to put it on this hot plate, and I'm going to leave it there for a little bit just to raise its temperature, then what we're doing is by raising its temperature, we're bringing it into a different part of pressure vs. temperature phase space, so it's in a different part of the phase diagram when I raise the temperature. I'm going to put it down here, and notice what happens if I change its pressure. That's biodiesel.

So there was a piston, doing mechanical work by moving up and down once I applied heat to the bottom of it. Now I could have added more heat, and it might have run a little faster, but it's also limited by the fact that that little piece of foam can't take high temperatures. So instead what I did instead is I put something colder on the top, and then, you saw that it got a lot faster! Same amount of heat was put in, just a bigger temperature difference across the engine. As I mentioned, even though we intuitively feel that hotter must be better, it's this difference that turns out to be the key in the efficiency of a heat engine, so making the cold side colder was just as effective.

Let's now dig into some thermo to figure out what's happening under the hood—no pun intended, I swear—in this heat engine. First, let's define more clearly what we mean by heat engine. This is important since the term heat engine is pretty generic, and as we'll see, there are a wide variety of heat engines, from the ideal case of Carnot, to the Stirling engine we just saw in the demo, to the Otto cycle commonly used in your car engine.

In general, the term heat engine refers simply to the conversion of thermal energy into useful work. The heat could come from a wide variety of sources, like gasoline, electricity, or the sun, to name a few. And the work is done for us by what is often called a working fluid; this could be an ideal gas, as in the cases we'll talk about today, but it could also be many other materials as well, like water, for example. In general, for a heat engine, we assume that there is an operating cycle. That means that the operation of the engine can be sustained and that the working substance, such as our gas, must be able follow a cycle that can be repeated over and over again.

Now, in a real engine, that would be, well, all the ones ever made in this world that do useful work for us, the processes are "non-ideal." As you may recall from other lectures, this means that they are irreversible. This means that for any engine, we always expect dissipative forces to be present. Now, the reason I'm telling you this is because I want to first talk about the very non-real case of an ideal engine. In this case, we assume that all processes are reversible, so there are no dissipative forces whatsoever. And just as we've done before, by first understanding the ideal case, we take our first step towards building the knowledge and formalism we'll need to understand the real case. Such an ideal engine is called a Carnot engine.

Here's the basic cycle of a Carnot engine, shown here in this image. First, we have some source of heat, which we call the "hot reservoir." You can think of this as some vast resource of thermal energy. In the car, it comes from making fire by burning gasoline. In the demo, it came from the hot plate that was plugged into the wall. This hot reservoir is held at some temperature, say T_h, and it is in contact with the working substance. For the demo, this temperature was set by setting the dial on the hotplate. As a result of being in contact with the thermal reservoir, the working substance absorbs some amount of heat.

As I mentioned, in my examples for the rest of today, I'll use an ideal gas as our working substance. So, let's call the amount of heat energy absorbed by the gas in a cycle Q_h, since it comes from the hot reservoir. Then, that ideal gas converts a part of the heat absorbed into useful work; let's call that amount of work, W. Any unused heat, that is, the heat that did not get converted into work, goes into the cold reservoir. Let's call that unused, or rejected heat Q_c, since it goes to a thermal reservoir at lower temperature. And yes, in our brilliant labeling scheme, we'll let the temperature of the cold reservoir be called T_c.

So, I'd like you to notice something very important here. How did I know that heat would flow from the hot reservoir to the working fluid and that any unused heat would flow to the cold reservoir? Well, if you've put on your second law hat, then you've got it, because we know from the second law that heat can only flow from a hot body to a cold body, and not the other way around.

The thermal efficiency of a heat engine can now be defined based on the simple process we just put forth. The efficiency will be the ratio of the net work done by the engine during one cycle to the energy input at the higher temperature during the cycle. So, efficiency equals W divided by Q_h. But, we also already said that Q_c is whatever heat was not used for work, so that means that W by definition is equal to Q_h minus Q_c. That means the efficiency is equal to Q_h minus Q_c divided by Q_h, which we can rewrite as 1 minus Q_c over Q_h.

So, the efficiency of a heat engine is simply the ratio of what we gain, which is the work done by the engine to what we give, which is energy transfer at the higher temperature. It turns out that this ratio is typically a lot less than 1, so most heat engines operate at much, much lower efficiencies than 100%. Did you know that your car engine only operates at an efficiency of about 25%? That means that from all of that precious ancient sunlight that we are extracting from the ground to put into our cars and trucks, we are only using a measly one fourth of it to make them move. The other three fourths is simply wasted as extra heat that's not used for anything.

Notice from our efficiency equation that in order to truly get to an efficiency of 100%, we would need for Q_c to be zero. This implies that a perfect engine cannot expel any energy whatsoever to the cold side, and all of the input energy would be converted to work, and the image would look more like this. But as I mentioned, this is not at all possible. In order to see why, we need to get a little bit more concrete here, beyond the abstract image of T's and Q's I've used so far. But anyway, by now I hope that you're just itching to get this process up onto a PV diagram. I know I am, especially since our working fluid is an ideal gas. And in order to make this all more concrete, I'll also bring back our piston from Lecture 5. remember how much fun we had with that, where we placed and removed the little pebbles on top to get a sense of the reversible mechanical work done by the gas inside? Well, here it is, back in all of its glory!

Remember, our system here is a cylinder fitted with a movable piston at one end, where the cylinder walls and the piston are thermally non-conducting, but, the bottom of the cylinder allows for thermal energy transfer. Now, in equilibrium the gas is exerting some pressure on the walls of the container, including on the piston, and the downward force from the weight of those pebbles exactly cancels this force from the gas pressure, leaving the piston at some given position. But this is not an engine yet. For that, we need a temperature difference. So let's take a look at what happens when we bring our two thermal reservoirs at temperatures T_h and T_c into contact with this system.

There are four key processes that we need to consider for the Carnot cycle. Two of them involve an isothermal path, so they occur at constant temperature. And the other two are adiabatic, so they occur with no heat transfer. Since this is the ideal case, we assume here that all processes are reversible.

So, we start at point A on the P-V phase diagram. Remember, this diagram is telling us what's happening to our ideal gas, or the working fluid of our engine. The piston at this point might look something like this. Now, we contact the gas from the bottom by our hot reservoir held at temperature T_h, and what happens is that an amount of energy Q_h is transferred to the gas. This is a process that makes the gas expand to a new volume, since we've raised its temperature, in turn, causing the piston to move upwards, as you can see in our drawing, and bringing us over to point B on the phase diagram.

So what's going on in the process A→B? Well, for one thing, it's isothermal, since the gas is in contact with the hot reservoir at a fixed temperature T_h. That means the gas is at this fixed temperature during the process. Also, during this process, the gas does work on the piston, since its volume increases when its temperature increases.

And what about the internal energy? Well, it's a good time to recall from our lecture on ideal gases, that the internal energy of an ideal gas is only a function of the temperature, well and the number of molecules, but certainly not a function of the volume and the pressure. And since we said this process is isothermal, and we're also assuming here that no gas escapes from the container, then we know that the internal energy does not change in going from A to B. That's pretty cool, since from the first law, we also know that the change in internal energy of a given process is equal to the heat plus work done by the system during that process, or delta-U equals Q plus W.

Let's remind ourselves of the sign convention we use; work done to the system is positive, and heat added to the system is positive. In this case, we have work being done by the system and heat added to the system, so Q will have a positive value and W will have a negative value. And since delta-U is zero for this process, that means that for this process, for this part of our ideal engine's cycle, the amount of heat added to the piston is exactly equal to the amount of work done by the system, since they must sum to zero. And as a reminder, the work done by the process is also equal to the area under this P-V curve.

Okay, so that was the first step. Now let's see what happens next. To get from point B to what we'll call point C, the base of the cylinder is removed from the hot thermal reservoir. And once it's removed, let's suppose that the bottom is made to be thermally insulating, just like we're assuming for the rest of the cylinder. Now what happens is that the gas keeps expanding, but now it expands adiabatically, since no thermal energy is allowed to enter or leave the system. So we have $Q = 0$, since it's an adiabatic process.

You can think of this process as one in which the gas is still hot, and we are removing little pebbles one at a time. And as we do, the hot gas can do work on the piston, increasing its volume while lowering both its pressure and temperature. In effect, what is happening in this idealized adiabatic process is that internal energy from the gas is being converted into mechanical work. As a result, more work is done by the piston, and the temperature goes down, let's say to T_c.

And next, we do the first process but in reverse. Of course, we have to compress the piston back if we want this to be a cyclic process. So, we're going to do this compression along another isotherm, in this case, at the lower temperature T_c. So we go back up the PV diagram over to what I'll call point D. Notice that for this process, since I want the temperature to be constant, I'll need to put back a reservoir in contact with the bottom. But, this is a different reservoir than in the first case, since it's at a lower temperature. Again, you can think of the compression as happening by adding pebbles one at a time, so we can picture the process as being slow and reversible. During this C→D part of the path, the bottom is no longer an insulator, but rather, in contact with a reservoir, so once again the gas can either give heat or or take heat to and from the reservoir. Since we're compressing the gas, it would have wanted to heat, up but instead, that thermal energy, we call it Q_c, is given over to the reservoir. The work in this case is not done by the gas, but rather, to the gas, by the piston, and it's equal to the area under this curve.

And finally, to get D back to A, once again we imagine the base of the cylinder being replaced by a thermally insulating wall. The gas is compressed further, but now it's compressed adiabatically, meaning there is no heat flow out of or in to the system. The temperature of the gas, therefore, increases back to T_h, and along the way, again, work is done to the gas by the piston pushing down on it, again, think about continuing to put those pebbles on to compress it, just up until the point when we're back to our original isotherm at T_h.

So, we made it through the Carnot cycle! And what does this get us? Here's the reason we're doing it. This is the absolute best we can ever do. It's an ideal heat engine that starts and ends in the same place, is based on reversible processes, and it gives us a total amount of work equal to the area enclosed by the path from A to B to C to D. The change in internal energy is zero for the full cycle, remember, internal energy is a state function, so, if we return to the same place in P-V space, it must have the same value. And since the internal energy change is zero, the net work done by the gas over one cycle must equal the net energy transferred into the system, which is the difference *Q_h* minus *Q_c*.

But the beauty of the Carnot cycle is that, since we now have a set of concrete processes as I outlined above, we can actually compute these Q's for each of the processes. And as a result, we can arrive at a very important result from here. I won't go through all of the math, but you've already learned what you need to know, since it just comes down to integrating the *P-dV* state functions over each process in order to find the work term.

Remember that the logarithm of the ratio of initial and final volumes that we found for isothermal expansion, way back in lecture 5, Well that comes back, among a few other things. But the point is that by computing explicitly the work and heat terms for each process, we find that for this cycle, the ratio of *Q_h* over *Q_c* is equal to the ratio of *T_h* over *T_c*. And that means that the efficiency of a heat engine, which remember we wrote down as 1 minus *Q_c* over *Q_h*, can be rewritten for a Carnot engine as one minus *T_c* over *T_h*.

Now, that is pretty cool; the efficiency of the ultimate, ideal-case heat engine is only dependent on the temperature difference between hot and cold reservoirs. Notice that if the temperature difference is zero, then we have one minus one, which is an efficiency of zero. We can understand this intuitively; no temperature difference means there cannot be heat flow, and if there's no heat flow, well, then basically nothing can happen! On the other hand, we can approach an efficiency of exactly one, or in other words, a perfect engine, if the temperature of the hot side tends towards infinity, or, the cold side tends towards zero, but you can see that neither of these limits is actually possible.

In our demo, the temperature difference was at first about 100°C from the hot plate, minus, say, 25°C for the air temperature. Of course, as we know, this is not the scale we should use when solving a thermo problem. For that, we have to use use Kelvin. That gives us an efficiency limit of 1 minus 298 over 373, or around 20%. Now, the Stirling engine in the demo is not quite as efficient as a Carnot engine for reasons I won't go into here, but I encourage you to go ahead and search for all the different types of engine cycles, and basically, they amount to variations of the four stages on the P-V path, which you're now well-equipped to understand.

But what Carnot does, is it tells us that absolute theoretical limit to the efficiency of making heat do work. At first, this value was 20%. When I put ice on the top side, I lowered *Tc* down from 298 to below 273 Kelvin. So, the theoretical limit of the heat engine just went up; that's why it sped way up. The actual engine, though, didn't operate anywhere near this efficiency, since it's in no way an ideal case.

Now, in most cases, like in our car or in a power plant, *Tc* is near room temperature. It would be highly impractical to maintain a cold reservoir of ice in real applications, because that alone would require an enormous amount of energy. So, since the cold side is stuck around 300 Kelvin for most cases, it's usually the goal in such engines to increase the efficiency by increasing *Th*.

There are many, many, losses in a real engines, so, I want to re-emphasize that the Carnot limit, based only temperature difference, is often not the actual operating efficiency of a heat engine. For example, in a real engine, say, the one in your car, there's a very short time interval for each cycle. I mean, at 1000 RPMs, those pistons are firing more than sixteen times per second. The result is that the gas inside of them, which for a car engine is simply air, doesn't quite make it all the way up to that high temperature of the hot reservoir, and similarly, there's not enough time for it to get all the way back down to the low temperature of the cold reservoir. That's just one example of the complications that arise when we move from the ideal case of Carnot to a real working engine.

So, in terms of how heat can be converted to work, we've made our way from perfect, to ideal, to real. And I hope that in this lecture you've gotten a good sense of one of the single most important thermodynamic cycles of human history. There are only a few other points I'd like to make here as I wrap up. First of all, I didn't have time to go into the details of heat pumps, which are different than the heat engines we talked about today. But, as you might imagine, a heat pump effectively works in the reverse of a heat engine, Instead of putting in heat to do work, we put in work to move heat.

And second, and this is really important, I want to be sure there's no confusion over the first and second laws and how they apply to this problem. The first law is a statement of energy conservation: delta-U equals Q plus W. The change in the internal energy of a system occurs because of energy transferred to or from the system by heat, or by work, or by both. Notice that the first law does not make any distinction between whatever happens because of heat or whatever happens because of work—either heat or work can cause a change in internal energy.

But here in this lecture, which is devoted to one of the most practical applications possible of thermodynamics, we find that there is a crucial distinction between heat and work, one that is not at all evident from the first law. In order to see the distinction, we needed the second law of thermodynamics.

From the point of view of engineering and practical applications, perhaps the single most important implication of the second law is the limited efficiency of heat engines. The second law states that a machine that operates in a cycle, taking in energy by heat and putting out an equal amount of energy by work, cannot be constructed.

More Engines—Materials at Work
Lecture 20

In this lecture, you will learn about the ways in which mechanical energy can be produced that are not as direct as making heat and then converting that heat into mechanical energy. There are many different ways to do this, but this lecture will focus on a few specific examples that expose different types of thermodynamic work. In some cases, the work being done is not likely to have much industrial relevance, although it demonstrates a wonderful example of the transformation of energy to motion, such as with the entropy and surface-tension engines. In other cases, such as the phase-change engine, there is great excitement for potential future advances and applications.

Magnetic Engines

- Our current ability to make seemingly endless and incredibly cheap heat will not last forever, so it's worth thinking about the following question: What other ways can we use heat to do useful work? What ways can we use non-fossil-fuel-based temperature differences?

- Magnetism is a special kind of configurational work that is very important in modern society. From data storage in computer hard drives to solenoids in car engines to various electric generators, magnetism helps the world run. The physical origin of magnetism arises from a combination of a particular type of interaction between electrons called spin and the induced magnetism that can come from external fields.

Magnetism makes it possible for computer hard drives to store data.

- There are different types of magnetism, but the one we will focus on is called ferromagnetism. Ferromagnets are the materials you probably think of as magnets from everyday experience. The key is that ferromagnets maintain a magnetization in the absence of an externally applied magnetic field—so, they can be magnetized.

- Very few ferromagnetic materials exist, and most of them contain either iron, cobalt, or nickel. Ferromagnetic materials are extremely important. In fact, the inner core of the Earth consists primarily of a liquid of iron and nickel.

- A key point is that if a ferromagnetic material gets hot enough, then it will lose the magnetism that it had—the alignment of all those spins goes away, and the material becomes what is called paramagnetic. This transition temperature is called the Curie temperature for any given magnetic material.

- We can actually take advantage of the temperature dependence of the magnetism of a material to do mechanical work. A system can use heat to create a change in a property—for example, magnetism—in order to perform mechanical work.

- In order to understand such a system from its thermodynamic viewpoint, we must write the work term due to magnetism, which is as follows: δW for magnetism equals the volume (V) of a material times the external magnetic field (H) times the change in the internal magnetic induced field (dB): $\delta W = VHdB$. So, the external magnetic field is the thermodynamic force, and the change in the induced field inside the material is the thermodynamic displacement.

Phase-Change Engines

- There's another type of engine we can build that relies on materials that change state abruptly depending on the temperature. These types of materials are called phase-change materials.

- It is true that all materials phase change in the sense that they can exist in a number of different phases, according to their phase diagram, but for phase-change materials, the actual shape of the material itself can be molded and "memorized" in two different phases so that when the material goes from one phase to the other, its shape changes as well. That's why we also call these materials shape-memory alloys.

- One particular phase-change material, known as nitinol, is an alloy of nickel and tin. The amazing thing about nitinol is that it can be trained to "remember" a shape by heat treatment under specific circumstances. After such training has occurred, the material can be bent, twisted, and pulled to any deformation, and it will return to its original shape when heated to its transformation temperature.

- This occurs because it undergoes a phase transformation, and in the higher-temperature phase, it always returns to its original shape. This transformation temperature can be tuned to be just a few degrees above room temperature. In addition, the material is extremely elastic, so it can be deformed quite a lot with very little effort, all the while being able to snap back to its original "trained" shape by a slight temperature increase.

- This is how a nitinol engine works. On one side, there is cold water so that the part of the wire immersed there can stretch and deform; on the other side, there is hot water, which causes that part of the wire to snap back to its original (more taut) shape.

- The result is a single wire constantly exerting a tension from one side, causing it to pull and turn, releasing the hot wire into the cold, allowing it to give; at the same time, it brings in more wire to the hot bath, which then goes through a phase change and pulls, and so forth. What we have in the end is a pretty efficient, metallic, single-piece solid-state engine that could theoretically run without maintenance for long periods of time.

Entropy Engines

- An entropy engine works entirely on the fact that entropy is changing in the material. Recall that the free energy (G) is equal to the enthalpy (H) minus the temperature (T) times the entropy (S): $G = H - TS$.

- Entropic effects are whatever contributions to the free energy come from the TS term because that is what has the entropy in it. And these entropic contributions to the free energy will carry more weight the larger the temperature.

- At room temperatures, for many, many materials—including metals, glasses, and ceramics—it is the term due to the enthalpy that entirely dominates the free energy. Because entropy dominates over enthalpy, the change in free energy can be written as approximately just the change in $-TS$.

- For a larger temperature, the effect of entropy is larger, which makes the system even more strongly "want" to be in the higher entropy state. That's actually counterintuitive. Usually, when we apply heat to something, it tends to become more elastic or more flexible; somehow, our intuition tells us that hotter stuff is "looser" in general. But in this case, with entropy in the driver's seat, the opposite is true: Higher temperature leads to a stronger, not weaker, tension.

Surface-Tension Engines

- There are many ways to make motion—and, of course, not just from heat. One way is to use electricity to turn a motor. This is, in fact, another form of magnetic work, and it's the kind we use all the time, including in a blender, a fan, a remote-controlled helicopter, or an electric car. But one approach that doesn't use heat that you may not have encountered is the one based on surface tension.

- How can we make something actually move using only the concept of surface tension? It's yet another form of thermodynamic work, and for this one, we use the symbol gamma (γ). That is called the surface energy, and it is defined as the energy associated with the creation of a unit area of surface.

- Similar to the pressure-volume work term PdV, the work term associated with creating (or, for that matter, removing) surface by some amount dA would be equal to γ times dA. The units for the surface energy gamma are joules per meter squared.

- For the case of a drop of water in air, the surface would be defined as the interface between the water and the air. In thermodynamics, we can classify different types of surfaces based on the type of interface involved—often this is between two different phases (such as a liquid and solid or a solid and gas) or a boundary between two different phases (such as two different solid phases) within the same material.

- In all cases, the presence of a surface requires energy in order to form. That is, it takes energy to introduce an interface into what would have otherwise been a pure and continuous material phase. This extra energy is called the surface energy.

- From an atomistic viewpoint, atoms in the bulk phase of a material—atoms just within the part of the material where there are no interfaces—are bonded in all directions and remain stable. However, the atoms at the surface are not ideally bonded in all directions. The bonds on one side have simply been cut away, making those atoms unstable, and as a result, they try to find a happier place, often by moving around in some way to lower their energy.

- This rearrangement of the atoms upon formation of a surface is called reconstruction, and it's what determines the shape at the surface. But as hard as those atoms try to be happy, their bonding energy is still not going to be as strong as it was when they were hanging out in the bulk region. That energy penalty is the gamma term—the surface energy.

- The result of having a surface energy is that if this energy is changed along some appropriate displacement, then thermodynamic work is performed. In this case, the displacement is area, which makes sense because we are talking about a surface. And gamma can then be thought of as the thermodynamic force.

- Why is it that you can fill a glass of water ever so slightly above the top of the glass and it still won't spill out? It's all due precisely to that force. Any change leading to an increase in the surface free energy is resisted by the surface, and this resisting force is called the surface tension. It's the same thing as gamma.

- The value for this surface tension depends on a whole lot of things. For a solid, it depends on the type of material as well as what particular orientation the surface cut was made. In a liquid, the orientation doesn't matter, but the material itself can have very different values for gamma.

Suggested Reading

Callister and Rethwisch, *Materials Science and Engineering*, Chap. 20.

Moran, Shapiro, Boettner, and Bailey, *Fundamentals of Engineering Thermodynamics*, Chap. 6.

Questions to Consider

1. List the different types of magnetism, and explain what the Curie temperature means.

2. How big of a difference in surface energy would be required to move a large boat, as opposed to the small ones used in the demo?

3. What are the pros and cons of the nitinol—or phase change—engine?

4. Why does a rubber band expand when cooled but contract when heated, unlike most materials?

More Engines—Materials at Work

Lecture 20—Transcript

Hi, and welcome to our second lecture on engines. Actually, I'm not quite comfortable calling what we'll be talking about today engines, since when I use that word most of us think about cars, trains, or planes. That type of engine, where we first make heat and then convert that heat directly into mechanical energy, that was the subject of the last lecture. Here in this lecture, I want to discuss other ways in which we can produce mechanical energy, ways that are not so direct.

There are many different ways to do this, as you might imagine, since there are so many different kinds of thermodynamic forces, but in this lecture I'll focus in on a few specific examples, ones that help us see different types of thermodynamic work. In some cases, the work being done is not likely to have much industrial relevance, although, it demonstrates a wonderful example of the transformation of energy to motion, like for the entropy and surface tension engines that I'll show you today. In other cases, like for the phase change engine, which I'll discuss in a little bit, there's a great excitement in the potential for future advances and applications.

So, I suppose the first question I should address in a lecture on engines other than heat engines should be, why would I want to make any other types of engines in the first place? I mean, the engines we have based on heat are nothing less than the foundation of all of our modern technological society. They provide us with nearly all of our electricity, all of our transportation, and do almost all of our work. Just keep feeding them heat, and they will keep on working for us in all these ways. Why even think about changing this all-encompassing heat-to-mechanical work framework?

Well, one thing that we know is that the conversion of heat into work is not that efficient. Remember the Carnot limit? That tells us that the ultimate limit of efficiency possible in converting heat into work is set by the temperature difference we are able to apply to the engine. And as we saw, given that in most practical applications we are stuck with the cold side being around 300 Kelvin—that's room temperature—and the hot side is limited both by how hot we can actually make it, as well as the limit of the materials being used to build the engine and transfer the heat.

This latter point is extremely important. If I build a piston for an engine out of aluminum, then the maximum temperature possible 659°C, since at that point aluminum melts, whereas, if I build it out of bronze, I can get up to 913°C, and Nickel takes me to 1452°C. And actually, the real operating temperature of an engine is typically far below the engine materials' melting points to avoid any risk of the materials' instability. So, the need for a large temperature difference in order to get high efficiency can be challenging. And even with a large temperature difference, Carnot tells us that the efficiency limit is still not very high. A car engine typically runs at about 25% efficiency.

But in this lecture I will still be talking about starting with some form of heat and then making something move in the end. So the Carnot limit still applies to everything I'll be telling you. However, an important point is that at least with fossil fuels we've been able to create large enough temperature differences that we do pretty well. Power plants do, after all, operate at around 50% efficiency. But the thing is, this ridiculously cheap source of high temperature heat will some day, not too far from now, be seen as a precious blip in our world's history.

Our current ability to make seemingly endless and incredibly cheap heat will not last forever. Nearly all of the heat we generate comes from burning fossil fuels, either coal, or oil, or natural gas. I like to call these resources "ancient sunlight, since their origins all lie in biomass that grew from the sun's energy and accumulated over many thousands of years, and then got buried for millions of years more. But such resources are quite finite. I know there are a lot of arguments around how much more oil we have, or how much more natural gas we can crack out of rocks miles underground, but to me, this is a very small point. In the long-term picture, it just doesn't matter. All that matters is that these resources will run out. It may be 50 years, or it may be 150 years, but at some point in the future, engines of the world may not run on high heat once we cannot produce heat at such high temperatures so easily.

So it's worth thinking about the following question. What other ways can we use heat to do useful work? What ways can we use non-fossil-fuel-based temperature differences? To take a look, let's run a few examples of some other types of engines, ones that are based on the unique properties of materials in different ways.

First, we'll make a magnetic engine, but perhaps not the kind you're used to. Magnetism is a special kind of configurational work that's very important in modern society. From data storage in computer hard drives to solenoids in car engines to various electric generators, magnetism helps the world run! And the physical origin of magnetism arises from a combination of a particular type of interactions between electrons, that's called spin, and the induced magnetism that can come from external fields.

There are different types of magnetism, but the one I want to focus on here is called ferromagnetism. These are the materials you probably think of as magnets from everyday existence. The key is that ferromagnets maintain a magnetization in the absence of an externally applied magnetic field. So they can be magnetized. Very few ferromagnetic materials exist, and most of them contain either iron, cobalt, or nickel. By the way, as few elements as there are that are ferromagnetic, they are extremely important! I'm sure you've seen all sorts of examples of these materials, not to mention the fact that the inner core of the earth itself consists primarily of a liquid of iron and nickel.

But a key point is that if a ferromagnet gets hot enough, then it will lose the magnetism that it had; the alignment of all those spins goes away, and the material becomes what is called paramagnetic. This transition temperature between ferromagnetic and paramagnetic is called the Curie temperature for any given magnetic material. Now, here's the thing; we can actually take advantage of the temperature dependence of the magnetism of a material—you guessed it—to do mechanical work! Let's take a look.

As we learned in the lecture, a ferromagnet has what is called a Curie temperature. That's a phase transition where the material goes from being magnetic to non magnetic above a certain temperature. So, in a setup I have here, there's a ferromagnet, that's this little loop right here sitting above the candle, and that's attached to a wire. And there's a permanent magnet over here, sitting on the edge. So that's drawing this loop over. You see how the permanent magnet is drawing the ferromagnet over. So it's attracted to that permanent magnet.

But, when I light the candle, the fire is going to heat up the ferromagnet and get it above the Curie temperature; let's watch what happens.

So there's that… Oh, and you can see that it comes away from the magnet because once the fire gets that ferromagnet, the little loop at the end, hot enough, it goes above the Curie temperature and completely loses its ability to be magnetized. So it's no longer going to be attracted to the permanent magnet on the end.

But, what's cool is that it also, when it starts cooling down, because it's away from the fire, it goes back below the Curie point. And so, it can be magnetized again. And that's why it goes back and forth the way that you see here. This is an engine that takes heat and makes mechanical motion. It's not exactly an efficient heat engine, but it's a really cool example of magnetic work and phases of materials.

So there you had a system that uses heat to create a change in a property, in this case, magnetism, in order to perform mechanical work. In order to understand such a system from its thermodynamic viewpoint, we must write the work term due to magnetism. And that is as follows: δW for magnetism equals the volume V of a material times the external magnetic field, written as H, times the change in the internal magnetic induced field written as dB. So the external magnetic field is our thermodynamic force, and the change in the induced field inside the material is the thermodynamic displacement.

The experiment that I showed you just now can be understood as follows. If we had a piece of iron initially at room temperature with an internal magnetic induction of B = 0 Tesla—those, by the way, are the units of magnetism—an external field is then applied to the system, say, with a field strength large enough to result in an induction in the material of B = 0.5 Tesla. The magnet is then heated to 750°C using the candle, which makes it lose its magnetism, since it went above the Curie temperature. So in that case, once it goes above that point, B goes back to 0. That, in turn, made it not attracted to the permanent magnet that was close-by, which takes it away from the heat source. And that, in turn, allows the magnetism to return, since it cools down, so it's once again attracted to the permanent magnet, and so, the mechanical work cycle continues.

That experiment shows an example of how we can take advantage of the temperature dependent properties of a material to make something move. It's both similar to and different from the P-V-T work of a typical heat engine. Similar, because for P-V-T, the temperature dependence of a property, in that case pressure, is also used. But very different, because, in the case of magnetism, the temperature dependent property is an on-off state, rather than a continuously changing variable.

In much the same on or off way, there's another type of engine we can build that relies on materials that change state abruptly depending on the temperature. These types of materials are called phase change materials, and they're really cool. Of course, right away you may be thinking, but aren't all materials phase change in the sense that they can exist in a number of different phases, according to their phase diagram? Well, absolutely that's the case. But, for phase change materials, the actual shape of the material itself can be molded and "memorized" in two different phases, so that when the material goes from one phase to the other, its shape changes as well. And that's why we also call these materials shape memory alloys. Let's take a look!

So in this demo we're going to make something move using a phase change. Nitinol was first discovered in 1962, and it has fascinating mechanical behavior. It's a shape memory alloy, like we discussed in the lecture. That means that it can be trained to take a certain shape, and when it goes through a phase transformation, it goes back to that shape.

Here, we'll use this concept, again, to make something move. So, what I have here is a single nitinol wire that has been trained to have a high tension state and a low tension state, depending on it's temperature. So it has two phases, one that has a higher tension than the other. The wire is wrapped around these pullies here, but it's just one single wire, okay?

So, the two phases happen right above and right below room temperature, which means, that if one side is placed in hot water and the other side in cold water, part of the wire contracts, pulling on the wire, but as it rotates out of the cold side, it goes through a phase change, and loosens, while the new part of the wire that rotates into the cold, contracts. And it continues in this way to power an engine.

Let's give it a shot. I'm going to pour some hot water into this container, so that's hot water. And I'm going to pour some cold water into this side, okay. And when I put this nitinol wire wrapped around these pullies inside, this middle part here, and I give it a little kick start, you get it going. You can see that it starts running right on its own. It's because the wire is undergoing a phase change depending on whether it's on a hot side or the cold side.

Now you notice that it's starting to slow down, and I bet you might be able to guess why. It's because the wire is heating the cold water, as it brings hot water over to that side, and it's cooling the hot water as it brings the cold water onto the hot side. Eventually, the water is simply, close enough to the same temperature, that we no longer have enough of the temperature differential, or temperature gradient, which we know that we need in order to keep doing mechanical work.

So there we had a moderate temperature difference, of say, 50 Kelvin, and with that, we were able to make this motor spin. There was no volume expansion of a gas or pressure change, rather, just a material that changes its shape back and forth with temperature. This particular phase-change material is known nitinol, and it is an alloy of nickel and tin.

Take a look at the phase diagram for these two materials. Hey, hold on a second. Let's just take in the fact that I just said that, and you knew what I meant. It's pretty cool to be able to read such a materials map, so I wanted to be sure to call that out. And this one in particular is pretty complicated. Notice that over here we have full miscibility between nickel and tin, while past about 3% tin, the alloy tends to phase separate into one of several different very stable, distinct nickel-to-tin ratios.

The reason I wanted you to check out the phase diagram for nickel and tin is because you can get a sense from this for how difficult it would be to precisely control the composition and phase of an alloy made from these elements. It's really only since, the 1990's that nitinol, which is a mixture of about 50% tin, and 50% nickel by atomic percent, can be made fairly routinely.

The amazing thing about nitinol is that it can be trained to "remember" a shape by heat treatment under specific circumstances. After such training has occurred, the material can be bent, twisted, and pulled to any deformation, and it will return to its original shape when heated to its transformation temperature. The reason? Well, because it undergoes a phase transformation, and in the higher temperature phase, it always returns to its original shape. And the cool thing is that this transformation temperature can be tuned to be just a few degrees above room temperature.

The other cool thing is that the material is extremely elastic, so it can be deformed quite a lot with very little effort and all the while being able to snap back to its original, "trained" shape by a slight temperature increase. And that is how the nitinol engine I just showed you works. On the one side, we have cold water so the part of the wire immersed there can stretch and deform, while on the other side we have hot water, which causes that part of the wire to snap back to its original, more taught, shape.

The result is a single wire that is constantly exerting a tension from one side, causing it to pull and turn, releasing the hot wire into the cold, allowing it to give, while at the same time, bringing in more wire to the hot bath, which then goes through a phase change and pulls, and so forth. What we have in the end is a pretty efficient, metallic, single-piece, solid state engine, that could run, theoretically, without maintenance for long periods of time.

Okay, so there we have yet another way to leverage a sharp change in the property of a material with temperature. It's not pressure, as in a traditional heat engine, and it's not a type of magnetism, as in the magnetic heat engine I showed a few minutes ago. In this case, the power comes from the material changing its phase from one solid form into another solid form, back and forth, with temperature.

Okay, so, I have two more engines to show you today. Our last engine based on heat, and then one that doesn't involve any heat at all. Now, this next one is cool because it works entirely on the fact that entropy is changing in the material. It's in fact, an example of an entropy engine, Let's take a look.

Now I know that you might be thinking, that sure did spin pretty slowly! And you're right, I wouldn't exactly win any awards if I were to race a car running on an entropy engine. Still, it's fascinating to consider the thermodynamics at play in this case.

Let's start with a rubber band. You can do this at home. Take the rubber band and hold it up against your skin, right above your upper lip. This is a good place because it's pretty sensitive to temperature. Now, while holding the rubber band in contact with your skin, stretch it quickly. You'll notice that it gets hotter when it's stretched. And now, start it in the stretched position, and again, while maintaining contact with your skin, allow it to contract back to its starting point. In that case, you'll notice that the rubber band gets colder; you can feel the cold pretty clearly on your lip. Stretching makes it warm, and recoiling makes it cold. Why is that the case?

Well, in order to understand this we turn to our good friend the free energy. Remember, that the free energy G is equal to the enthalpy H minus the temperature times the entropy, or T times S. What are called entropic effects are whatever contributions to the free energy come from the TS term, since that is what has the entropy in it. And as we've discussed in previous lectures, these entropic contributions to the free energy will carry more weight the larger the temperature. And, that's kind of obvious since it has a factor of T in it. But at room temperatures, for many, many materials that we know and love, like metals, glasses, ceramics, for example, it's the term due to the enthalpy that entirely dominates the free energy.

But for polymers, and well, these are different. A rubber band is made out of polymers. Polymers are long little strands of some repeating molecular sequence, and often, when they come together, the result is a kind of spaghetti of these strands. This is where we can see the role of entropy. Take a look at these three different states of a single polymer strand. On the left, it's all tangled up into a ball, while on the right, it's fully stretched out into a straight line, and in the middle it's somewhere in between.

Remember our statistical definition of entropy. It's the function of possible microstates for a given macrostate. So, in the case on the right, where the strand is fully stretched out in a line, how many different ways can we arrange it to be that way? The answer, is only one; only one possible microstate gives us this arrangement.

On the other hand, when the strand is all tangled up, like on the left here, there are many, many different ways the tangling could occur and give the exact, same energy. That means that there are many different microstates that represent the same macrostate, so in that case, the entropy is very high.

In fact, since these spaghetti strands are quite weakly bonded together in a rubber band, we can ignore the enthalpy term, and at least, at room temperature, changes in the free energy are entirely dominated by changes in entropy. When the rubber band is stretched, the strands straighten out and the entropy goes way down, while when it is unstretched, the strands are coiled around, and the entropy skyrockets up again.

So that gives us the connection we need. Turning back to our free energy expression, since I already told you that entropy dominates over enthalpy, the change in free energy can be written as approximately just the change in minus TS. And for a larger temperature, the effect of entropy is larger, which makes the system even more strongly "want" to be in the higher entropy state. That is the state where the rubber band is fully unstretched

So, in effect, what happens when we put heat onto the rubber band is that it makes the rubber band have a strong force of contraction. If you think about it, that's actually pretty cool, since it's counterintuitive. Usually, when we apply heat to something, it tends to get more elastic, or more flexible; somehow our intuition tells us that hotter stuff is "looser" in general. But in this case, with entropy in the driver's seat, the opposite is true; higher temperature leads to a stronger, not weaker, tension in the rubber band.

In the entropy wheel, the spokes are replaced by rubber bands, as you saw in the demo. If only one of the sides of the wheel is heated, the rubber bands in the proximity contract and pull the rim closer to the center. This, in turn, causes a shift in the center of mass and gravity of the wheel, which causes the wheel to rotate, and now the rubber bands on either side swap positions. Those originally heated rubber bands now cool down, and the other set of rubber bands get heated up. So the same contraction happens, and the wheel continues to rotate. The entire driving force for the rotation arises from a change in entropy.

So there we had three different ways to make something move using heat, all covering different aspects of thermodynamics. In the original heat engine I discussed in the last lecture, the increase of pressure from heat did the trick. Here, so far, we've seen that we can also get mechanical motion from the temperature dependence of the magnetic properties of a material, from the temperature dependence of the phase of a material, and just now from the temperature dependence of the role of entropy in a material.

There are, indeed, many more ways to make motion, and of course, not just from heat. One way that you'll all be aware of is to use electricity to turn a motor. This is, in fact, another form of magnetic work, and it's the kind we use all the time, whether in a blender, a fan, a remote controlled helicopter, or an electric car. But I'll bet that, in the category of "making things move in a way that doesn't use heat," one approach you may not have encountered is the one based on surface tension. Here, let's take a look!

So here we're going to make something move using another kind of force, that is, surface tension. When there's a change in the surface tension, it pulls the surface to the region of greater surface tension. And this can be used to make something move. So let's take a look.

First, we're going to sprinkle some oregano onto some water, okay? And what you can see is that the oregano floats on the top. So the oregano pieces are very strongly connected to the surface of the water. Now, watch what happens when I dip my finger into water. You can see, really, not much of anything. But now, I'm going to put my finger in some soap. So I have some soap here, and I'm going to put my finger in the soap, like that, and I'm going to dip it into the water again.

And now you can see when I did that, the oregano moved away from my finger. Why? Well, because the soap in the water makes the water have a different surface tension than the water without the soap. And so you have a different surface tension on one side of the oregano, compared to the other side. And that makes it move.

We can have a little more fun with this, if we want, by making little, tiny boats and doing the same thing. So here I have a whole lot of little boats that I've pre-printed, and they're floating on top of water. And watch what happens when I put some alcohol into the little chamber in the boat. So there's a little chamber inside each boat. When I put some alcohol in it, you see that it starts to move.

Why? Because as the alcohol leaves the back of the boat, it creates a change in surface tension between the back of the boat and the front of the boat. So I'm going to activate a bunch of these by dripping alcohol into those chambers. And so we have a surface tension difference between the front and backs of these boats that makes them move around on the surface of the water. … and… a little bit of a log jam here. Let's clear that up.

And that is another kind of thermodynamic force that allows us to make something move—surface tension.

So how exactly did that work? How can I make something actually move using only the concept of surface tension? Well, it's yet another form of thermodynamic work, and for this one, we use the symbol gamma. That is called the surface energy, and it is defined as the energy associated with the creation of a unit area of surface.

Similar to our *P*-*V* work term, PdV, the work term associated with creating, or for that matter, removing, surface by some amount dA would be equal to γ times dA. As you can see, the units for the surface energy gamma are Joules per meter squared. So, how should we even think about this thing I just defined for you, the surface energy? Well, first of all, what is a surface?

For the case of a drop of water in air, it would be defined as the interface between the water and the air. And, that's the key word—interface.

It turns out that in thermodynamics we can classify different types of surfaces based on the type of interface involved; often this is between two different phases, like a liquid and a solid, or a solid and gas, or, it's a boundary between two different phases, like two different solids within the same material. In all cases, the key point is that the presence of a surface requires energy in order to form. That is, it takes energy to introduce an interface into what would have otherwise been a pure and continuous material phase. This extra energy is called the surface energy.

From an atomistic viewpoint, atoms in the bulk phase of a material, that means just within the part of the material where there are no interfaces, those atoms are bonded in all directions and remain stable. However, the atoms at the surface are not ideally bonded in all directions, since as you can see, the bonds on one side have simply been cut away. This makes those atoms unstable, and as a result, they try find a happier place, often by moving around in some way to lower their energy.

And this rearrangement of the atoms upon formation of a surface is called reconstruction, and it's what determines the shape at those surfaces. But as hard as those atoms try to be happy, their bonding energy is still not going to be as strong as it was when they were hanging out here in the bulk region. And that energy penalty? That's our gamma term— our surface energy.

Now, the result of having a surface energy is that, if this energy is changed along some appropriate displacement, then the thermodynamic work is performed. In this case the displacement is area, which makes sense; we are, after all, talking about a surface. And gamma can then be thought of as our thermodynamic force.

Why is it that you can fill a glass of water ever so slightly above the top of the glass and it won't spill out? Well, it's all due precisely to that force. Any change leading to an increase in the surface free energy is resisted by the surface, and this resisting force is called the surface tension. It's the same thing as gamma.

Now, the value for this surface tension depends on a whole lot of things. For a solid, it depends on the type of material, as well as what particular orientation the surface cut was made. In a liquid, the orientation doesn't matter, but the material itself can have very different values for gamma. That's how the oregano magically floated away and how those boats zipped around. In the case of the oregano, when I introduced a soapy liquid into the water with my finger, that soapy stuff immediately started diffusing outward. But soap water has a much lower surface tension than just pure water. As a result, the little flakes of oregano felt a different surface tension on one side—the side where the soap was—than on the other side, where the soap hadn't reached yet in its diffusion outward. And this difference in surface tension led to a net force, pushing the oregano out in a wave.

The same thing happened for the boats. The alcohol inside has a very different surface tension than the water, so, as it trickled out the back, a net force formed due to the difference in surface tensions between the front and back sides of the boat. The boat rode along because of this net force. And of course, this isn't something we could use to power a real, large size boat. I mean, unless you carry thousands of gallons of alcohol with you. So powering the world with surface tension differences may not be a very practical direction. But still, you see in this example yet another way of making something go and how the thermodynamic work term associated with it.

In this case, it had nothing to do with heat. So, where did the energy come from? Well, one way to think of it is that it took energy to get the chemicals that had different surface tensions in the first place. Notice that once the alcohol and water are mixed together, it's pretty hard to re-use that liquid. I could keep adding alcohol to those boats and eventually they wouldn't run anymore, since the liquid they're sitting in would be a mixture of water and alcohol with too similar of a surface tension to make any difference.

So, there you have it—four other engines that run on four quite different concepts and materials properties. And as I said, there are many, many more examples I could give. Just consider that I haven't even included in today's discussion any force or work terms involving light or electricity.

But today I just wanted to show you a few examples, which I hope gave you a sense of how we can take advantage of materials in different ways to do work. The possibilities and numbers of work terms in thermodynamics is incredibly vast!

The Electrochemical Potential
Lecture 21

This lecture will introduce a new thermodynamic potential, one that combines chemistry with electricity, called the electrochemical potential. It's a potential of enormous importance because it describes the operation of something we use every day—the battery. A battery converts chemical energy into electrical energy, and if it's rechargeable, it can go the other way, converting electrical energy back into stored chemical energy. The electrochemical potential not only allows us to explain how a battery operates, but it also describes the underlying thermodynamics of any system where chemical reactions are occurring together with charged particles.

Batteries

- The analysis of a battery introduces another new form of energy—it's a form of internal energy for the system called the electrochemical potential. The electrochemical potential is similar to the chemical potential in the sense that it is another thermodynamic energy potential, one that once you know how to write down can help to find the equilibrium conditions for a given system.

- In a battery, reactions occur that both create and consume electrons, which in turn provide the electrical current when the battery is connected to a circuit. Electrons are by definition charged species that carry a charge of −1. That's a fundamental unit of charge itself.

- But atoms and molecules can also be charged. This state of charge that an atom or a molecule can take on is known as the oxidation state. Atoms tend to have natural oxidation states. In fact, some atoms will readily lose an electron or two or even three, while others will readily take on additional electrons. And then there are still others (noble gases) that simply don't ever want to deviate from their neutral state.

- So, if a battery is going to involve having chemistry occur that also moves charges around, then the materials involved in the operation of the battery should be able to either take on or give away charges.

- This is why in batteries we often see elements like zinc, magnesium, lithium, sodium, nickel, lead, and manganese. All of these are elements that have nonzero oxidation states and are happy to give away and then take back some charge. When they're in a state of net positive charge, so they've given some electrons away, we refer to them as ions.

Lithium is a common element for batteries to contain.

The Electrostatic Potential Energy

- The operation of a battery involves the trafficking of charged components, including both electrons and ions. Because the reactions that occur in a battery generate charged components (ions in a solution and electrons in the electrodes), we have to introduce a new source of internal energy for the system: the electrostatic potential energy. This is energy that arises from the electrostatic "pull" of opposite charges.

- A positive charge and a negative charge will be attracted to one another by a force equal to a constant times each of the two charges divided by the distance between them squared. This relationship was rigorously determined by the French physicist Charles Coulomb in

1785, and it is therefore known as Coulomb's law, and the constant in the equation is known as Coulomb's constant.

- We can compute work done from the Coulomb force. The displacement in this case is the change in the distance between the two charged species. You can think of the Coulomb force of repulsion between the two species as the two particles having energy stored in the electrostatic field, and you can think of this energy as being potentially available whenever it is needed. You can think of the potential energy for an electrostatic force as being defined as the capacity for doing work that arises from the position or configuration of the charges.

- Any collection of charges will have a potential energy associated with it. So, the general expression for the potential energy of a given charge (q) that comes near another fixed charge (Q) is equal to the following work expression: $V_{\text{electrostatic}} = kqQ/r^2$, where k is Coulomb's constant.

- The electrostatic potential energy per unit of charge is units of joules per coulomb, so we often divide the above expression by that test charge q to arrive at the following: Electrostatic potential per unit charge is equal to Coulomb's constant (k) times the charge of the fixed particle Q divided by the distance it is from that particle (r): kQ/r.

- Because the charge of the test particle has been divided out, the electrostatic potential is a property that is related only to the electric field itself and not the test particle. For this version, we usually use the symbol Φ, and in this case, the work done to bring a particle with charge q to any point in space will equal the electrostatic potential Φ at that point times the charge q.

- The electrostatic potential is the key variable we need for our thermodynamics. And the thermodynamic work expression for a charged particle is equal to the charge of the particle times Φ, where Φ depends on the position of the particle.

The Electrochemical Potential

- The electrochemical potential will become a new term in the fundamental equation for the internal energy—a term that accounts for the new form of energy that arises due to the separation of charges in a system.

- For a closed system where only mechanical work is possible, the fundamental equation is written as $dU = TdS - PdV$, assuming a reversible process. However, there can be many other forms of work, and they simply get added to the right-hand side of this equation if they are present in the system.

- So, for the case of chemical work, we have an addition term—the sum over all chemical components in the system of the chemical potential of each component times the change in the total number of moles of that component: $\sum_{i=1}^{n} \mu_i dN_i$.

- For electrostatic work, we add another term to the right-hand side, namely a sum over each component of the electrostatic potential for that component times the change in the total charge of that component. This is the electrostatic work term, which can be incorporated into the same sum over the different components.

- The electrochemical potential is the combination of those two terms in the summation over the number of components: the chemical work and the electrostatic work. However, wouldn't it be nice if we could rewrite this as a single potential? In fact, we can. To do so, all we need to do is rewrite that change in the total charge of a given component in the system as a change in the number of moles of that same component. This will bring us to our final definition of the electrochemical potential.

- To do this transformation, we introduce new terms. The total charge of a given component in the system is the valency (Z) of that component times the charge of an electron times Avogadro's number times the number of moles of that component.

- The valency is simply the number of electrons that the component typically gives up. For zinc, for example, this number is +2 because it likes to lose 2 electrons. For sodium, this number would be +1 because it likes to lose only a single electron.

- The number of electrons a given component likes to give up times the charge of an electron is equal to the charge each component (this could be an atom or a molecule) likes to give up. Avogadro's number times the number of moles of that component is the total number of them. A mole of zinc atoms would be 6×10^{23} atoms. So, when we multiply this all together, it is exactly equal to the total charge that the given component has in the system.

- We now have a description of the electrostatic potential that depends on the change in the number of moles of a component. This allows us to group it together with the chemical potential term. And we can use Faraday's constant (F) instead of the charge of an electron times Avogadro's number. Finally, we have the fundamental equation expression, which now has $dU = TdS - PdV + \sum_{i=1}^{n} \mu_i dN_i$.

- We write the electrochemical potential with the same symbol as the chemical potential (μ), but to signify that it's been "electrified," we add a small line on top of it ($\overline{\mu}$). And $\overline{\mu}$ is equal to μ plus the valency (Z) times Faraday's constant (F) times the electrostatic potential Φ: $\overline{\mu} = \mu + ZF\Phi$.

- The electrochemical potential contains the intrinsic chemical potential of each component plus the electrostatic potential acting on charged species. Just as the chemical potential of a species must be the same in all phases at equilibrium, the electrochemical potential of charged components must be the same in all phases to reach electrochemical equilibrium.

- A battery operates by extracting electrical work from a chemical reaction. And just as with chemical reactions, electrochemical reactions can be analyzed using the Gibbs free energy and the chemical potential in order to determine the equilibrium state at constant temperature and pressure. The Nernst equation relates the electrostatic potential energy change to the change in free energy of the overall reaction for the system divided by 2 times Faraday's constant.

Suggested Reading

Callen, *Thermodynamics and an Introduction to Thermostatistics*, Chap. 8.

DeHoff, *Thermodynamics in Materials Science*, Chap. 15.

Questions to Consider

1. Where is the energy in a potato battery coming from?

2. How long will this battery last?

The Electrochemical Potential
Lecture 21—Transcript

In today's lecture, we're going to introduce a new thermodynamic potential, one that combines chemistry with electricity. It's a potential of enormous importance, since it describes the operation of something we all use every single day. I'm using one right now in this microphone. It's likely you're using one too, whether in your phones, computers, watches, TV remote, power tools, e-reader, Nintendo DS, night light, calculator, or car. I'm talking, of course about the battery.

A battery converts chemical energy into electrical energy, and if it's rechargeable, it can go the other way, converting electrical energy back into stored chemical energy. So we're combining electrical energy and chemical energy, well, you know by now as exciting and amazing as thermodynamics is, it's not exactly super-creative when it comes to terminology. This critical new thermodynamic potential is called the electrochemical potential. It not only allows us to explain how a battery operates, but it describes the underlying thermodynamics of any system where chemical reactions are occurring, together with charged particles.

For example, a technology of enormous importance is called electroplating, which is widely used in industry for coating metal objects with a thin layer of a different metal, just think about all those chrome-plated faucets. And another example of a process that involves the chemical reaction of charged particles is corrosion, which, well, you probably already know what that is; just take a look at those rusty spokes on your bike, especially if you've left it out in the rain a few too many times.

Now, for today's lecture, I'll be focusing on the battery as an example of the electrochemical potential. And just so we get ourselves in the battery mood, let's be sure we have a feeling for the energy scales we're talking about. Did you know that your cell phone runs on just a few watts of power? And as a reminder, a watt is a joule per second. So, let's say, if a cell phone runs at 5 watts, then, if it has a battery life of 5 hours, it means the battery stores an amount of energy of 90,000 joules. A typical incandescent light bulb might run at 100 watts, so to keep a kitchen with 6 of those on lit for 2 hours, well, that would take 720,000 joules of energy.

Now, for batteries, we often don't use joules as our units of energy, but rather, watt-hours. That way we can think more closely about the connection to the power being used. A watt-hour is just what it sounds like; it's the amount of energy needed to provide one watt of power for one hour, so, that's 3600 Joules of energy. For reference, a typical AAA battery can store roughly 1.5 watt-hours, and the energy needed to drive a car 200 miles is about 100 kilowatt-hours. The typical maximum power in your house is around 40 kilowatts, and the world, well, it's increasing by the minute, but in 2013, it uses about 13 terawatts of power, give or take a few watts. So that means that with a mere ten thousand billion Triple-A batteries, you could power the world for about an hour! But my main point here is that I'd like you to be sure that you're familiar with these different energy scales and comfortable with the difference between power and energy and that you can go back and forth between the two.

Now, all this talk of power and energy makes me want to go power something. Did you know that you can charge your phone with a potato? Or power a light with a tomato? Let's go take a look!

So here we have a battery, which is a way of storing electrochemical energy that a lot of us are very familiar with. I take this battery, and I hook it up to wires. And as you can see, I can power something; in this case, I'm turning on a clock.

But, did you know that from just two metal wires and a couple of potatoes I can also get electricity? Here I have a piece of zinc, and here I have a piece of copper. What I'm going to do is I'm going to wire these up through the potato and hook them back up to the clock.

Now, for this demo, just to up the current a little bit, I'm actually going to use two potatoes; I'm going to put a bridge in between them. So I've go copper wire going into this one, and a zinc wire going into that one. And then I'm going to have my zinc electrode coming out of this potato, and my copper electrode is going to come out of this one.

And now, when I put these wires into contact with my electrodes, you can see that, again, the clock turns on. Now, notice that it doesn't matter where I put the electrode. I could take this out and put it in a different place, maybe over here, and you can see that you're still going to get the clock turning on.

So what's going on here? Well, charged particles are leaving one of the metals, and charged particles are sticking on to the other metal. This is the basis for the Daniell cell, which is one of the very first and simplest batteries that was ever made. And notice, you don't actually need the potato. You could use anything. You could use an onion; you could use salty water, or just about any other vegetable or fruit you pick. All that's doing is it's acting as the electrolyte in the battery, which is something you'll learn about in the lecture. It's really the action of the charged particles leaving and coming on to these metal electrodes that causes the current to flow.

Okay, so that's pretty cool. But there's a really crucial point here, and it's kind of a punch line that I'm going to give you up front, and then, we'll do some thermo to help us understand it. You see, the real deal is that the power does not at all come from the potato. The potato serves an important role in allowing the electricity to flow, but it does not provide the power itself. In order to see just what does, let's now turn to thermodynamics to shed light on this!

As I already mentioned in the beginning, the analysis of a battery introduces another, new form of energy; it's a form of internal energy for the system called the electrochemical potential. We've already discussed the chemical potential extensively back in Lecture 9. The electrochemical potential is similar, in the sense that it's another thermodynamic energy potential, and, once you know how to write it down, it can help you find the equilibrium conditions for a given system.

In a battery, reactions are occurring that both create and consume electrons, which in turn provides the electrical current when the battery is connected to a circuit. Now, electrons are, by definition, charged species that carry a charge of minus one. That's a fundamental unit of charge itself.

But atoms and molecules can also be charged. As an example, let's look at a Zinc atom, as shown here. As you can see in this image, in the zinc atom, we have a core of protons and nuclei and a bunch of electrons here forming the outer shell of the atom. Now, for zinc, we have an atomic number of 30, so in its neutral state, this atom would have 30 such electrons zipping around out here in their various energy levels.

However, zinc also is happy to lose a few of those electrons, for example, two of those 30 electrons, such that the atom only has 28 total electrons. In that case, and I should say that the core hasn't changed here, we have an atom with a net charge of plus 2. That's because overall, it has two more protons than it does electrons. This state of charge that an atom or a molecule can take on is known as the oxidation state. Atoms tend to have natural oxidation states. In fact, some atoms will readily lose an electron or two, or even three, while others will readily take on additional electrons. And then, there are still others that simply don't ever want to deviate from their neutral state—noble gases, I'm thinking of you.

So you can imagine that, if a battery is going to involve having chemistry occur that also moves charges around, then the materials involved in the operation of the battery should be able to either take on or give away charges. This is why, in batteries, we often see elements like zinc, magnesium, lithium, sodium, nickel, lead, and manganese. All of these are elements that have a non-zero oxidation state and are happy to give away and then take back some charge. When they're in a state of net positive charge, so they've given some electrons away, we refer to them as ions. And that allows me to describe the true essence of a battery: The operation of a battery involves the trafficking of charged components, including both electrons and ions.

Because the reactions that occur in a battery generate charged components, as we will see, these will be ions in a solution, and electrons in the electrodes, we have to introduce a new source of internal energy for the system—the electrostatic potential energy. This is energy that arises from the electrostatic pull of opposite charges.

Imagine that I have a positive charge over here and a negative charge over there. These charges will be attracted to one another by a force equal to a constant, times each of the two charges, divided by the distance between them squared. This relationship was rigorously determined by the French physicist Charles Coulomb in 1785, and therefore, it's known as Coulomb's law, and the constant in the equation is known as Coulomb's constant.

You might be thinking, aha! we have a force, and with a force can come work, and as we've learned, in thermodynamics we can include any and all relevant work terms into the first law. So what is it that makes force into work? Well, it's displacement. It's the basic and very fundamental nature of thermodynamic work—a force times a displacement. And we've seen this already in many forms—pressure times change in volume, surface tension times change in area, chemical potential times change in number of moles, applied magnetic field times change in magnetic induction, and so on and so on. And since I'm mentioning this all, let's not forget that we can even consider heat as a thermodynamic force times a thermodynamic displacement; in that case, the force is temperature, and the displacement is entropy.

So here, we have introduced a new force, the Coulomb force, and from it we can compute work done. The displacement in this case is the change in the distance between the two charged species. Let's consider how much work it would take a positive and negative charge that are originally sitting at some distance apart, call it r_0, and bring them infinitely far away.

In order to compute that work, it's the integral over the force times distance, so, integral over Coulomb's constant, times the two charges, divided by their separation squared, from r_0 to infinity. We'll assume for now that the charges don't vary, so we can pull them out of the integral, and we're left with the total work done to rip these charges apart from one another is equal to Coulomb's constant, times the two charges, times negative 1 over r, evaluated at infinity (which is simply zero), minus the same thing evaluated at r_0. The two minus signs cancel, and we're left with the electrostatic work to tear the charges apart equals Coulomb's constant times the charges divided by r_0.

By the way, if the two charges were the same, say, either both positive or both negative, then this value would simply be the work needed in order to bring them to a distance of r_0 apart; or another way to think about it would be that, if you let them go, and again now, we're assuming that they are charges of the same sign, then they will fly apart and when far enough away, they'll have a combined kinetic energy equal to that value.

Now, I gave you this last example of them flying away apart from each other, because I want you to see the next concept in an intuitive way. You see, once the charges have been brought to some distance r_0 apart, the fact is, that they are very much feeling that Coulomb force of repulsion between them. You can think of this as the two particles having energy stored in the electrostatic field. And you can think of this energy as being potentially available whenever it's needed.

Notice the word "potentially." That was purposeful, since it helps us think of the potential energy for an electrostatic force. We can think of the potential here as being defined as the capacity for doing work, which arises from the position or configuration of the charges. Any collection of charges will have a potential energy associated with it. So the general expression for the potential energy of a given charge, say, little *q*, that comes near another fixed charge, say, big *Q*, is equal to that work expression we derived: *V*_ electrostatic = the Coulomb constant, times little *q* times big *Q* divided by *r*.

Okay, so, that's good. Now, where do we go from here? Well, we are usually interested in the electrostatic potential energy per unit of charge. And that would be units of Joules per Coulomb, so, we often divide the above expression by that test charge, little *q*, to arrive at the following: electrostatic potential per unit charge is equal to Coulomb's constant times the charge of the fixed particle, big *Q*, divided by the distance it is from that particle.

Since the charge of the test particle has been divided out, the electrostatic potential is a property that's related only to the electric field itself and not the test particle. And for this version we usually use the symbol capital Phi, and you can see that in this case the work done to bring a particle with charge *q* to any point in space will equal the electrostatic potential Phi at that point times the charge *q*.

Now, I want you to feel your oneness with the concept of the electrostatic potential, that capital Phi thing, since that is the key variable we need for our thermo. And the thermodynamic work expression for a charged particle, as we saw, is equal to the charge of the particle times capital Phi, where Phi depends on the position of the particle.

And now we've reached a beautiful moment, since we're ready to introduce the electrochemical potential. This will become a new term in the fundamental equation for the internal energy, a term that accounts for this new form of energy that arises due to the separation of charges in a system. Remember that for a closed system where only mechanical work is possible, the fundamental equation is written as follows, *dU* equals *TdS* minus *PdV*, assuming a reversible process. But, as we just discussed, there can be many other forms of work, and they simply get added to the right-hand side of this equality if they are present in the system.

So for the case of chemical work, we had an additional term, the sum over all chemical components in the system of the chemical potential of each component, times the change in the total number of moles of that component. For electrostatic work, we now add another term to the right hand side, namely, a sum over each component of the electrostatic potential for that component, times the change in the total charge of that component. This is the electrostatic work term, and as you can see, it can be incorporated into the same sum over the different components.

By the way, if you're a little unsure about the fundamental equation, or for that matter, even what I mean by the term components, please do go ahead and take a quick refresher look at some of the earlier lectures, like, on the basic concepts or the first law.

Okay, so, as you can imagine, the electrochemical potential is the combination of those two terms in the summation over number of components—the chemical work and the electrostatic work. However, we're still not quite where we want to be. Wouldn't it be nice if we could rewrite this as a single potential? In fact, we can. To do so, all we need to do is be able to rewrite that change in the total charge of a given component in the system as a change in the number of moles of that same component. This will bring us to our final definition of the electrochemical potential.

So, how do we do the transformation? Well, as is typical in such cases, we introduce new terms. The total charge of a given component in the system can be written as what is called the valency of that component, times the charge of an electron, times Avogadro's number, times the number of moles of that component. Now the valency is simply the number of electrons that the component typically gives up. Remember that for Zinc, this number is +2, since it likes to lose 2 electrons. For say, the element sodium, this number would be +1, since it, on the other hand, likes to lose only a single electron.

So you can see why this all works; the number of electrons a given component likes to give up, times the charge of an electron is equal to the charge each component (remember, this could be an atom or a molecule) likes to give up. Avogadro's number times the number of moles of that component, well, that's the total number of them. A mole of zinc atoms would be 6 times 10 to the 23rd atoms. So, when we multiply this all together, it's exactly equal to the total charge that the given component has in the system.

And notice that I've achieved my goal, namely, I have an expression for the electrostatic potential that depends on the change in the number of moles of a component. And this allows me to group it together with the chemical potential term, as shown here. And so, we don't have to tire our hands out by writing e times Avogadro's number over and over, we can use Faraday's constant instead, written as the letter F. That's exactly equal to the charge of an electron times Avogadro's number, which, in case you're interested, happens to be 96,485 Coulombs per mole.

Alight, so finally, after all of that, we have the fundamental equation expression, which now has $dU = TdS$ minus PdV plus the sum over all components over the electrochemical potential for that component, times the change in its number of moles. The electrochemical potential, we write with the same symbol as the chemical potential, so, a lower-case Greek mu, but now, to signify that it's been, let's say, "electrified," we add a little tilde on top of it. And mu-tilde is equal to mu plus the valency Z, times Faraday's constant F, times the electrostatic potential Phi.

Alright, and now, we're ready to rock and roll with our thermodynamic analysis of a battery! I'll use one of the simplest kinds of batteries in our example; it's called a Daniell cell, and it was invented in 1836 by the British chemist John Daniell. In this type of battery we have two different liquids, as shown in this image. There are a number of different liquids that will work, but let's get specific for this example. On the left, we have a solution of copper sulfate, and on the right, we have a solution of zinc sulfate. A sulfate group is SO_4. And in between the two containers, there is a connecting tube, and it's a bit special since it only allows certain species to pass. I'll come back to this in a moment, but for now, you can think of at least one crucial aspect of this connecting tube being the fact that it allows there to be a complete circuit.

And speaking of a circuit, if we now connect the zinc and copper electrodes with a wire, current will flow from the zinc electrode to the copper electrode. Of course, that's not very interesting by itself, but when electrons are moving through a wire, we have an opportunity to make them do work for us. So for example, if I were to put a light bulb here in the middle of this wire, it would turn on, making at least some of those electrons create photons and give off light.

The key point here is that reactions are happening in each of these containers. And they are reactions that cause stuff to get charged or uncharged, which in turn, creates a difference between the charge on the left and the charge on the right, which is what sets up a driving force for electrons to move from left (the zinc) to the right (the copper).

Think about it this way; the chemistry that's happening in these two containers, makes it so that there is an effective "hill" for any type of charge. This hill makes charges want to roll down it, going from a state of higher energy to lower energy. And while those precious electrons are rolling down a hill, well, we can catch them along the way and make the fact that they have some drive to them—also known as a current—do something useful for us.

Let's look at what's happening in detail, and we'll just take one of those electrodes, the zinc side. When an electron leaves this piece of metal to go down that hill, the only way for the metal to stay charge neutral—and that's what it wants to do—is for a zinc ion to leave the metal. Remember how zinc likes to lose two electrons? Well, that's in fact exactly what happens. So a zinc atom inside this metal electrode gives up two electrons to the hill. As soon as those electrons leave the vicinity, the zinc atom, which now has a charge of plus 2, very strongly wants to leave the metal. So zinc two plus leaves down here, while two electrons go up there. And then, another zinc atom can give up two electrons to the hill, and leave the metal as a charged ion. And again, and again.

However, this whole fun process—think of it like a big mountain that the electrons get to take a ski lift up and then ski down. This whole process would have completely stopped after the very first zinc atom left the metal and went down the slopes. The reason is that negative charge cannot build up on the right hand side. Basically, in general, charge cannot build up anywhere around here. Instead, it has to be neutralized. So when the electrons carrying a charge of minus 2 from that zinc atom go through the wire and over to the copper side, well, there has to be something over there with a charge of plus 2 that can cancel the negative charge, or rather, grab the charge of the electrons and be neutral over all. And that's exactly what happens.

On the copper side, in the copper sulfate solution a copper atom separates from a sulfate group, so in the solution we get a $CuSo_4$ group that becomes Cu and SO_4, but the kicker here is that they don't come away from each other neutral, instead, the SO_4 group grabs two electrons from the copper. And so we have a Cu_2+ floating around that can attach itself to the copper electrode. Now, no other Cu_2+ atoms would come along until that one Cu_2+ atom that attached can get neutralized. But it just so happens that it can, because it can grab those electrons from the zinc atom. And this is the process of the battery. Zinc atoms come off of the zinc electrode, while copper atoms add on to the copper electrode.

That is, in fact, a mass transfer. It's a chemical reaction that changes the number of moles of each of these metals. So for that, there's a chemical potential associated with each side. But notice that in each case the transfer also involves a charge. That's why, instead of just the usual chemical potential, we must use the electrochemical potential to describe the thermodynamics at play.

Now, what I'll do is show you how we would set up such a problem. So, the electrochemical potential contains the intrinsic chemical potential of each component, plus the electrostatic potential acting on the charged species.

And just as the chemical potential of a species must be the same in all phases at equilibrium, well, the electrochemical potential of charged components must be the same in all phases in order to reach electrochemical equilibrium. Now, as I've described, a battery operates by extracting electrical work from a chemical reaction. And just as with chemical reactions, electrochemical reactions can be analyzed using the Gibbs free energy and the chemical potential in order to determine the equilibrium state at constant pressure and temperature.

And by the way, that fact that the Gibbs free energy must be a minimum for any possible process? Just as a reminder, that comes from the second law, and we use Gibbs because it's relevant for the experimental conditions, a system at constant temperature and pressure, which, to a good approximation, we can assume we have here.

Now, what we really want to know at the end of our thermodynamic analysis is the potential difference between the two electrodes. That's because that potential difference, call it delta-Phi, is what determines the current that will flow when the battery is placed in service, and the amount of work that is obtained as the system equilibrates. So for each part of the cell, we break down the reactions at play, and there are four of them: first, the zinc electrode with the zinc sulfate solution; second the copper electrode with the copper sulfate solution; third, the two solutions with each other connected by that special bridge, and fourth the two metal electrodes with each other connected by the wire. As you can see, this leads to a lot of fun reaction equations! And for each reaction, we write down the condition of equilibrium, which is the change in the free energy must be zero. This allows us to write balance equations between the different electrochemical potentials, as you can see here.

By taking these expressions and doing a series of substitutions, we arrive at a very important equation, called the Nernst equation, which relates the electrostatic potential energy change to the change in free energy of the overall reaction for the system, divided by 2 times Faraday's constant. The 2, by the way, comes from the fact that we have 2 electrons traveling through the wire for each reaction cycle that occurs.

And now, after all of this fun, I hope that you can see why saying "potato power" makes little sense. You see, the real power that came from our demo did not come from the potato. That simply served as our bridge! Instead, the power came from the two pieces of metal that I stuck into the potato. As soon as I did and you saw the light come on, the zinc electrode was losing atoms while the copper electrode was gaining atoms. And it was that gain and loss of charged species in the electrodes that gave us the power!

Okay, so I know that was a lot of discussion about a battery, but it is, after al, the prototypical system to think about electrochemical potentials, and it also happens to be an immense part of our daily lives. I hope that the next time you turn on your cell phone or open up your laptop when it's not plugged in, you think about all those electrons going downhill skiing, and how the chemical reactions in the battery are leading to charged metal atoms and the electrons they leave or take, are going up and down those potential lifts.

And while you're thinking about that, be reassured by the fact that you are now armed with the proper thermodynamic potential to understand and handle such cases, where the number of moles of something is a changing quantity, and as it changes so does the charge.

Chemical Reactions—Getting to Equilibrium

Lecture 22

In this lecture, you will learn about chemical reactions and how the concepts of thermodynamics can be used to tell us when a reaction will occur and when it will not. Chemical reactions are fundamentally part of everything we do. In many processes, there is work being done, and it is a direct result of substances changing their nature. This is called chemical work, and it is based on a process wherein the number of moles of a component in the system changes. That chemical potential can be considered the thermodynamic force in a chemical reaction, and the change in number of moles is the thermodynamic displacement.

Chemical Reactions

- Chemical reactions happen when, through some process, a set of one or more chemical species become a set of one or more different chemical species that were not in the original state.

- Chemical reactions come in a variety of forms, including the following.
  - Decomposition: A single chemical species becomes two or more species upon the reaction occurring.

  - Synthesis: Two or more chemical species become a single species upon the reaction occurring.

  - Displacement: The same number of chemical species is found before and after reaction, but the atomic species in the chemicals switch.

- Formally, we say that chemical reactions can be either exothermic or endothermic. In general, an endothermic reaction is one in which energy input is required for the reaction to occur, while an exothermic reaction is one that releases heat energy upon reacting.

- Endothermic and exothermic refer to the process of the reaction itself; they do not refer to any energy required to overcome a barrier to "get it going."

- A few examples of endothermic reactions are melting an ice cube or evaporating water, baking bread or cooking anything, melting a solid, the way in which a chemical ice pack (such as one in a first aid kit) works, and the photosynthesis part of plants growing.

- A few examples of exothermic reactions are burning a candle, making an ice cube from water, rusting iron, the way in which a chemical hot pack works, burning gas on your stove, and nuclear fission.

- In terms of thermodynamics, the most important topic of interest is to find an answer to the following question: Which direction will a chemical reaction occur? This, of course, depends on the current equilibrium conditions of the system.

Burning a candle is an example of an exothermic reaction.

- For standard chemical reactions in a lab, the control variables are typically temperature and pressure. When we have those conditions of constant temperature and pressure, there is a particular thermodynamic variable, a potential energy variable, that is extremely useful, and it will be the one we use to analyze reactions—the Gibbs free energy.

- As a reminder, the Gibbs free energy is equal to the internal energy minus temperature times entropy plus pressure times volume: $G = U - TS + PV$. (This can also be written as $G = H + PV$ to incorporate enthalpy).

- By considering the first law and our definition of entropy in a reversible process, the differential form of the Gibbs free energy can be written as follows: $dG = -SdT + VdP + \sum_{i=1}^{n} \mu_i dN_i$.

- This type of equation is called a fundamental equation, because such an equation completely specifies all changes that can occur in the given thermodynamic system. Fundamental equations form the basis for applying the second law of thermodynamics to identify the equilibrium state of a given system.

- If a chemical reaction occurs at constant temperature and pressure, the change in Gibbs free energy in a chemical reaction will tell us whether a reaction will go in the forward or in the reverse direction.

- If the change in G for a given reaction (ΔG_{rxn}) is less than zero, then the forward reaction is favored. If it's equal to zero, then the reaction is in equilibrium, and if it's greater than zero, then the reverse reaction is favored. This makes sense when considering the second law of thermodynamics, which states that the entropy of the universe for any spontaneous process must increase, and the definition for the change in the Gibb's free energy.

- From this equation for the change in the Gibbs free energy, for a system at constant temperature and pressure, dG is equal to a sum over all the components in the system of the chemical potential of that component times its change in number of moles.

Balancing Chemical Reactions

- The chemical potential dictates mass flow. It is what we consider to understand how molecules in a reaction are driven from one phase into another in closed systems or how the addition or subtraction of molecules occurs in open systems. We add up the chemical potentials of each side of a reaction, and the heavier side—that is, the side with the largest chemical potential—is the side that we can consider to be the reactants.

- To find the equilibrium condition for any given reaction, we find the point at which the change in the Gibbs free energy (dG) is zero. For a given reaction, this will equal the sum over those chemical potentials times number of moles, where we use a positive sign for the species on the right and a negative sign for the ones on the left. This gives us the net change in G for that possible process.

- However, we cannot simply create something out of nothing, so we must make sure that the reaction is balanced. In thermodynamics, while chemical species may change from one form to another in a given reaction, the total number of atoms is conserved. So, the two sides of the reaction have to be balanced.

- For example, we can't write $H_2O \rightarrow H_2 + O_2$, even though we know that the two diatomic molecules are what form. Instead, to find the balance, we write this as $2H_2O \rightarrow 2H_2 + O_2$ so that the number of hydrogen and oxygen molecules, although they've taken on new meaning by decomposing, is fully conserved.

- Those coefficients that end up balancing the reaction—and we like to use integers instead of fractions—are called the stoichiometric coefficients of the reaction. These numbers are very important because if we "feed" a reaction with a certain number of moles of reactant, then the stoichiometric coefficient tells us something about how many moles of reactant we will get. In fact, it tells us how many moles of what needs to be mixed with how many moles of something else in the first place.

The Haber Reaction

- The following reaction is arguably the single most important reaction developed in human history: $N_2 + 3H_2 \rightarrow 2NH_3$. Notice that this is balanced and that we have produced ammonia from nitrogen and hydrogen.

- This reaction is important because it was the first time humans could make nitrogen how plants like it, and that ability led to a new way of growing food, leading to the industrial revolution of food. This process takes nitrogen from an unusable form and makes it usable in the form of ammonia. This process is called "fixing" the nitrogen, which just means converting it into some biofriendly form.

- There are two important things we do to quantify this aspect of reactions. First, we want to know the change in free energy per mole of reaction. Second, we only really want to know this for the amount of the reaction that has occurred to completion—or, as it is often called, the extent of the reaction.

- This extent concept can be understood as considering the reaction occurring in terms of the change in number of moles of a given species divided by the stoichiometric coefficient for that species. The change in free energy for the reaction is then divided by this "scaled" change in number of moles, which leads to what we want—namely, the change in the Gibbs free energy per mole of reaction that actually occurs.

- A thermodynamic analysis of reactions only tells us about the equilibrium case—that is, eventually what the reaction will do given enough time. The ideas of metastability, of barriers to reactions occurring, and of the timescale over which a reaction occurs do not come from our thermodynamic analysis.

- For the Haber reaction, the key discovery was not that the reaction occurs but, rather, how to make it occur much more efficiently by introducing catalysts that create intermediate processes and by varying the pressure to allow for large scales, high output, and low costs.

- The simple question of whether a reaction will occur or not, given that the system can find a way to its happy place, is crucial because it forms the basis of even how to think about ultimately what can be made from what.

Suggested Reading

Callen, *Thermodynamics and an Introduction to Thermostatistics*, Chap. 8 and 13.

DeHoff, *Thermodynamics in Materials Science*, Chaps. 8 and 11.

Moran, Shapiro, Boettner, and Bailey, *Fundamentals of Engineering Thermodynamics*, Chap. 13.

Questions to Consider

1. What are the different types of reactions that can occur?

2. Can thermodynamics tell us anything about how fast a reaction is expected to happen?

3. Why were so many bubbles produced in the demo?

4. Estimate the change in entropy of the universe after the reaction from the demo.

Chemical Reactions—Getting to Equilibrium

Lecture 22—Transcript

Today we're going to talk about chemical reactions, and we'll learn how the concepts of thermodynamics can be used to tell us when a reaction will occur, and when it will not. And why am I so excited? Well, because chemical reactions are fundamentally part of everything we do. And I'm not just talking about the candle example I discussed in an earlier lecture, where we learned about the chemical potential. I'm talking about understanding what particular process is going to make a set of materials do precisely what we want; or how to get the most energy out of a given type of combustion process; or for that matter, what our bodies do with the food we eat to literally maintain life itself!

And these are examples of processes that are different in a crucial way compared to some of the other types of processes we've covered in this class, like a gas expanding to do mechanical work; or a dye diffusing into a liquid to create a new mixture; or, for that matter, magnetic, surface tension, electric, or entropic work. In those cases, work is done, systems change, lots of fun stuff happens; however, nothing happens to the inherent building blocks of the system.

I could make a mixture of water and ethanol, and as we know, the volume is not simply equal to the original volumes of each one separately. So, an interesting thing is happening due to the fact that I made a mixture, and we talked about that in Lecture 9 and how it can be described by the concept of partial molar quantities. However, in that case, nothing changed about the inherent amount of components in the system. If I started with 1 mole of water and 1 mole of ethanol, then I ended with 6 times 10 to the 23rd molecules of water and 6 times 10 to the 23rd molecules of ethanol, that is, the same number. In other words, nothing was consumed or altered chemically at the level of the component itself during the process. That kind of process is not a chemical reaction.

In the examples I first mentioned, like combustion for energy conversion, materials chemistry, and, well, life itself, among so many other processes, there is work being done, and it is a direct result of substances changing their nature. This is called chemical work, and, it's based on a process wherein the number of moles of a component in the system changes. And we've already learned about this type of thermodynamic work. Remember how we learned about the chemical potential in Lecture 9? Well, as a reminder, that chemical potential can be considered the thermodynamic force in a chemical reaction, and the change in number of moles is the thermodynamic displacement.

We showed back then that a candle burning is undergoing a chemical reaction, one that turns paraffin wax plus oxygen in the air into CO_2 plus water. And I mentioned that a reaction will always occur in the direction from high chemical potential to low chemical potential. Our thermodynamic analysis showed that this reaction is heavily directed towards the candle burning.

Now, as we set the stage to go deeper into the thermodynamics of chemical reactions, remember also how I discussed the fact that thermo only tells us which way a chemical reaction wants to go. It does not tell us whether it can go. For that, we need to understand the kinetics of the problem, and remember that thermodynamics only tells us about the equilibrium properties of materials. We take an "eventually it will happen" approach to the world.

So, for the case of the candle burning, this point is illustrated by the fact that as badly as that reaction wants to happen—and our thermodynamic analysis tells us that it will—it doesn't happen, not until we light it with an external source of heat. This is the kinetic part of the problem. Sure, the reaction is very favorable, but there's a barrier for it to occur. Without some heat input into the system, it cannot get over that barrier and so it remains stuck in what we call a metastable state. That means there's a more stable state it wants to be in, in this case, CO_2 plus $H_2 2O$, but it is separated from that state by a barrier.

Again, I'll repeat because it's so important. Thermodynamics does not speak to such barriers; it only tells us how to find equilibrium states, not what the road looks like along the way. That's also embedded in the very concept of state functions, which remember, have no dependence on the path taken.

Okay, so, enough setting the stage, let's do a really fun chemical reaction. Have you ever mixed vinegar with baking soda? What you'll find is that you can get quite a bit of bubbles forming when these two chemicals are mixed. In fact, many kids throughout time have built play volcanoes where the eruption that occurs is made from this reaction. Sodium bi-carbonate reacts with acetic acid to form sodium acetate plus carbonic acid, which then decomposes into water plus CO_2. And it's all that CO_2 that causes the bubbles. I like this particular reaction so much that I wanted to take it to another level here for you. Let's take a look!

So here we're going to see a really fun reaction. In this bowl, we're going to watch a reaction of hydrogen peroxide, which has the formula H_2O_2, decompose into O_2 gas plus water. Now, I'm going to add a little bit of sodium iodide to help the reaction go more quickly. And, it's a pretty fast reaction, so, I'm going to make sure I wear my safety goggles.

So here's the sodium iodide, and I'm going to add this to the H_2O_2.

You can see a whole lot of fizzling happening, which is all those O_2 molecules, all those oxygen bubbles getting released, and also, you might have noticed that this container got really hot. A lot of steam actually came out. And that's because the reaction is highly exothermic; it really wants to go forward as it releases a whole lot of heat as it goes downhill in the reaction.

Now, the chemical energy is stored in the hydrogen peroxide, and it's really on display as you see it dissolve into O_2 plus H_2O. But now, let's do this again, only this time I've added a little bit of liquid detergent to the mix, so I've got in here sodium iodide, plus some liquid detergent. That way, when the O_2 gas forms from the reaction, it's going to be trapped inside of bubbles, because it's soapy now, instead of just being released into the air.

So, here I have the same hydrogen peroxide, H_2O_2, in this flask, and I'm going to add the sodium iodide into it.

Now that was a lot of fun! And it's a great example of the strength of the driving force for this reaction to occur. Another way of thinking about it is that what we just witnessed is a very large chemical potential difference for a reaction, which is the fundamental thermodynamics going on inside of that flask.

Now that was a lot of fun! In that reaction, we had hydrogen peroxide mixing with sodium iodide, and as you could see, it was a very favorable reaction! Notice another point here; unlike the example of a candle burning, in this case, there was no barrier to the reaction occurring. The second I mixed the two chemicals together, the reaction went forward. I did not need any added heat, and in fact, a whole lot of heat came out during the reaction. I'll come back to this heat thing in a moment, but first, let's turn to one of our favorite things to do, and that is, define important terms!

So, for starters, let's put forth a basic definition for what I mean by a chemical reaction. Okay, chemical reactions happen when through some process, a set of one or more chemical species become a set of one or more different chemical species that was not in the original state. Chemical reactions come in a variety of forms, including, first, decomposition: in this case, a single chemical species becomes two or more species upon the reaction occurring; second, synthesis: here, two or more chemical species become a single species upon the reaction occurring; and third, displacement: in this case, the same number of chemical species is found before and after reaction, but the atomic species in the chemicals switch.

So let's see, for case one, decomposition, an example could be the reaction of water to form hydrogen and oxygen, or $2H_2O$ goes to $2H_2$ plus O_2. For case two, synthesis, an example could be the formation of carbon dioxide, so, C plus O_2 goes to CO_2. And for case three, displacement, how about Cl_2 plus 2NaBr goes to Br_2 plus 2NaCl, which is a simple way to form pure bromine from its naturally occurring salt form. In writing these examples down, you may have noticed that the numbers all work out so that the reactions are balanced. I'll come back to this in a bit.

Now, in addition to these basic categories, a really important one relates back to the heat I just mentioned. Formally, we say that chemical reactions can be either exothermic or endothermic. In general, an endothermic reaction is one in which energy input is required for the reaction to occur, while an exothermic reaction is one that releases heat energy upon reacting. Now, there's a very important point here related to that barrier concept; endothermic and exothermic refer to the process of the reaction itself; they do not refer to any energy required to overcome a barrier to "get it going." For example, I just mentioned a few moments ago that burning a candle requires heat to start the reaction, so you may be tempted to consider that an endothermic process, since energy input was needed. But that's not how we should think about the use of the word endothermic here, since that refers only to the reaction barrier, not to the thermodynamic process of the reaction itself. For that, as we know, by just putting our finger near the flame, the burning of a candle is clearly exothermic.

A few examples of endothermic reactions would be, melting an ice cube; or evaporating water; or baking bread; or cooking pretty much anything; melting a solid; a chemical ice pack, like in a first aid kit; and plants growing, well, the photosynthesis part anyway. A few exothermic reaction examples would be, well, in addition to the burning candle, it could be making an ice cube from water; rusting iron; a chemical hot pack;, burning gas on your stove; and sure, let's just throw this in for good measure, nuclear fission.

Now, the first and really most important topic of interest, in terms of thermodynamics, is to find an answer to the simple question, which direction will a chemical reaction occur? This, of course, depends upon the current equilibrium conditions of the system. For standard chemical reactions in a lab, the control variables are typically temperature and pressure. And I hope that as I say that, that you're getting an idea, or at least having a good dose of déjà vu. See, when we have those conditions of constant temperature and pressure, there is a particular thermodynamic variable, a potential energy variable, that's extremely useful, and it will be the one we use to analyze reactions. Yes, you knew it was coming—it's the Gibbs Free Energy!

As a reminder, the Gibbs free energy is equal to the internal energy minus temperature times entropy plus pressure times volume. So G equals U minus TS plus PV, or, if you're a fan of enthalpy, this can also be written as G equals H plus TS. And as we saw back in lectures 10 and 11, by considering the first law and our definition of entropy in a reversible process, the differential form of the Gibbs free energy can be written as follows, dG equals minus S

times dT plus V times dP plus a sum over all the components of the chemical potential of each component times the change in number of moles of that component.

Remember, this type of equation is called a fundamental equation, because such an equation completely specifies all changes that can occur in the given thermodynamic system. Fundamental equations form the basis for applying the second law of thermodynamics to identify the equilibrium state of a given system, and that's what we'll do now for the case of a system undergoing chemical work, or more simply put, for the case of reactions.

So if a chemical reaction occurs at constant temperature and pressure, the change in the Gibbs free energy in a chemical reaction will tell us whether a reaction will go in the forward or in the reverse direction. If the change in G for a given reaction, or, delta-G sub RXN, as we like to write it, if this is less than zero, then the forward reaction is favored. If it's equal to zero, then the reaction is in equilibrium, and, if it's greater than zero, then the reverse reaction is favored. I won't go through the detailed math for how those definitions come about, but it's quite straightforward by considering the second law of thermodynamics, which states that the entropy of the universe for any spontaneous process must increase, and the definition for the change in the Gibb's free energy that I just wrote down for you.

So, from this equation for the change in the Gibb's free energy, you can see that for a system at constant temperature and pressure, we can write dG as equal to a sum over all the components in the system of the chemical potential of that component times its change in number of moles. Now, as I mentioned, we already saw back in Lecture 9 that the chemical potential dictates mass flow. It's what we consider to understand how molecules in a reaction are driven from one phase into another in closed systems, or how the addition or subtraction of molecules occurs in open systems. Remember, I used the analogy of a see saw, where we add up the chemical potentials of each side, and the heavier side, that is, the side with the largest chemical potential, wins. That means that that's the side that pushes the seesaw down and the side that we can consider to be the reactants.

Now, suppose that we want to find the equilibrium condition for a given reaction. Well, the way we find equilibrium for any process, is to find the point at which the change in the Gibb's free energy is zero, so where does dG equal zero? For a given reaction, this will equal the sum over those chemical potentials times number of moles, where we use a positive sign for the species on the right, and a negative sign for the ones on the left. This gives us the net change in G for that possible process. But, there's another really important point here, namely, that we cannot simply create something out of nothing, so we have to make sure that the reaction is balanced.

As an aside, in processes such as nuclear fusion, it does, in fact, seem that we are creating something completely new, but that's a bit more complicated case that we're not going to go into in this course. And even in nuclear reactions, if we count all of the atoms and electrons and energy involved in the process correctly, we'll find that nothing is created or destroyed. But in any case, for the cases that we're interested in thermodynamics, while, chemical species may change from one form to another in a given reaction, the total number of atoms is conserved. So I have to balance the two sides of the reaction. As an example, I couldn't write down H_2O goes to H_2 plus O_2, even though I know the two diatomic molecules are what form. Instead, to find the balance, I would write this as $2H_2O$ goes to $2H_2$ plus O_2, and there you see that the number of hydrogen and oxygen molecules, although they've taken on new meaning by decomposing, is fully conserved.

Those coefficients that wind up balancing the reaction—and we like to use integers here instead of fractions—those are called the stoichiometric coefficients of the reaction. And if you think about it for a bit, you can see that these numbers are, in fact, very important. That's because if I "feed" a reaction with a certain number of moles of reactant, then the stoichiometric coefficient tells me something about how many moles of product I'll get. Actually, it's even more than that; it tells me how many moles of what needs to be mixed with how many moles of something else in the first place.

For what is arguably the single most important reaction developed in human history, we have N_2 plus $3H_2$ goes to $2NH_3$. Notice that this is balanced. Also notice, that we have produced ammonia from nitrogen and hydrogen. So, why is this so important? Well, because it was the first time humans could make nitrogen in a way that plants like. And as you can guess, that ability led to a new way of growing food, leading to nothing less than the industrial revolution of food itself.

Nitrogen is a crucial ingredient in plant growth, and you might think, well, but air is 78% nitrogen, so what's the big deal? Well, the thing is, nitrogen in the air is not suitable for plants, because it's in the form of N_2, which is two nitrogen atoms connected by a triple bond. This is one of the strongest bonds in nature, and plants simply cannot break it. So to them, this type of nitrogen is useless. Of course, carbon and oxygen are also crucial for plants, but unlike nitrogen, these are easily obtained by the plants from soil and air.

So what the Haber—or sometimes as it's known, the Haber-Bosch—process does, is it takes nitrogen from that unusable form and makes it usable in the form of ammonia. This process is called fixing the nitrogen, which just means converting it into some bio friendly form. Fritz Haber developed the process in this reaction in the early 1900's, and it completely changed the world. If you want to get a sense of just how much, consider this, today, 500 million tons of nitrogen for fertilizer is produced this way each year, and about half of the protein in all human beings on this planet comes from nitrogen that was fixed by this process. Take a look at this human population curve and notice where the Haber reaction was developed; it's no accident that after that process, the population was able to, and did, grow at a much faster pace.

Okay, so that's the background, now let's go back to the thermodynamics of the reaction itself. What if I were to start this reaction with a mole of N_2 and a mole of H_2, what would I get? Well, in that case, since the stoichiometric coefficients are $3H_2$ per N_2, then you can see that for each N_2 I need $3H_2$ to react, so, I would consume the full mole of H_2, but only $^1/_3$ of a mole of N_2 in the process, and in the end, I'd get out $^2/_3$ of a mole of NH_3. If I had wanted to consume all of the N_2 to begin with, I should have started with a number of moles of each reactant in line with the stoichiometric coefficients, so in this case 1 mole of N_2 and 3 moles of H_2 would have been good, and I would have wound up with 2 moles of ammonia.

Now, there are two important things we do to quantify this aspect of reactions. First, we want to know the change in free energy per mole of reaction. Second, we only really want to know this for the amount of the reaction that has occurred to completion, or, as it's often called, for the extent of the reaction. This extent concept can be understood as considering the reaction occurring in terms of the change in number of moles of a given species divided by the stoichiometric coefficient for that species. The change in free energy for the reaction is then divided by this scaled change in number of moles, which leads to what we want, namely, the change in the Gibbs free energy per mole of reaction that actually occurs. Don't worry if that's a little confusing at first. The main point is that this version of the change in free energy is the one we want for a reaction, since it combines all that stuff I just mentioned about how much each species is able to react dictated by those stoichiometric coefficients.

In the end, our expression for delta-*G_RXN* per mole is equal to a sum over the number of components in the reaction of the chemical potential of the component times the stoichiometric coefficient of that component. And the same conditions as before apply, but now, since it's per mole, remember we need to write a bar on top of the variable as per our convention.

So for the Haber reaction, we have that delta-*G* with a bar on top for the reaction is equal to 2 times the chemical potential of NH_3 minus 3 times the chemical potential of H_2 minus 1 times the chemical potential of N_2. And as before, if this quantity is less than zero, the reaction occurs spontaneously in the forward direction. Equal to zero means it is in equilibrium; and greater than zero means it occurs spontaneously in the reverse direction.

Okay, now, for standard temperature and pressure conditions, and for the given phases of these materials (all gas phases in this case), we can look up the tabulated values for the chemical potentials. And what we find is that the change in free energy for this reaction at standard conditions—so that's temperature equals 298 Kelvin and pressure is one atmosphere—that's equal to minus 33 kilojoules.

As a reminder, when we're talking about standard state conditions, we usually put a little naught symbol above the variable to specify this. And by the way, the actual absolute value for the chemical potential is not important, and it would be a real pain to work with absolute values. I mean, after all, we don't measure mountain peaks by their distance to the center of the earth, but rather, only by something more meaningful to us, namely, their distance from sea level. For chemical potentials, all that we care about is their difference from one side of the reaction to another. So it makes no difference how we define them as long as we are consistent. And as a matter of practice, the convention is to take the chemical potential of each element as zero for reference. This implies that the state of matter in which substances are decomposed into the elements under standard conditions for the scale of the chemical potential represents our analogy to sea level for elevations. Here are just a few examples of standard state chemical potentials. Notice that the pure elements are zero, and also, that the chemical potential of a pure substance depends on its state and crystal structure. For example, liquid water and water vapor have different chemical potentials at the same temperature and pressure.

Okay, so, let's go back to our reaction. Now, we just derived the fact that this reaction has a tendency to go from the point of view of its free energy per mole of reaction. But that was only at standard conditions. How do we predict what happens when these conditions vary? Well, first, what about temperature? In so many cases of industrial importance, we know that the temperature of a given reaction is considerably higher than the standard 298 Kelvin, or in other words, room temperature. Car engines run at temperatures more like 500 Kelvin, a natural-gas-fired power plant runs above 900 Kelvin, steel is made over a range of temperature from 1200 to 2000 Kelvin, and glass is made at a whopping 2700 Kelvin, just to name a few examples.

So how do we take into account the role of temperature in a reaction? In order to answer that question, we can go back to our original definition of the change in molar Gibbs free energy for a reaction, namely, that it is equal to the change in molar enthalpy minus temperature times the change in molar entropy. And here we see the role of temperature. The simplest thing to do is just fix the delta-*H* and delta-*S* at their values for standard conditions and then see that the effects of temperature from the *T*-delta-*S* term.

Let's take a look at how this works for the Haber reaction. The change in enthalpy for this reaction at standard conditions, which we write as delta-*H* with the little naught symbol up there, is equal to negative 92.38 kilojoules. That means that if the reaction were to occur at zero temperature, then 92.38 kilojoules of energy would be released in the form of heat by the reaction as it occurs from left to right. Remember that at constant pressure, enthalpy change is in fact equal to heat.

But what about entropy? Well, it turns out that, for this reaction, the entropy change at standard conditions decreases as we form ammonia from N_2 and H_2. This actually makes sense; remember, the entropy is equal to the number of degrees of freedom that the system has available to it. If we reduce the number of gas molecules that make up the system, we can see how the number of degrees of freedom would also reduce, and therefore, the entropy. In this particular case, delta-*S* naught equals minus 198.3 joules per kelvin, so indeed, that's quite a negative value.

Now, at room temperature, delta-*G*-naught equals negative 92.38 kJ minus 298 kelvin, times minus 198.3 joules per kelvin, which equals about minus 33 kJ, just what we had before from our analysis of the chemical potentials. This tells us the reaction is spontaneous at room temperature. But, now, let's increase the temperature. So, for example, at 500°C, or 773 Kelvin, the value for delta-*G*-naught now becomes plus 61 kJ, so the reaction no longer occurs spontaneously! Now, this is perhaps counterintuitive, since we're kind of used to thinking of reactions happening more favorably the higher the temperature, but that is simply not the case for a reaction where entropy decreases.

Okay, now, a reaction can also deviate from the case of standard conditions in ways that don't just involve temperature. This gets a bit more complicated, and it's beyond the level of detail I'll be going into for this lecture, but I'll tell you the punch line in case you're interested. For reactions that occur under non-standard state conditions, we simply write the change in molar free energy of the reaction as equal to the change in molar free energy under standard conditions, plus the ideal gas constant *R* times the temperature, times the natural logarithm of what is called the reaction quotient. Again, I won't go into detail here, but that reaction quotient is related to the partial pressure if it's a gas, or the concentrations, if it's a liquid or solid

Now, let me conclude by going back to the concept of kinetics. As I mentioned, a thermodynamic analysis of reactions only tells us about the equilibrium case, that is, eventually, what the reaction will do given enough time. That means that, if we apply our analysis to diamond, it tells us that it should, in fact, all react to graphite. But we know that this doesn't happen, at least not within any time frame we're living in.

So, the ideas of metastability, or of barriers to reactions occurring, and of the timescale over which a reaction occurs do not come from our thermodynamic analysis. In fact, for the Haber reaction, the key discovery was not that the reaction occurs, but rather, how to make it occur much more efficiently by introducing catalysts that create intermediate processes and by varying the pressure to allow for large scales, high output, and low costs.

Still, even without all of the kinetics, the simple question is, will a reaction occur or not, given that the system can find a way to its happy place? That question is crucial, since it forms the basis of even how to think about, ultimately, what can be made from what!

The Chemical Reaction Quotient

Lecture 23

In this lecture, you will learn about a crucial concept in the application of thermodynamics to reactions—namely, the activity coefficient and how it relates to concentration and the chemical driving force during a reaction. This, in turn, will lead you to a way to compute this driving force at any concentration and to an equation to find the equilibrium of a reaction. This is of fundamental importance when the ability of a reaction to occur depends on the concentrations of the reactants and products, which are changing during the reaction.

The Activity Coefficient

- What we want to know, and what thermodynamics can tell us, is which way reactions will go when they occur. However, thermodynamics does not tell us whether they will go or not; it only tells us if the energy gradient favors the reaction.

- There often are kinetic barriers to the reaction occurring. We can figure out if a reaction will occur by adding up the values of the standard-state molar free energies for each compound, and from those, we can calculate ΔG of the reaction.

- Chemical reactions of any kind can proceed initially in one direction or the other, dictated by the difference in free energies between the reactants and products. Either way it goes, the reaction can then stop for two reasons: One of the reactants is simply used up, or the reaction reaches an equilibrium state—with all products and reactants still present but now in some sort of new balanced condition.

- The way we describe the second case, where reactant and product concentrations in solution are changing, is by considering the change in the chemical potential as a function of concentration for each component involved in the reaction.

- There is always additional entropy caused by mixing, and it only depends on the concentrations of the mixtures. As the mole fraction of one of the components varies, the molar entropy of mixing increases as two components are mixed and reaches a maximum for a 50-50 mixture.

- From this expression for the entropy of mixing, we can also write down expressions for the free-energy change upon mixing. We can write the chemical potential of either one of the species in the mixture as having changed due to the concentration of that component changing.

- In the example of two components *A* and *B* mixing, the chemical potential of *A* is equal to the standard-state chemical potential of *A* (μ^o_A) plus the gas constant (R) times the temperature (T) times the natural log of the mole fraction of component *A* ($\ln X_A$): $\mu_A = \mu^o_A + RT\ln X_A$.

- This is a powerful equation that tells us how the Gibbs free energy varies with changes in concentration of something in solution. However, this form of the equation only applies to ideal solutions, meaning that we have ignored interactions, just like in the ideal gas law. So, the equation only considers changes in the free energy—or chemical potential if we take the molar free energy—due to the entropy of mixing.

- But most reactions are not ideal because interactions between components can be very different from the interactions of one component with itself. For the components *A* and *B*, this would mean that *A* and *B* have a very different bonding character than either *A* with *A* or *B* with *B*. This kind of effect is not taken into account in the ideal-solution model.

- So, we need to expand the range of applicability of this relationship. However, this form of the equation, where we take the standard-state chemical potential and add a single term that goes as the logarithm of something, is a very convenient mathematical form, so we'd really like to retain it. We do this by defining something called the activity (a).

- The chemical potential for a given component—for example, component i—at any concentration is equal to the chemical potential for that component at standard-state conditions (in other words, when it is in its pure form and not reacting) plus RT times the natural logarithm of the activity of that component: $\mu_i = \mu^o_i + RT\ln a_i$.

- Quite simply, we have replaced the mole fraction with the variable a_i. But it is a very powerful switch, because now we can essentially include any changes that happen to the chemical potential of the reacting species during the reaction. At the same time, we retain the same mathematical form—namely, a natural logarithm.

- It's nice to write the activity in terms of the activity coefficient, for which we use the symbol gamma (γ), times the mole fraction of that component: $\mu_i = \mu^o_i + RT\ln(\gamma)X_i$. By writing it this way, we can see how far away from the ideal-solution model the system is.

- For example, if γ for a component is equal to 1, then the activity of that component becomes the mole fraction of the component. We get back the same equation for the chemical potential that we had when we ignored interactions and just had the ideal-solution model.

- The ideal-solution model means that the chemical potential will change upon mixing things together, but only because the entropy of the mixture increases. The activity, which includes the interactions, can tell us how important the interactions were in the first place. Interactions of components can lead to a huge difference in the effects of mixing, even fully reversing the qualitative effects of mixing.

Raoult's Law and Henry's Law

- The deviation of the activity coefficient from 1 is a measure of how far away from ideal the system is. And in so many cases for reactions, the activity coefficient can vary quite a bit as a function of the mole fraction of each species in the solution. The factor by which the activity coefficient deviates from 1 has an enormous impact on the chemical behavior of the system, which is driven by the chemical potential.

- As the concentration of the mixture changes, the activity coefficient also changes. Once the activity coefficient converges to around 1, the solution behaves pretty much like an ideal solution. This limit, where the solution starts to behave like an ideal one because the concentration is approaching 1, is called Raoultian behavior. Raoultian behavior simply means ideal behavior, so we're back to the chemical potential of mixing depending on the natural log of just the molar concentration.

- François-Marie Raoult was a French physicist who was experimenting with vapor pressures around 1882. He noticed that if he had a solution of two liquids—for example, A and B—then as long as no other gases were present, the total vapor pressure above the solution was equal to the weighted sum of the "pure" vapor pressures of the two components.

- He found that the total vapor pressure (P) is equal to the mole fraction X_A times the partial vapor pressure P_A plus X_B times P_B: $P = X_A P_A + X_B P_B$. Although we're talking about pressures, this is equivalent to the same ideal-solution-model behavior we found for the free energy and chemical potential.

- For low concentrations, just as with the high-concentration regime, we call the behavior of the mixture Henrian. William Henry formulated Henry's law in 1803, and like Raoult, he was also playing around with liquids and gases and measuring pressures.

- Henry's law states that at a constant temperature, the amount of gas that dissolves in a liquid is proportional to the partial pressure of that gas in equilibrium with the liquid. For a dilute solution, Henry's law tells us that the pressure goes as a constant (called Henry's constant, k_H) times the molar concentration of the solute (C): $P = k_H C$.

Physician and chemist William Henry (1775–1836) proposed what is now known as Henry's law.

- It is interesting to compare this to Raoult's law, which states that the pressure goes as the partial pressure of that component in its pure state times its mole fraction, valid for the other extreme—at high concentration.

- Both of these laws appear to give something similar: a dependence of a constant times a mole fraction. This is actually true if the two components being mixed have a similar bonding behavior with each other as they do with themselves. But that's just for the ideal-solution case.

- If interactions are strong and different between the two components, then in the case of Henrian behavior, we are looking at a regime where these differences are in a sense maximized, because each solute is surrounded primarily by only solvent, as opposed to other solute molecules.

The Equilibrium Constant

- The change in the chemical potential in a reaction is equal to the difference in the standard-state chemical potentials plus RT times the natural log of what is called the reaction quotient (Q): $\Delta G = \Delta G^o + RT\ln Q$. This reaction quotient is the ratio of the activities of the products multiplied together, divided by the activities of the reactants multiplied together.

- The activities in the Q change during the reaction as the system strives to reach equilibrium, and these activities are highly dependent on concentration. The concentrations of these components are changing as they react, so the activities will change as well.

- This reaction quotient allows us to keep being able to calculate the chemical driving force for the reaction at any point, as long as we also know what the activities are for the reactants and products.

- At the special point at which this driving force, this change in chemical potential due to a reaction occurring, is zero, there is no longer a way for the system to lower its free energy, so it is in equilibrium.

- We can find this point simply by plugging in zero for the change in chemical potential. We find that the equilibrium point for the reaction is determined when the RT natural log term is equal to those differences in standard-state chemical potentials.

- This point, where we have equilibrium, is so important that we call the reaction quotient at that point the equilibrium constant. It's still just the same reaction quotient, but it only refers to the equilibrium point in the reaction, so many textbooks give that its own variable (often K).

- Among those who do a lot of chemistry, this equation for the equilibrium of a reaction has been considered the most useful in thermodynamics. Most important is the fact that the activity product ratio that makes up the equilibrium constant is independent of variations in the system composition. The equilibrium constant K is controlled completely by the difference in standard chemical potentials.

- This means that it is only a function of the temperature and pressure. And for any given temperature and pressure, it's a constant. As soon as the substances reacting are chosen, then the value for the equilibrium constant is fixed. It is the activity product ratio that would be observed if the system reaches equilibrium.

- The equilibrium constant is a very useful piece of information to have about a reaction. If K is very large, it shows that a reaction will tend to go "to completion"—that is, mostly products will be present at equilibrium—and if K is small, it shows that the reaction hardly goes at all before enough products are formed to stop it.

- The equilibrium constant will not tell you whether reactants will actually react or at what rate; instead, it indicates the activity product ratio at equilibrium, not whether equilibrium is easily achievable. The reaction continues until the ratios of the activities of the products and reactants equal this equilibrium constant.

Suggested Reading

Anderson, *Thermodynamics of Natural Systems*, Chap. 9.

Callen, *Thermodynamics and an Introduction to Thermostatistics*, Chap. 8.

Moran, Shapiro, Boettner, and Bailey, *Fundamentals of Engineering Thermodynamics*, Chap. 14.

Questions to Consider

1. What is the molar Gibbs free energy of formation for an element?

2. What is the difference between the equilibrium constant and the reaction quotient?

3. What is the difference between an exothermic and an endothermic reaction?

The Chemical Reaction Quotient
Lecture 23—Transcript

In the last lecture, we talked about chemical reactions. But as you may remember, at the end, I kind of teased you a bit. I talked about how temperature plays a role in reactions through the entropy term in the free energy. But, then I also told you that reactions can deviate from the case of standard conditions in ways that don't just involve temperature. I briefly mentioned that in such cases what we do is to add an additional term to the molar free energy change expression, the logarithm of something called a reaction quotient. Today, I'm here to tell you that we will, indeed, learn much more about the reaction quotient.

Some scientists, especially chemists and chemical engineers, will tell you that the expression for the free energy change during a reaction that includes this reaction quotient thing, they'd tell you that that equation is the single most important equation in all of thermodynamics! And while I'm not sure I would say such a sweeping, dramatic thing, I can certainly understand why some would say this.

In this lecture, I want to be sure we understand what the reaction quotient is, and how it is applied, and why it's so important. So, I want to contrast two different types of reactions to help understand the key reason for why we need the reaction quotient. In the first reaction, we will have pure compounds reacting together. In the second reaction, we will have dissolved compounds reacting together.

As an example of the first type of reaction—pure compounds reacting together—let's consider a really cool metal called Titanium. Titanium is really strong; it has a very high melting point, and it's highly corrosion resistant, which means, it doesn't damage much at all over time. It's used a lot in the aerospace industry, as well as other areas, like medical technologies, or, for pipes in the oil and gas industry.

Now, even though it's the 10th most abundant element in the Earth's crust, pure titanium is actually pretty hard to extract. See, Titanium comes naturally as an oxide, TiO_2. But we just want the Ti part by itself, and it turns out that getting rid of the oxygen for this particular metal is not easy. So what is done is first, TiO_2 is converted into $TiCl_4$ using the reaction $TiO_2 + 2Cl_2 + 2C \rightarrow TiCl_4 + 2CO$. This takes quite a bit of energy and needs to be done very carefully. But then, once we have the $TiCl_4$ we can use sodium to make pure titanium. The reaction for that is $TiCl_4 + 4Na \rightarrow Ti + 4NaCl$.

Okay, so there are two reasons I'm using this particular example. First, this is a really important aspect of materials science and a fantastic example of how knowing and controlling a chemical reaction is crucial; we use pure metals all of the time in our lives, but they almost never come to us in their pure form in the earth, so, we need to understand how to extract them.

But the other reason I'm using this example is because it's a nice, simple reaction that involves pure substances. As a reminder of what we discussed in the last lecture, what we most badly want to know, and what thermodynamics can tell us, is which way the reactions will go. Remember also, that thermodynamics does not tell us whether it will go or not; it tells us only if the energy gradient favors it. There are often kinetic barriers to the reaction occurring. We can figure out if this reaction will occur by adding up the values of the standard state molar free energies for each compound, and from those, calculate delta-G of the reaction.

In this case, that delta-G is equal to minus 600 kilojoules per mole at 900°C. Remember that from the second law, if the change in G for a reaction is negative, then the reaction will proceed. This means that the reaction, as written, is spontaneous, which means that it wants to occur.

So, what happens is that, at that temperature, Titanium Chloride and sodium get used up during the reaction. But here's the key point, while being used up, they do not change their Gibbs free energies. The reaction will proceed as long as any reactants are left. When either the sodium or the titanium chloride is completely used up, the reaction must stop. It just runs out of stuff to react.

We can think about this reaction in a very simple graphical manner as shown here. The y-axis is the total molar free energy, and the x-axis is just used to place my reactants and products. I plot bars to represent the magnitude of the combined molar Gibbs free energies of the products and of the reactants. The difference in the height of the bars is minus 600 kilojoules per mole, and just to be complete, I put a bar in between them that I call energy barrier. This may or may not be higher than the reactants and products; in this case I've drawn it higher, which indicates that the reaction would need some sort boost to get it going. The middle bar in that represents an activation energy barrier that prevents the reaction from occurring.

The point here is that the size of these bars does not change during the reaction, because none of the products or reactants changes in any way, only the amounts of those change. The value of delta-*G* never goes to zero. This is the key point, so let me just emphasize once more. For this reaction of pure compounds, all of the molar free energy differences are based on the values of the molar free energy of the products and reactants in their reference states, and these values do not change during the reaction.

Okay, now, I want to contrast this case with another reaction example. In the second case, I'll use a reaction in which all products and reactants are dissolved in water. Let's consider carbonic acid, or H_2CO_3 in solution. This is a pretty important chemical, since when CO_2 mixes with water, part of it reacts with water to form H_2CO_3. And since H_2CO_3 is a bit on the acidic side, that means that when CO_2 mixes with water, the water becomes just a bit more acidic. You could, in fact, take a straw and blow bubbles into a glass of water, and after just a little bit, the water would go from a pH of 7 to a pH of about 5. The term pH is a measure of the acidity of a solution, and going from 7 to 5 means the solution becomes 100 times more acidic.

So the reaction that occurs is H_2CO_3 (aq) goes to HCO_3 minus (aq), plus H plus (aq). And note that as it's written, I've put the letters "aq" in parentheses to help remind us that these are all dissolved in solution. That just means there's water all around. Those H plus atoms are what make things acidic.

So, here's the critical point. This reaction is very different than the first one I showed you, because now, the products and reactants are capable of changing their concentration during the reaction! At the very start of this dissolved-compounds reaction, we begin in the same way, by looking up the values of the molar free energy at standard conditions for each species.

We find a value of delta-G equals plus 36.3 kilojoules per mole. So, this is a positive value, which means that the reaction goes spontaneously to the left, as it is written. But that's fine, but as soon as the reaction starts, the concentrations of H plus and HCO3 minus start to decrease, the concentration of H_2CO_3 starts to increase, and in this case, the Gibbs free energies of all three change.

Let's plot this in the same we that we did for the titanium extraction reaction, between pure compounds, by looking at the molar free energies for the reaction. On the left, I'm showing the starting conditions, represented by the numbers I just gave you, which come from looking up molar free energies in tables.

But note, that we include here a superscript so that it reads mu-naught, which indicates that standard temperature and pressure conditions were used for all products and reactants. The reactants, which in this case are H plus and H_2CO_3 minus have a greater net molar free energy than the product, which is H_2CO_3. In the middle, I'm plotting the free energies after the reaction has proceeded for some time. The point is, that as shown by the arrows on the plot, the molar Gibbs free energy of the reactants decreases and that of the products increases during the reaction!

This is in very stark contrast to the titanium chloride plus sodium reaction between pure compounds, where the free energies remained constant throughout. Sooner or later for the case of carbonic acid, as the reaction between dissolved compounds continues, a state of equilibrium is reached, and the molar Gibbs free energies of reactants and products are equal, as shown on the right here. The reason why this reaction is so different is that the solutes can change their concentrations during the reaction, which in turn, changes their molar Gibbs free energies, quite different from the all-pure-substances situation.

So, what do we do for such a case? Well, there's no need to worry. This situation can easily be described using what is known as the equilibrium constant. But before we get to that "most important equation in thermodynamics," as some people call it, let's summarize the key point of where we are so far. So, chemical reactions of any kind can proceed initially in one direction or the other, dictated by the difference in free energies between the reactants and products. And here's the thing, either way it goes, the reaction can then stop for two reasons, first, one of the reactants is simply used up, which would be what stops the titanium extraction reaction; or second, the reaction reaches an equilibrium state, with all products and reactants still present, but now in some sort of new balanced condition.

The way we describe the second case, where reactant and product concentrations in solution are changing, is by considering the change in the chemical potential as a function of concentration for each component involved in the reaction. Remember in our lecture on mixing and osmosis, how we derived an expression for the change in entropy upon mixing? We found that there is always additional entropy caused by mixing, and it only depends on the concentrations of the mixtures.

Here's a plot of the molar entropy of mixing. As the mole fraction of one of the components varies, the molar entropy of mixing increases as the two components are mixed, and it reaches a maximum for a 50/50 mixture. The function you're seeing here is the one we derived from a simple statistical model; it's a plot of the natural logarithm of the mole fraction being mixed. Now, it turns out that from this expression for the entropy of mixing, we can also write down expressions for the free energy change upon mixing. That's something we talked about back in Lectures 17 and 18 on binary phase diagrams.

We can write the chemical potential of either one of the species in the mixture as having changed due to the concentration of that component changing. In the example of two components, A and B, mixing, the chemical potential of A is equal to the standard state chemical potential of A, which we write as mu-naught-sub-A, plus the gas constant R times the temperature T times the natural log of the mole fraction of component A.

This is a powerful equation that tells us how the Gibbs free energy varies with changes in concentration of something in solution. That's exactly the case for our carbonic acid reaction. However, the form I'm showing you only applies to ideal solutions. Remember that ideal here means that we have ignored interactions, just like in the ideal gas law. So the equation only considers changes in the free energy—or chemical potential if we take the molar free energy—due to the entropy of mixing.

But as you may have guessed, most reactions are not ideal, since interactions between components can be very different from the interactions of one component with itself. For the components A and B, this would mean that A and B have a very different bonding character than either A with A or B with B. And this kind of effect is not taken into account in the ideal solution model.

So, we need to expand the range of applicability of this relationship. But this form of the equation, where we take the standard state chemical potential and add a single term that goes as the logarithm of something, that's a very convenient mathematical form, so we'd really like to retain it. We do this by defining something called the activity. The chemical potential of a component at any concentration is equal to the chemical potential at standard state conditions, plus RT times the natural logarithm of the activity of that component.

So, what have we done? Well, quite simply, we've replaced the mole fraction with this other variable, and that's it! But, it is a very powerful switch, because now we can essentially include any changes that happen to the chemical potential of the reacting species during the reaction. At the same time, we retain the same mathematical form, namely, a natural logarithm. So, we write the chemical potential for a given component—say component i—as the standard state chemical potential for that component (in other words, when it is in its pure form and not reacting), plus RT times the natural log of the activity of that component, a_sub_i.

It's nice to write the activity in terms of a coefficient, for which we use the symbol gamma and is called the activity coefficient, times the mole fraction of that component. By writing activity it this way, we can really clearly see how far away from the ideal solution model the system is. For example, you can see that if gamma for a component is equal to 1, then the activity of that component becomes the mole fraction of the component. We get back the same equation for the chemical potential that we had when we ignored interactions and just had the ideal solution model. The deviation of the activity coefficient from 1 is a measure of how far away from ideal the system is. And, as we were discussing, in so many cases for reactions, the activity coefficient can vary quite bit as a function of the mole fraction of each species in the solution.

Take a look at these plots of both the activity and the activity coefficient for a mixture of carbon disulfide and acetone. On the left is a plot of the activity vs. concentration for this solution. So here, if the graph showed a straight line from (0, 0) to (1, 1) then we would simply have ideal behavior, since the activity would just be equal to the mole fraction at any concentration. But as you can see, when we compare the true, experimentally measured activity of the system, it deviates considerably from ideality.

Now take a look at the graph on the right. This is a plot of the activity coefficient for the same solution. And here, in the behavior of the activity coefficient, we can clearly see a few things. First of all, there is a very strong dependence on concentration. Remember that if the activity coefficient is 1, then the activity is just equal to the mole fraction, so we have the ideal solution model.

And just to be sure we're on the right page, as a reminder, the ideal solution model means that the chemical potential will change upon mixing things together, but only because the entropy of the mixture increases. In the ideal solution model, we have ignored the effects of interactions. So again, the activity, which does include those effects, can tell us how important they were in the first place!

Okay, back to the activity coefficient plot. So, you can see that, at low concentrations, the coefficient deviates a lot from 1, by a factor of more than 3. This factor has an enormous impact on the chemical behavior of the system, which again, remember, is driven by the chemical potential. Just think about it. At a mole fraction of, say, 0.2, then, in the ideal solution model we would have a change in the chemical potential due to mixing of R times T times the natural log of 0.2. But if I now have that factor of 3 coming from the activity coefficient, then we would have a natural log of 0.6, instead of 0.2. And that gives a result that's about three times smaller.

So it's not just, say, a small change, like 10% or 20%, but rather, interactions that can lead to huge difference in the effects of mixing! At higher concentrations, a factor of 2 or 3 in the activity coefficient could even lead to a complete change in the sign of that logarithm term, which would fully reverse the qualitative effects of mixing. So a key point here is that as the concentration changes, the activity coefficient also changes. In the case of carbon disulfide mixed with acetone, it gets smaller with increased concentration of CS_2. This means that the system behaves more and more like an ideal solution as more and more CS_2 is mixed in.

If we think about what this means at a molecular scale, it means the following. As we add more CS_2 to acetone, the difference in how the CS_2 and acetone interact with each other, compared to how CS_2 interacts with itself, and how acetone interacts with itself, becomes less and less. And once the concentration of CS_2 reaches about 90% mole fraction, the activity coefficient has effectively converged to around 1, so the solution behaves pretty much like an ideal solution.

Actually, this limit where the solution starts to behave like an ideal one because the concentration is approaching 1, that limit has a special name; it's called Raoultian behavior. Mathematically, this behavior is written that as follows: As the concentration X_CS2 → 1, then the activity a_CS2 → X_CS2. Raoultian behavior simply means ideal behavior. It's the same thing. So we're back to the chemical potential of mixing depending on the natural log of just the molar concentration.

François-Marie Raoult was a French physicist who was experimenting with vapor pressures back in the day, that would be around 1882. He saw that if he had a solution of two liquids, say A and B, then, as long as no other gases were present, the total vapor pressure above the solution was equal to the weighted sum of the "pure" vapor pressures of the two components. So, he found that the total vapor pressure *P* is equal to the mole fraction X_A times the partial vapor pressure p_A plus X_B times p_B. Although here we're talking about pressures, this is equivalent to the same ideal solution model behavior we found for the free energy and chemical potential.

In our example, as the concentration of CS_2 becomes high enough, each CS_2 effectively will "see," for the most part, other CS_2's, as opposed to acetone molecules. This is why the behavior of a given component is guaranteed to become more and more ideal as the concentration of that component increases. It simply starts to feel less and less like it's in a mixture in the first place!

Now, let's take a look at the other extreme. Over in the other limit, where we have very low concentrations of CS_2, you can see from the plot of the activity that the behavior is in fact quite non-ideal. This is where we saw the largest deviation from 1 in the activity coefficient plot.

Over at low concentrations, where the CS_2 is highly dilute in the solution with acetone, each CS_2 molecule effectively "sees," for the most part, only acetone molecules. So, in this limit, we can approximate that there is a strong change from ideality, especially if the molecules interact strongly with each other differently than they do with themselves. For low concentrations, just as with the high concentration regime, we also have a special name, We call the behavior of the mixture Henrian for this case. William Henry formulated Henry's law back in 1803, and like Raoult, he was also playing around with liquids and gases and measuring pressures. Henry's law says that at a constant temperature, the amount of gas that dissolves in a liquid is proportional to the partial pressure of that gas in equilibrium with the liquid. For a dilute solution, Henry's law tells us that the pressure goes as a constant (It's called Henry's constant) times the molar concentration of the solute.

It is interesting to compare this to Raoult's law, which states that the pressure goes as the partial pressure of that component in its pure state times its mole fraction, valid for the other extreme at high concentration. Both these laws appear to give something similar, a dependence of a constant times a mole fraction. This is actually true if the two components being mixed have a similar bonding behavior with each other as they do with themselves. But that's just for the ideal solution case.

If interactions are strong and different between the two components, then in the case of Henrian behavior, we are looking at a regime where these differences are, in a sense, maximized, since again, each solute is surrounded primarily by only solvent, as opposed to other solute molecules. That's why in the activity coefficient plot we saw the largest deviation from 1 occurring right near zero. This is where the mixture is said to have Henrian behavior. For a narrow concentration range near zero, the deviation from ideality is roughly a constant, and that's Henry's law.

Just to ground Henry's law in an example, consider a can of some carbonated beverage. Before you open it, there's CO_2 gas inside, sitting above the liquid drink. And that CO_2 is pressurized; it's not too high of a pressure, but usually about twice the atmospheric pressure. Now, the liquid also has CO_2 inside of it, dissolved in the liquid. We can think of this as none other than our two-component solution. Here, component A would be the liquid, and component B would be the CO_2. In the unopened can, the CO_2 that's mixed in a solution with the liquid is at fairly low concentrations, so the behavior of the chemical potential of CO_2 in the solution will be Henrian. That means that the amount of CO_2 that is actually dissolved in the liquid is proportional to the partial pressure of the CO_2 above the liquid in the can. If the gas inside the can is essentially all made up of pure CO_2, which is usually the case, then this partial pressure above the liquid is just the total pressure of the gas.

In a typical can of, say, soda, the amount of CO_2 dissolved in the liquid is around 0.2% mole fraction. This is pretty low, so we know we're in the Henrian limit. This means that we might expect that if I double the partial pressure of CO_2 inside the can, then I could get up to 0.4% CO_2 dissolved in the beverage. And in fact, within a certain range of concentrations, this is the case; the amount of carbonation is driven simply by the pressure of CO_2 above the liquid. But, as the concentration of CO_2 continues to increase, then the behavior begins to deviate from Henrian, and this type of dependence will no longer be observed.

Note that when the can is opened, all of that CO, that was under pressure goes away—That's the pop sound we hear when we open it. Now the concentration of CO_2 dissolved is much higher than it wants to be, given that the partial pressure of CO_2 above the liquid has just been dramatically reduced. That's why the drink bubbles; the concentration of the CO_2 in the liquid must decrease until it's suitably proportional to the partial pressure, which, when the can is opened is extremely low.

And this is a perfect opportunity to return to our original example of a reaction where the reactants are dissolved in water, the one involving carbonic acid. That's because, as I mentioned a bit earlier in this lecture, CO_2 dissolved in water at low concentrations leads to the formation of carbonic acid, or H_2CO_3. And then remember that we talked about how carbonic acid decomposes into HCO_3 minus and H plus, with the reaction that I showed before, $H_2CO_3 \rightarrow HCO_3$ minus plus H plus This reaction is what causes carbonated water to have a slightly more sour taste, since it becomes a little bit acidic. That's also why sometimes alkaline salts, like sodium bicarbonate, are added to soda water. They reduce the acidity, which reduces the sourness.

And now that we have all this activity stuff under our belt, we arrive at an exciting place, that would be the point where we are ready to learn the thermodynamic equation that governs the carbonic acid reaction. And it goes as follows. The change in the chemical potential in a reaction is equal to the difference in the standard state chemical potentials plus R times T times the natural log of what is called the reaction quotient, or Q. This reaction quotient is the ratio of the activities of the products multiplied together, divided by the activities of the reactants multiplied together.

So, for the carbonic acid reaction, this Q would equal the activity of HCO_3 minus times the activity of H plus divided by the activity of H_2CO_3. And the chemical potential change occurring during the reaction is written as the standard state chemical potential of HCO_3 minus plus the standard state chemical potential of H plus minus the standard state chemical potential of H_2CO_3, then add to that RT times the natural log of Q, which is the natural log of those activities as shown.

The activities in the Q change during the reaction as the system strives to reach equilibrium. From our discussion a moment ago about the activity of CS_2 or CO_2 in a solution, and the different limits of Raoultian and Henrian, one thing we saw for sure is that these activities are highly dependent on concentration. And see, that's just the thing, the concentrations of these components are changing as they react, so the activities will change as well.

This reaction quotient allows us to keep being able to calculate the chemical driving force for the reaction at any point, as long as I also happen to know what the activities are for the reactants and products. And then, there's that very special point, where in fact, this driving force, this change in chemical potential due to a reaction occurring, when that is zero. At that point, there is no longer a way for the system to lower its free energy, and so it's in equilibrium.

We can find this point simply by plugging in zero for the change in chemical potential. We find that the equilibrium point for the reaction is determined when the RT natural log term is equal to those differences in standard state chemical potentials. This point, where we have equilibrium, is so important that we call the reaction quotient at that point the equilibrium constant. It's still just the same reaction quotient, but it only refers to the equilibrium point in the reaction, so, many textbooks will give that its own variable, often K is used for the equilibrium constant.

So, certainly among those who do a lot of chemistry, this equation for the equilibrium of a reaction has been called the most useful in thermodynamics. But, most important, is the fact that the activity product ratio that makes up the equilibrium constant on the right-hand side, is independent of variations in the system composition. As you can see, the equilibrium constant K is controlled completely by the difference in standard chemical potentials. This means that it's only a function of the temperature and pressure. And for any given temperature and pressure, it's a constant. As soon as the substances reacting are chosen, then the value for the equilibrium constant is fixed. It's the activity product ratio that would be observed if the system were able to reach equilibrium.

The equilibrium constant is a very useful piece of information to have about a reaction. If K is very large, it shows the reaction will tend to go "to completion," that is, mostly products will be present at equilibrium, and if K is small, it shows that the reaction hardly goes at all before enough products are formed to stop it. If you're trying to design a process to produce some kind of new chemical, then you can imagine that how important it would be to know to what extent the reactions involved in this process should theoretically proceed.

The equilibrium constant will not tell you whether reactants will actually react, or at what rate. Remember, in thermodynamics, we take an "eventually it will happen" approach, just as a candle does not light itself, even though the burning of a candle is a highly favorable reaction. There can be any number of reasons for reaction kinetics to be very slow, or even completely hindered. The equilibrium constant indicates the activity product ratio at equilibrium, not whether equilibrium is easily achievable.

So, finally, let's apply this equation to our carbonic acid reaction. So what we really want to figure out is how much of this reaction will actually occur before it reaches equilibrium. We can now do that by equating plus 36.3 kJ/mol to minus RT times the natural log of the ratio of activities. Plugging in the value of R, which is 8.3 Joules per mole-Kelvin and room temperature, which is 298 Kelvin, we can solve for K. We find that the natural log of K must be equal to minus 14.66, and then solving for K, we get the value, K equals 10 to the minus 6.37. This means that the ratio of activities for this reaction equals 10 to the minus 6.37 when the reaction has reached equilibrium.

And this is the answer to our question of what happens to the reaction after it starts! The reaction continues until the ratios of the activities of the products and reactants equals this equilibrium constant, which, in this case, is a very small number. That means that very little of the reactants have formed products at equilibrium. Basically, not all that much happens! The physical result of this is that carbonic acid is a weak acid.

So, there you have it. In this lecture, we talked about a crucial concept in the application of thermodynamics to reactions, namely, the activity coefficient and how it relates to concentration and the chemical driving force during a reaction. This in turn led us to a way to compute this driving force at any concentration and to an equation to find the equilibrium of a reaction. This is of fundamental importance when the ability of a reaction to occur depends on the concentrations of the reactants and products, which are changing during the reaction.

The Greatest Processes in the World

Lecture 24

The laws of thermodynamics, and the variables they govern, ultimately describe how energy moves. They describe why energy transforms from one form into another and how matter is affected during these transformations. The technologies we still need to invent in order to address the world's future energy—and water—needs will rely on the design of new materials, and that can only be done with an intimate knowledge of thermodynamics. By learning thermodynamics, we learn how to understand, describe, predict, and design the greatest processes in the world.

The Value of Energy

- Fossil fuels have been able to completely change the world in essentially every way imaginable in what is by any measure quite a short time period. And scientists and engineers have been at the forefront of this revolution: whether by helping to design new ways to find and extract the energy from the ground, or new ways to use this energy in our factories, in our farms, in our transportation, and in our homes.

- The ability to burn carbon-based materials is what powers our industry, growth, and comfort. But all of these massive amounts of ultra-cheap energy have also led our world to quite a strained position. Concerns about sustainability, climate, and national security are requiring that we reexamine our reliance on carbon burning.

- Among the many thought-provoking questions these topics have inspired, one entirely practical question is critical: Is the burning of carbonaceous materials the only way in which our planet can be tapped for energy?

- Currently, there is enough ancient sunlight trapped in carbonaceous forms—whether from conventional or unconventional reservoirs—to last at least another generation, maybe more. But as our 200-year blip of seemingly limitless fossil fuel energy begins to decline, we will have to rely on alternative and sustainable energy resources. Fossil fuels are finite; eventually, and not too far from now, they will run out.

- The energy we get from fossil fuels is so dense and such a precious resource that it is a daunting task to even think about how we might someday be able to fully replace them. Unlike many other revolutions that have occurred, the revolution that must occur in energy and water over the next century will almost entirely center on our ability to carefully design and manufacture completely new materials.

- Ironically, it is precisely because of the foundational science and engineering discoveries enabled by thermodynamics that we have arrived at this global tipping point. Now that we are where we are in human civilization, there is no starting over. We need to look to the future, and we need to use the science and technology that we have learned over the past centuries to place humanity in a more sustainable position with respect to the planet.

- So, instead of asking what would happen if we had to begin again, we need to ask the following: What happens now that we cannot begin again? Thermodynamics is the answer to this critical question. Scientists and engineers have used subjects like thermodynamics to harness energy to build our modern society, and they are also the ones who must create a new way to get that energy.

- The work to invent and develop and optimize alternatives to fossil fuels must be done now, because it is long-term work that might take decades. Scientific breakthroughs cannot be predicted, but it is easy to predict that if we don't put any people and resources onto a problem, then no breakthroughs can occur. We need to work now on this problem so that we are ready for it later.

Solar Thermal Fuels

- An example of a new type of fuel that is being developed is called a solar thermal fuel, which is a material that absorbs sunlight really well, and when it's exposed to the Sun, it undergoes a transformation at the molecular level. As light shines on the molecule, it changes its shape. In its new, twisted shape, the molecule stores energy—a kind of energy that can be released later, whenever you need it, in the form of heat. Once the energy is released, the fuel is ready to be recharged by the Sun.

- Temperature is related to vibrational motion, and that the heat capacity of a material depends on how many internal degrees of freedom are available to store the energy. This case has a special atomic arrangement where it can store even more internal energy, but it's not taking in thermal energy to get there—instead, it's converting sunlight energy into this specially strained internal energy state.

- When the molecule is triggered, it goes back to its original position, and as it does, the motion of the molecule is so intense that it's like a little tiny explosion. And because the motion of atoms is related to temperature, the temperature of the molecule goes way up, and thermal energy is released.

- It still all comes down to the thermodynamics of the energy pathways. In this case, we have sunlight to chemical energy to thermal energy. One way to think of a solar thermal fuel is that it's like a rechargeable heat battery. It can be charged over and over again by the Sun and deliver heat on demand when needed.

- This type of fuel could completely change the way we make and use energy. First, unlike fossil fuels, it's a material that operates in a manner that leads to no by-products. The fuel does not combust but, rather, simply cycles back and forth between the charged and the uncharged states.

- Second, it's chargeable by the Sun, so it is based on a resource that is effectively infinite. Third, it represents the only way to use the Sun's energy in the form of heat that is distributable and transportable. That's because other solar thermal technologies do not store the energy indefinitely until it is purposefully released but, rather, simply get hot and then immediately begin losing that stored heat.

- Because of their potential advantages, it's perhaps not surprising that solar thermal fuels are not new. They were heavily explored in the 1970s, although they were widely abandoned by the mid-1980s because the materials under consideration back then simply could not withstand the cycling. They would degrade a little bit each time they charged. After a mere 10 or 20 cycles, the amount of useful fuel left was miniscule.

- By optimizing the thermodynamic free energy of this system, we have invented an entirely new platform for making solar thermal fuels, one that leads to materials that can be recharged over and over again with absolutely no degradation for tens of thousands of cycles and that have energy densities even higher than the best-performing batteries on the market today. This represents the first sunlight rechargeable thermal fuel that provides safe, recyclable, high-density, transportable, and distributable heat on demand. And it's entirely possible because of thermodynamics.

- One example of how we're using solar thermal fuel already is in solar cooking. Solar cookers exist, but a key limitation is that until now, you could only cook with them when the Sun was shining. With solar thermal fuels, we've created a new cooker. You can charge it up during the day by leaving it out in the Sun, and then when you want to cook, you can simply flow the charged-up fuel the other way and trigger it to release the stored heat.

Water and Energy

- Water and energy are intimately related to one another. That's because producing and transporting freshwater takes massive amounts of energy. And it's not just about energy required to get water; it's also about the water required to get energy. The energy security of the United States remains precariously reliant on large volumes of freshwater for cooling.

- But the biggest challenge with water is in the availability of it. Even though 70 percent of the planet is covered with water, less than 3 percent of that is freshwater, most of which is locked up as ice in polar caps and glaciers. This leaves less than 1 percent of all of Earth's water as drinkable.

- One way to solve this challenge would be to invent a way to take the salt out of ocean water in order to make it drinkable. This process is called desalination, and there are many ways to do it, but the problem is that none of these ways is cheap enough to become a global solution to freshwater, at least not currently.

Most of the less than 3 percent of water that is freshwater is locked up as ice in polar caps and glaciers, meaning that it is undrinkable.

- The most efficient technique we have today is called reverse osmosis, which is based on massive pumps to push saltwater across specialized membranes that, at high enough pressures, will allow the water to pass but not the salt.

- To desalinate water, we need to overcome the entropy of mixing, which is quite powerful. We have to get to pretty high pressures to force water to be unmixed and move to lower entropy states. The pressure that drives the water to higher entropy is called the osmotic pressure. The cost of the energy required to overcome the osmotic pressure is enormous.

- Most of the ancillary processes involved in desalination have already been intensively optimized over the past several decades, but the membrane itself—the key material involved in the process—has been based on the same technology for the past 40 years. If we want to turn desalination into an environmentally benign and economically affordable solution for the global water supply, we need to completely rethink the membrane.

- In the new age of materials design and with a solid understanding of thermodynamics, we have used thermodynamics to design the ultimate membrane for water desalination. At only a single atom thick, it is the thinnest membrane that one could ever make. The material that allows us to make such a membrane is called graphene, and it's a sheet of carbon atoms that looks like a honeycomb lattice.

- However, we cannot desalinate water with this single-atom-thick material in its natural form. For that, we need to make holes in the graphene that are just the right size. They need to be big enough to let water pass through but small enough so they block the salt from going through.

- This is where thermodynamics enters the picture: The key to understanding and therefore designing just the right pores lies in controlling the entropy of the water molecules as they pass through the membrane.

- By tailoring the pore in just the right way, we are able to maximize the entropy of the water while keeping the pore small enough to completely reject the salt. The result is a completely new membrane for desalinating water.

Suggested Reading

Anderson, *Thermodynamics of Natural Systems*, Chap. 1.

Callister and Rethwisch, *Materials Science and Engineering*, Chaps. 1 and 22.

DeHoff, Thermodynamics in Materials Science, Chap. 1.

Questions to Consider

1. Name your favorite thermodynamic processes and variables, and think about how you could use these to process a material in a particular manner.

2. At its core, what is thermodynamics about?

3. How does thermodynamics enter into your daily life?

The Greatest Processes in the World
Lecture 24—Transcript

Hi and welcome to our very last lecture of this course. I've had a wonderful time taking you through all of the foundational topics of thermodynamics. I hope that you have a good sense for why we decided to call this course the "Four Laws That Move the Universe." Because the laws of thermodynamics, and the variables they govern, ultimately describe how energy moves. They give us a clear picture of why energy transforms from one form into another, and how matter is affected during these transformations. In short, they allow us to understand movement of any kind and any form, whether from matter, heat, motion, or any other form of energy we know. This is why "the laws that move the universe" truly captures the essence of thermodynamics.

Starting from the meaning of temperature itself, we've covered quite a massive amount of scientific knowledge. We set forth the crucial variables that allow us to describe any system and any process. Sometimes these variables felt rather intuitive, like pressure and volume, and sometimes perhaps, not so much, like entropy, or the chemical potential. But either way, we felt the meaning of these crucial variables all the way down to the scale of the atom and all the way up to our technologies, our buildings, and the planet itself. And we didn't just learn about variables by themselves, we made connections between them.

We saw the ideal gas law in action, one example of a state function. And we developed the relationship between heat and work. We learned that internal energy represents a kind of energy that is stored inside of a material, and that it can tell us how that material holds onto energy from heat and responds to changes in temperature. And speaking of temperature, we saw that a change in it represents a crucial driving force for heat to flow.

We felt our oneness with entropy and learned that it's the missing link between temperature and thermal energy. And we saw that the entropy of the universe must increase for any process to occur. And speaking of processes, we covered all types in this course. But the ones where temperature and pressure can be controlled, well those are so important we introduced a new variable, the free energy, to discuss such conditions. By minimizing the free energy, we learned that it would be possible to find the happy place for any material—That's also known as equilibrium—and that this place could be plotted in a phase diagram, which is the ultimate materials map.

Having a phase diagram for a given material is like adding GPS navigation to your car; it helps show how to get from one point to another in the highly interconnected property, structure, and performance of the materials. Through these maps we learned all sorts of beautiful phenomena, like how materials change their phase between gas, liquid, and solid in many different ways, and sometimes, even possess completely new phases, like in the case of a supercritical fluid. We saw that when two different materials are mixed together, these phase diagrams can take on immense complexity and show new behaviors, like a dramatic lowering in the melting point, or, a complete separation of the material into two distinct concentrations. And when things are mixed, we learned that partial molar quantities are the way to describe them. We saw that the partial molar value of the free energy is so special that it gets its own name, the chemical potential, which is the chemical driving force for any change in the amount of a substance to occur.

We learned about heat engines, which formed the basis of the industrial revolution. We learned that while work can make heat with 100 percent efficiency, this does not go the other way; heat cannot do work with 100 percent efficiency, and the Carnot limit tells us that the absolute best a heat engine can ever achieve is dependent on a temperature difference. We saw how other types of work, like magnetism, surface tension, phase change, and entropy itself, could be used to make things move.

And beyond these highlights, we learned many other foundational thermodynamic concepts, which are critical pillars of the science and engineering of modern times, like eutectic melting, osmotic pressure, the electrochemical cell, and that beautiful moment for a material where all three phases coexist, called the triple point, along with many, many other important phenomena.

And speaking of modern times, in the very first lecture, in order to motivate you to think about the crucial role of thermodynamics in our lives, I asked a question. I asked you to imagine what would happen if we had to start all over. What if we had to begin again, to rebuild the world that we know today? What science would we need? What variables would matter? How should we think about these variables in relationship to one another? And perhaps most importantly, what forms of energy do we have, and how could we use energy in these different forms to move the variables around and achieve some desirable function?

Over the past 23 lectures, we have learned the foundational science that answers these questions. We have learned how to relate heat to motion and to materials and energy in all of its forms. We have learned thermodynamics. Here, in this last lecture, I'd like to talk about the topic that is near and dear to my heart, that would be energy. What gets me the most excited about the research that I do, what literally keeps me up at night sometimes because of how exciting it is, is the fact that technologies we still need to invent in order to address the world's future energy and water needs, those technologies will rely on the design of new materials. And that can only be done within an intimate knowledge of thermodynamics.

Let's get a sense of what energy means to us. What is the value of energy? Well, think about the following comparison. For about one dollar today, I can buy a liter of gasoline in the U.S., and that liter of gasoline contains roughly 32 million joules of energy in it. A typical human can perform work at a rate of about 60 Watts sustainably, that is, without any rest for at least some number of hours on end. Sixty Watts means 60 joules per second. So, that means that if a person did some kind of physical labor continuously for eight hours each day, then it would take four weeks for that person to have worked as much as 32 million joules. One liter of gasoline, in terms of energy, is equivalent to four full 40-hour weeks of labor from a person. On the one hand, we paid a dollar for that energy, while, for the person, well it depends of course, but even at just $10 per hour, we'd have had to pay 1600 times as much as the gasoline.

I love that comparison, because I feel that it very clearly and simply illustrates why fossil fuels— almost free labor when compared to humans—have been able to completely change the world in essentially every imaginable way, and in what is by any measure, it's changed that world in an incredibly short time period. And scientists and engineers have been at the forefront of this revolution, whether by helping to design new ways to find and extract the energy from the ground, or new ways to use this energy in our factories, in our farms, in our transportation, and in our homes.

When we turn the knob on a lamp or switch on the TV, we rarely think about how the energy comes to our appliances, that is, that the power traveling through wires into our sockets is made possible because carbonaceous materials, like oil, natural gas, or coal, are being burned hundreds of miles away. We may not think about it, yet this ability to burn carbon-based materials is what powers our industry, our growth, and our comfort. But all of these massive amounts of ultra-cheap energy have also led our world to quite a strained position.

What I mean by the word “strained” here is a collective effect. It’s the culmination of this energy boom that has had, to name only a few examples, led to wildly rapid population growth, the release into our biosphere of unprecedented amounts of toxic chemicals, the precarious lack of diversity in our approach to agriculture, and the many conflicts that have already occurred and many more that likely will occur due to the heterogeneous global distribution of this precious resource. Because of this, we have arrived at quite an interesting point in our world history. Concerns about sustainability, climate, and national security are requiring that we reexamine our reliance on carbon burning.

And among the many thought-provoking questions these topics have inspired, one entirely practical question is critical. Is the burning of carbonaceous materials the only way in which our planet can be tapped for energy? Currently, there is enough ancient sunlight trapped in carbonaceous forms, whether from conventional or unconventional reservoirs, to last at least another generation, maybe more. But as our 200-year blip of seemingly limitless fossil fuel energy begins to decline, we will have to rely on alternative and sustainable energy resources. Fossil fuels are finite. They may not run out in the next 30 years, or even the next 50, or 100 years. But eventually, and not too far from now, they will.

It’s not just that we need a “slightly” better version of the materials we already have. You see, that doesn’t capture the challenge we face. Rather, we need complete game changers. We need materials that have factors of 5 or 10 better efficiencies, or lower costs, compared to current ones. And we can't just suddenly make the switch. The energy we get from fossil fuels is so dense and such a precious resource that it's a daunting task to even think about how we might someday be able to fully replace it.

So, unlike many other revolutions that have occurred, the revolution that must occur in energy and water over the next century, that revolution will center almost entirely around our ability to carefully design and manufacture completely new materials. Ironically, it's precisely because of the foundational science and engineering discoveries enabled to a large extent by thermodynamics, that we have arrived at the global tipping point I just described. The fact that I even asked the question, "What if we had to begin again?" in that first lecture, speaks volumes in itself. We have advanced the human civilization so much, and changed the world so much, and improved our well-being so much, that just that question alone is awe-inspiring to think about. That's why I asked it!

But now, I want to ask yet another, different question. Now that we are where we are in human civilization, there is no starting over. We need to look to the future, and we need to use the science and technology we have learned over the past centuries to place humanity in a more sustainable position with respect to the planet. So, instead of asking what would happen if we had to begin again, I ask, "What happens now that we cannot begin again?" The answer to our first question was the topic of thermodynamics. I want to convey to you now, that thermodynamics is also the answer to this second, critical question.

When I launched my research program in clean energy and water, I was well aware that there was one topic in particular that I absolutely needed to become intimately and truly knowledgeable about, a topic I needed to know inside and out, in such a way that I could make an impact on energy problems and try to change the game. You probably guessed it; it was, of course, the topic of thermodynamics. We scientists and engineers have used subjects like thermodynamics to harness energy to build our modern society. And we are also the ones who must create a new way to get that energy—a clean, safe, renewable way that could last not just another fifty or hundred years, but as long as the sun itself, which, by the way, is another seven billion years, give or take.

But here's the catch. We can't just wake up one day, when we need it later on, and say, ok, now let's do the sun-powered thing. It won't work that way, because all of the clean energy technologies in place are far too expensive to be able to scale up to the world's energy needs, which, by the way, are continuing to grow without bounds. The work to invent and develop and optimize these alternatives to fossil fuels must be done now, since it's long-term work that may take decades. Scientific breakthroughs cannot be predicted, but it's easy to predict that if we don't put any people and resources onto a problem, then no breakthroughs will occur. We need to work now on this problem so that later, we are ready for it.

So let's turn now to the research and to how materials design and thermodynamics can do something about these challenges. I'll first give you an example of a new type of fuel we're developing. It's called a solar thermal fuel. This is a material that absorbs sunlight really well, and when it's exposed to the sun, it undergoes a transformation at the molecular level.

Take a look at this illustration. As light shines on the molecule, it changes its shape, going through a kind of twisty turn. And here's the key. In that new, twisted shape, the molecule is storing energy. It's a kind of energy that can be released later, whenever you need it, in the form of heat. And then, once the energy is released, the fuel is ready to be recharged by the sun again. Now, remember how we learned that temperature is related to vibrational motion and that the heat capacity of a material depends on how many internal degrees of freedom are available to store the energy? Well, you can think of this case as having a special atomic arrangement where it can store even much more internal energy, but it's not taking in thermal energy to get there. Instead, it's converting sunlight energy into this specially strained internal energy state.

When the molecule is triggered, it goes back to it's original position, and as it does, the motion of the molecule is so intense that it's like a little tiny explosion, and since, as we know, the motion of atoms is related to temperature, the temperature of the molecule goes way up, and thermal energy is released. It still all comes down to the thermodynamics of the energy pathways. In this case, we have sunlight-to-chemical energy, to thermal energy. One way to think of a solar thermal fuel is that it's like a rechargeable heat battery. It can be charged over and over again by the sun and deliver heat on demand when needed.

As you can imagine, this type of fuel could completely change the way we make and use energy. First, unlike fossil fuels, it's a material that operates in a manner that leads to no byproducts. The fuel does not combust, but rather simply cycles back and forth between the charged and the uncharged states. And second, it's chargeable by the sun, so it's based on a resource that's, effectively, infinite. Third, it represents the only way to use the sun's energy in the form of heat that is distributable and transportable. That's because other solar thermal technologies don't store the energy indefinitely until it is purposefully released, but rather, simply get hot and then immediately begin losing that stored heat.

So, because of their potential advantages, it's maybe not surprising that solar thermal fuels are not new. In fact, they were heavily explored in the 1970's, although widely abandoned by the mid 1980's. Why? Because the materials under consideration back then simply could not withstand the cycling. They would degrade a little bit each time they charged. The result? After a mere 10 or 20 cycles, the amount of useful fuel left was miniscule. It would be like buying a rechargeable battery that can only charge 10 times, not very useful.!

Enter the new age of materials design and a solid understanding of thermodynamics. What we have done in my research group is to re-imagine the way a solar thermal fuel can be made. You see, there are many, many molecules that undergo some sort of structural change when light shines on them. These are, in general, called photoswitches, because, well, because they do just that; they switch when exposed to light. However, none of these are good energy materials, since they don't store very much energy when switched, and often, they don't even stay in the switched state very long, like, for just seconds or minutes. So that's not a very useful fuel.

But, by using the concepts of thermodynamics, we have invented a completely new chemical platform that turns essentially any photoswitch into a good solar thermal fuel. The trick lies in the fact that we combine these photoswitches, these tiny, little molecular solar storage devices, with a separate material that holds them at a fixed distance apart.

To understand why this works, let's consider the thermodynamics at play. Remember how we learned that temperature is a driving force? We used the zeroth and then the second laws to show that heat will always flow from hot to cold? Okay, so, for most technologies that absorb sunlight and then attempt to use that energy as thermal energy, like say, for hot water, this temperature driving force is the single biggest challenge. Since a container cannot be made to be perfectly adiabatic, there will always be some loss of heat from the material. It turns out that solar hot water heaters made today are pretty good, but within about 12 hours, the water temperature will usually be back down to that of its surroundings.

But, for a solar thermal fuel, we have something quite different. In this case, the technology allows the material to hold onto all of the heat it absorbs until just the point at which it's needed. That's because it operates much more as a phase change material than just an absorber of heat. More specifically, it's a phase change material, but in this case, the phase change is caused by light instead of heat. We learned in our lectures on the phases of matter that a lot of thermal energy is needed to get a material to change from one phase to another, say, a gas to a liquid, and that this energy, called the latent heat, is released when the material goes back to the original phase.

For the solar thermal fuel, the two phases correspond not to gas, liquid, or solid, but rather, to those two different molecular structures, and the phase change is caused by absorbing light energy as opposed to thermal energy. Then, as we coax the material to go back to its original state, it gives back all of that energy it had absorbed from light to phase change, but now it gives it back not as light, but rather, as heat, that same latent heat that we learned about.

It was from this thermodynamics understanding and perspective that we came up with the idea to template the molecules onto another material. What we wanted to do was increase by as much as possible that latent heat stored between the two phases, while, at the same time, block the pathways between the two phases. This way, we increase the storage density, while also increasing the lifetime of the fuel when it's charged. The hybrid material allows us to do this. Putting the molecules onto a template opens up completely new chemistries between one photoswitch and another—chemistry that is different depending on whether the molecule is in the switched state or not.

And how could we tell whether a particular chemistry would work well or not? Well, by minimizing that very important thermodynamic variable we have come to know and love throughout this course, the free energy of the system. By balancing contributions from our good friends enthalpy, temperature, and entropy, we were able to develop new, stable and powerful hybrid materials for solar thermal fuels. This approach led to a tripling of the energy stored per molecule, while at the same time increasing the storage lifetime from seconds to years.

Take a look at this diagram of what's happening at the molecular scale. On the left, you can see what a single molecule alone would want to do when light shines on it. But over here on the right, by packing these molecules in tightly onto a surface, you can see that the shape and movement of the molecule is now highly constrained. This constraint leads to a dramatic change in the entropy, as well as the internal energy of the material. And it's this new constraint that we can take advantage of to do new chemistry that leads to improved bonding configurations—that would be the enthalpy term.

In fact, by optimizing the thermodynamic free energy of this system, we have invented an entirely new platform for making solar thermal fuels, one that leads to materials that can be recharged over and over again with absolutely no degradation for tens of thousands of cycles, and that have energy densities even higher than the best performing batteries on the market today. This represents the first-ever sunlight rechargeable thermal fuel that provides safe, recyclable, high density, transportable, and distributable heat-on-demand. And it's entirely possible because of thermodynamics.

One example of how we're using this fuel already is in solar cooking. Some of you may know this already, but wood-fired stoves, especially in the third world, are responsible for more deaths than malaria and AIDS combined. A wood-fired stove inside a home releases the equivalent amount of carcinogens as 1000 cigarettes being smoked every hour. And up to seven hours per day and 50 percent of a family's income in some of the poorest nations is spent simply on getting the wood to cook with.

Solar cookers exist, but a key limitation is that until now, you could only cook with them when the sun was shining. But this is not when most families want to, or in many cases are even able to, do the cooking. With our solar thermal fuels, we've created a new cooker that charges up during the day by leaving it out in the sun, and then, when you want to cook, you can simply flow the charged up fuel the other way and trigger it to release the stored heat. A five-hour charge under the sun allows the cooker to provide a 300°C hotplate for more than an hour, plenty of time to make a nice meal.

Okay, so, another example of research we're doing has to do with what you know is one of my favorite molecules, that would be, water. Water and energy are intimately related to one another. That's because producing and transporting fresh water takes massive amounts of energy. The energy required for purifying, transporting, and treating water accounts for a staggering 30 to 40 percent of energy consumption in municipalities. And it's not just about energy required to get water, it's about the water required to get energy. The energy security of the United States remains precariously reliant on large volumes of fresh water for cooling. For example, several power plants in Texas have routinely been forced to shut down lately due to droughts, causing brownouts and widespread instabilities on the power grid. Water and energy cannot be decoupled from one another.

But, the biggest challenge with water is in the availability of it. Even though 70 percent of the planet is covered with water, only less than 3 percent of that is fresh water, most of that is locked up as ice in polar caps and glaciers. So this leaves less than 1 percent of all of earth's water as drinkable. And today, it's estimated that more than two billion people globally do not have regular access to clean drinking water. That number is expected to double to more than four billion before the year 2050. So this is a huge problem.

Now, one way to solve this challenge would be to invent a way to take the salt out of ocean water in order to make it drinkable. That way, there would be plenty of water to start with, since about 97 percent of all the water on earth is saltwater. This process is called desalination, and there are many ways to do it, but the problem is that none of these ways is cheap enough to become a global solution to fresh water, at least not currently. The most efficient technique we have today is called reverse osmosis, which is based on massive pumps to push saltwater across specialized membranes that, at high enough pressures, will allow the water to pass, but not the salt.

We saw this first hand in our lecture on osmosis, where we watched water pass back and forth across the membrane of the egg. When the egg was put into salty or sugary water, the water inside the egg had a driving force to push through the barrier and leave. When the egg was then placed back in fresh water, the water had a driving force to go back across the membrane the other way, inside the egg. We learned that the cause of this driving force was due to the entropy of mixing inside the water. When water is mixed with something else, like salt or sugar, its entropy goes way up. In fact, one of the earliest examples of entropy that I gave was salt dissolving in your soup. It's not because of energy, but rather entropy that drives that dissolution.

So why is this important? Well, to desalinate water, we need to overcome this entropy of mixing, which is quite powerful. That's why we need those massive pumps; we have to get to pretty high pressures to force water to be unmixed and move to lower entropy states. As we've learned, this pressure, the one that drives the water to higher entropy, is called the osmotic pressure. It's why the process is called reverse osmosis, because literally, we are reversing the natural direction of the osmotic pressure of the water caused by the difference in entropy between fresh and salt water. And the cost of the energy required to overcome the osmotic pressure is enormous. Fully half of the cost for the whole process lies in the energy it takes to drive those pumps.

Now, the thing is that most of the ancillary processes involved in desalination have already been intensively optimized over the past several decades; but, the membrane itself, the key material involved in the process, has been based on the same technology for the past 40 years. If we want to turn desalination into an environmentally benign and economically affordable solution for the global water supply, we need to completely rethink the membrane. Enter the new age of materials design and a solid understanding of thermodynamics.

We have used thermodynamics to design the ultimate membrane for water desalination. Why do I call it the ultimate membrane? Well, because it's the thinnest membrane that one could ever make. It's only a single atom thick! The material that allows us to make such a membrane is called graphene, and it's a sheet of carbon atoms that look a lot like a honeycomb lattice. But as amazing as this single-atom-thick material is, I cannot desalinate water with it in its natural form. For that, I need to make holes in the graphene that are just the right size. They need to be big enough to let water pass through, but small enough so they block the salt from going through.

And that's where thermodynamics enters the picture. You see, what we discovered is that the key to understanding, and therefore designing just the right pores, lay in controlling the entropy of the water molecules as they pass through the pores of the membrane. Remember, entropy is a measure of the number of possible states that the system can be in. Well, as the water molecules pass through the pores in graphene, they can be restricted in terms of how and what geometry they are allowed to take as they pass through. It's a combination of the pore size and shape, as well as the specific chemistry around the pore edge that dictates whether the water has a smaller or larger number of possible ways it can squeeze through. The fewer possibilities, the lower the entropy. The greater the number, the greater the entropy.

So, by tailoring the pore in just the right way, we were able to maximize the entropy of the water while keeping the pore small enough to completely reject the salt. The result? A completely new membrane, in a sense, the ultimate membrane for desalinating water. One that maintains 100 percent salt rejection while providing a permeability to water that is 500 times larger than the best RO membranes available today! Permeability, by the way, is a measure of the volume of water that can pass through the membrane, per area of the membrane, per time and pressure applied. As you can imagine, higher permeability means a lot more throughput is possible for the same membrane.

Okay, so there I just gave you two distinct examples from my own research where I've used the basic concepts of thermodynamics that we've learned in this course to design new technologies for energy and water. Whether looking at possible future technologies such as these, or learning about the past discoveries that allowed for the framework of thermodynamics to be created, I hope that throughout this course you have shared in my excitement as we've gone through this wonderful subject.

Understanding Thermodynamics is to understand the interactions and reactions that power the manufacturing of steel for a building, the development of new medicines designed to attack a harmful disease, the stars in the universe, or in the plants that grow. It's the principles of thermodynamics that make it possible for us to understand some of the greatest processes in the world.

That last phrase, "greatest processes of the world," is a wonderful place to close this lecture and this course. Because by learning thermodynamics, we have learned how to understand, describe, predict, and design the greatest processes of the world. Many of these processes we have already learned, understood, and mastered to build the world we live in today. And many more of these processes are still to be discovered as we build the world of our future.

Bibliography

Anderson, Greg. *Thermodynamics of Natural Systems*. 2nd ed. Cambridge: Cambridge University Press, 2009.

Callen, Herbert B. *Thermodynamics and an Introduction to Thermostatistics*. 2nd ed. Hoboken, NJ: Wiley, 1985.

Callister, William D., and David G. Rethwisch. *Materials Science and Engineering: An Introduction*. 8th ed. Hoboken, NJ: Wiley, 2009.

DeHoff, Robert. *Thermodynamics in Materials Science*. 2nd ed. Boca Raton, FL: CRC Press, 2006.

Halliday, David, Robert Resnick, and Jearl Walker. *Fundamentals of Physics*. 9th ed. Hoboken, NJ: Wiley, 2011.

Lambert, Frank L. "Shuffled Cards, Messy Desks, and Disorderly Dorm Rooms—Examples of Entropy Increase? Nonsense!" *Journal of Chemical Education* 76, no. 10 (1999): 1385.

Moran, Michael J., Howard N. Shapiro, Daisie D. Boettner, and Margaret B. Bailey. *Fundamentals of Engineering Thermodynamics*. 7th ed. Hoboken, NJ: Wiley, 2010.

Porter, David A., and Kenneth E. Easterling. *Phase Transformations in Metals and Alloys*. 2nd ed. Boca Raton, FL: CRC Press, 1992.

Serway, Raymond A., and John W. Jewett. *Physics for Scientists and Engineers*. 9th ed. Chaps. 19–22. Independence, KY: Cengage Learning, 2013.

Notes